INTRODUCTION TO ECOLOGY

R. Dajoz

Translated by

A. South
Department of Biological Sciences,
City of London Polytechnic

CRANE, RUSSAK & COMPANY, INC.

NEW YORK

Published in the United States by
Crane, Russak & Company, Inc.
347 Madison Avenue, New York N.Y. 10017

LCC 76–27620
ISBN 0–8448–1008–8

First published 1969 by Dunod, Paris
Second edition 1971
First printed in Great Britain 1976

Printed in Great Britain

TRANSLATOR'S FOREWORD

The translation of this book has been a demanding but rewarding task. Dajoz's approach to ecology is a comprehensive one and it is unusual to find such a wide range of information in an introductory text. There is an increasing need for this broad-based approach to the subject with the development of interdisciplinary environmental studies in sixth forms and at first year undergraduate level. I believe that Dajoz's synthesis of ecology, using levels of organisation of increasing complexity, although traditional, has much to commend itself to the newcomer in ecology.

English language references have been substituted for their French equivalents, where appropriate, especially where practical details are involved. In many instances the former were the definitive works. Some small revisions have been incorporated in the text to allow for advances in the subject which have taken place since the second edition appeared in 1971, including an updated section on useful text books for further reading in the bibliography.

The third edition of Précis d'écologie appeared after this translation had been prepared for publication. While it has not been possible to make a complete revision of the text, the section on population ecology (chapters 6, 7 and 8) has been partly rewritten.

I should like to acknowledge the help and advice that I have received from my wife and from several of my colleagues in the Department of Biological Sciences, City of London Polytechnic.

AUTHOR'S PREFACE

The science of ecology is developing rapidly at the present time. Neglected for many years in France, it is now taking its proper place in research institutions as well as in secondary and higher education. However, although there are ecological texts published in other languages, and especially in the English language, France has tended to fall behind in this field.

The present book seeks to fill this gap, although it would be impossible to cover the whole subject or to enter into details within the framework of a review. In particular, material relating solely to plant ecology has been omitted. Those aspects of plant ecology which relate to ecology in general have, however, been included. The extensive vocabulary of the subject has been simplified and technical terms reduced to a minimum.

Ecology makes use of information from a variety of disciplines, and this material has been combined with data obtained from original research. It enables the naturalist to view particular aspects of biology in a fresh light, and provides a new approach for pure and applied research.

The importance of the issues raised by recent ecological research introduces the question of whether ecology should be included in every person's education. This is the author's conviction, and it is for this reason that the book has been written.

This second edition has been enlarged by about fifty pages, thanks to the co-operation of the publisher. This has permitted increased coverage of some subjects which were referred to only briefly in the first edition. In addition, more examples from plant ecology have been included. An attempt has been made, both in the text and in the bibliography, to include information from recent ecological publications. However, some selection was unavoidable owing to the increasing numbers of papers which are now appearing, the latter being a symptom of the growing interest in ecology.

CONTENTS

CONTENTS

CONTENTS

INTRODUCTION

'Our surroundings and everything that our senses perceive, continually present us with a wide range of varied sensations. The ordinary man remains unaware of these things, especially if they are familiar, but the genuinely philosophical man cannot observe these things without displaying some interest. There is an amazing amount of activity throughout the universe and this does not appear to diminish for any reason. Every part of creation is exposed to continuous change'
(Lamarck. Recherches sur les causes des principaux faits physiques. Discours préliminaire.)

Although the word ecology was virtually unknown to the general public when the first edition of this book appeared in 1970, it has become more widely used today. The appearance of this word in common usage, as in many other instances, has broadened its meaning and, at the same time detracted from its original meaning. It is thus expedient here to clarify the meaning of ecology. It belongs to the biological sciences, that is those sciences concerned with the study of living organisms. Thus ecology cannot be regarded as appertaining to geography, economics, politics or any other branch of human knowledge as some authors would lead one to believe. This change in meaning, which is completely unjustified from the scientific point of view, may be explained, in part, by the considerable influence that the consequences of the disregard of ecological principles have had, and continue to have, for man in the twentieth century.

The biologist is concerned with the study of living organisms at various levels of organisation, each level being distinguished by properties which appear as the result of the development of new structures. The molecular biologist works at the macromolecular level; the cytologist is interested in the cell; the physiologist investigates the functions of organs. Morphology, anatomy and systematics are concerned with the study of whole organisms. Ecology is concerned with higher levels of organisation, including populations, communities, ecosystems and even the entire biosphere.

The term ecology was introduced by the German biologist E. Haeckel in 1866 and is derived from two Greek words: οικος meaning

1

house and λόγος meaning reason and signifying science. Thus ecology literally stands for 'science of the habitat'. The study of ecology is therefore concerned with the conditions necessary for the survival of living organisms together with, on the one hand, the relationship between various living organisms, and, on the other hand, the interactions between living organisms and their physical environment. The ecologist employs the techniques, concepts and results of other biological sciences and even mathematics, physics and chemistry. It would be wrong, however, to suggest that ecology was not a distinct science but simply a particular 'viewpoint' or 'way of thinking'. Many of the concepts, problems and methods of ecology are specific to the subject. While research in autecology can often be related to physiology or biogeography, the study of populations and ecosystems is specific to ecology.

THE SIGNIFICANCE OF ECOLOGICAL INTERACTIONS

In the course of his research, the ecologist cannot consider living creatures in isolation from their environment but rather studies the whole complex. In this, he is distinguished from, for example, the physiologist who works on an animal or plant under artificial conditions and then draws conclusions from a synthesis of the information obtained. The ecologist studies a living organism as a whole and 'not in an artificial, constant environment but rather under natural conditions where the forces that operate are continually changing'. As Prenant (1934) stated 'it is unlikely that a biologist would dispute this obvious difference and yet, there is a tendency to describe the natural causes of mortality solely in terms of physical factors studied in the laboratory'. Prenant illustrates his point by reference to examples of the sea urchin *Echinaster sepositus* and hermit-crab *Eupagurus prideauxi*. These animals show a narrow stenohaline distribution in nature and yet, in the laboratory, are able to survive considerable, even rapid, changes in salinity. Similar comparisons can be made for plants. The small herb *Spergula arvensis*, belonging to the family Caryophyllaceae, shows optimum growth in pure culture at a pH of between 6.0 and 6.5. However, the optimum pH is lowered to 4.0 if the plant is grown in mixed culture with the crucifer *Raphanus raphanistrum*. There are numerous other examples to demonstrate the difference between data obtained from laboratory experiments and that obtained from the field. There is, for example, little similarity between metabolic measurements made on birds and mammals maintained in cages and fed on cultivated plants with a high nutritional value and measurements made for the same animals living under natural conditions and which are more active, having to search for their food, which is often of poor nutritive value. This does not mean that laboratory studies can be ignored. On the contrary, they are invaluable but it is rarely possible to extrapolate from laboratory data into a field situation. The many interactions between living organisms and their environment make this impossible. 'The dynamic approach is fundamental to modern biology. It is not enough to study individual components and processes in isolation. The particular problems of the way in which the components are organised and coordinated still remain to be solved. These problems are the consequence of dynamic interactions between the components. Their behaviour will differ according to whether they are studied in isolation or as part of the whole system' (Von Bertalanffy). For this reason the introduction into ecology of such concepts as cybernetics and systems theory should certainly prove useful in the future.

A more recent example shows how difficult it is to understand the nature of these interactions. In Africa the Rwindi-Rutshuru plain on the south side of Lake Edward has been incorporated in the Albert National Park since 1929. The fauna of large mammals began to show changes after this date and these have been studied. The establishment of this national park was accompanied by the prohibition of scrub burning, previously carried out by the natives to prevent the regeneration of trees. The short grass savannah, which formed an open habitat suitable for several species of ungulate, gradually disappeared. At the same time, stocks of two antelope, the topi (*Damaliscus korrigum ugandae*) and Buffon's kob (*Adenota kob neumanni*), were reduced from 10 000 to 1200, and from 15 000 to 3000, respectively over an area of 800 km² between 1931 and 1940. The large carnivore predators of these ungulates showed similar reductions in their populations, numbers of lions, for example, falling from 250 to about a hundred. The regeneration of trees attracted elephants which invaded the reserve, causing considerable damage. Herds increased from 150 to 500 individuals between 1931 and 1940 over an area of 1200 km². Buffalo were similarly affected by the new environmental conditions that developed in the reserve. The stocks of Buffon's kob and topi have, however, begun to increase since 1950. Bourlière and Verschuren (1960) believed that these changes in population were not related to the cessation of scrub fires, but were simply more or less regular fluctuations in numbers, about a mean value, determined by the limiting capacity of the open savannah habitat. This shows how even studies made over several decades are unable to provide a conclusive explanation of these changes in ungulate populations in the Albert Park.

THE HISTORY OF ECOLOGY

It is fairly obvious that ecological ideas are often used without this being realised. The angler, for example, knows that trout are to be taken from fast flowing streams and other well-oxygenated waters, while coarse fish, such as the roach and carp, are found in slow flowing rivers.

Primitive man already had a basic knowledge of the ecological requirements of many living organisms as the result of his hunting for game and edible plants and his search for shelter. Some understanding of ecological principles is to be found in the works of scholars from the Middle Ages and even earlier. Darwin can be regarded as an early ecologist as a result of his theories on the evolution of species described in his book *The Origin of Species* (1859), and of his work on earthworms (*The formation of vegetable mould*, 1881). The ideas of Malthus (*An essay on the principle of population*, 1798), that numbers of organisms increase by geometrical progression while food supplies only increase in an arithmetical progression, had a significant effect on Darwin and formed the basis for his ideas on the 'struggle for existence'.

The number of ecological works, especially in the botanical field, increased after about 1850. Humboldt, de Candolle, Engler and Gray laid the foundations of plant geography. Forbes, as a result of his studies on the fauna of the Aegean Sea (1844) showed that each of the different vertical zones in the sea had its own characteristic species, and thus provided evidence of the dynamic relationship between organism and environment. Haeckel introduced the term ecology in 1866 and Mobius the term biocenose in 1877 as the result of his studies on oyster beds. Published work, mainly of an analytical nature, tended to develop somewhat haphazardly until about 1930. During this period, studies were made of the responses of animals to various environmental factors, the limiting values of these factors, faunal succession and synecology. Authors during this period included names like Shelford, Adams, Davenport and Chapman. The Swiss Forel laid the foundations of limnology with his studies on Lake Leman which appeared between 1892 and 1904. Murray and Hjort established oceanography on a sound footing. The work of Uvarov expanded ideas relating to the effects of physical environmental factors. Population dynamics first appeared with the mainly theoretical mathematical studies of Volterra, Gause and Lotka. Active discussion in France during this period resulted in a book by Prenant.

The development of modern ecology dates from about 1930 and there have been considerable developments in the subject in a number of countries. It must be realised, however, that progress in ecology is about fifty years behind that of the laboratory-based subjects like embryology and genetics, which developed more rapidly because of their greater popularity with biologists. There are several reasons for this lack of interest.

The first seems to be a desire to generalise and describe laws applicable to living organisms as a whole. The capacity for generalization from observations made on a single species or several species undoubtedly explains the success of genetics, embryology, cytology and molecular biology. Ecology, on the other hand, is still largely at the analytical stage. It is not possible, when studying interactions between different organisms and between these organisms and their environment, to ignore the great diversity of the plant and animal kingdoms; and comprehensive laws, where they do exist, can only begin to be formulated when sufficient data has accumulated to make a tentative synthesis possible. There is one noteworthy exception to this statement. Investigations into the more important biogeochemical cycles were all completed during the second half of the nineteenth century as a result of the progress made in chemistry. In this case it was possible to ignore specific examples, at least initially. There is little doubt that the study of these cycles was stimulated by agricultural interests.

The second reason for the delay in the development of ecology is undoubtedly due to the influence of the positivism of Auguste Comte. For a long time his ideas led philosophers to regard natural phenomena as if they were isolated and independent from one another. The third reason was the lack of direction which ecology suffered until the 1930's. Ecology appeared to have hardly any practical applications in comparison with the laboratory-based aspects of biology. However, in the nineteenth century, especially at the turn of the century, and even up to the present time, 'ecological blunders' have had unforeseen and often catastrophic consequences. In the U.S.A., for example, the continuous cropping of cotton in Florida without any attempt at rotation, has resulted in the deterioration and erosion of the soil because the interactions between soil and plant have been ignored. The widespread use of D.D.T. has often resulted in an increase in numbers of the very insect pests that it was intended to control, due largely to the destruction of the natural balance by the killing of beneficial insects which proved to be more sensitive to insecticides than the pest species. There are many examples of this kind of error and others will be cited in later chapters. They have one useful effect: they serve to emphasise the urgent need for ecological studies to be carried out before man interferes with nature. Ecological mishaps are not restricted to the present day: they can be traced back through history. The transformation of the fertile valleys of the Tigris and Euphrates into desert was due to erosion and to the accumulation of salts caused by poorly designed irrigation in ancient Mesopotamia. Intensive cultivation of the potentially unstable soils of tropical rain forests was at least partly responsible for the decline of the Maya civilisation in Central America. However, even if the consequences of these errors appear to have been relatively unimportant at a time when the world population was rather smaller, they become increasingly important today when the world population is expanding rapidly and when the methods of interference have a potentially greater effect and are carried out with little planning.

Thus ecology teaches us a lesson in humility, showing that man cannot continue to ignore the laws of nature with impunity without eventually suffering consequences which even technology will be unable to overcome. 'Nature to be commanded must be obeyed' (F. Bacon, *Novum Organum*). Ecological investigations lead one to challenge those anthropocentric ideas which regard nature as existing for man to use or abuse at will and to exploit at his convenience.

THE SUB-DIVISIONS OF ECOLOGY

Three main areas are traditionally distinguished in ecology: autecology, population ecology and synecology. These divisions are rather arbitrary but form a convenient outline for the subject.

Autecology describes the relationship between a single species and its environment. It is mainly concerned with limits of tolerance and preferenda of species for various environmental factors, and it examines the effect of the environment on the morphology, physiology and behaviour of organisms. It does not take interactions with other species into account. Defined in this way, autecology is clearly related to physiology and morphology. However, it has, in addition, characteristics of its own. The temperature preferenda for a species, for example, makes it possible to explain, partly at least, the local and geographical distribution, abundance and activity of that species.

Population ecology is concerned with qualitative and quantitative properties of populations. It describes the changes that take place in the abundance of various species; and attempts to explain their causes and to predict them (population dynamics). Although the genetical aspects (population genetics) and ecological aspects of populations have tended to develop independently from one another in the past, they are now beginning to converge to give a clearer understanding of population ecology.

Synecology analyses the relationships that exist between individuals of different species in a community, and also between them and their environment. The term *biocoenotics* is virtually synonymous with synecology. Synecology can be approached from two different viewpoints:

(i) The static approach (descriptive synecology) involves the description of communities of living organisms found together in a specific habitat. This provides information on the specific composition of these communities, together with the abundance and frequency of the component species.

(ii) The dynamic approach (functional synecology) has two aspects, one being the evolution or development of associations and the examination of factors that determine the succession of communities in a particular habitat. The other aspect of functional ecology examines the transfer of materials and energy between the various parts of an ecosystem, and includes such concepts as food chains, ecological pyramids, productivity and efficiency. This aspect is often called *quantitative synecology*.

Other ways of sub-dividing ecology are based on the nature of the habitat and correspond to the three main divisions of the biosphere, i.e. marine, terrestrial and freshwater ecology. The types of organisms and methods of study are as a rule very different for each of the three environments although, in most instances, the same general principles can be applied.

Much of the autecological research at the present time is directed towards an understanding of the mode of action of environmental factors at the physiological and biochemical levels. There is also considerable interest in the way in which various species adapt to their environment and, conversely, how their responses affect the structure and function of ecosystems. Since this requires a better understanding of the mechanisms of adaptation and evolution, it has led to renewed interest in natural selection, adaptation, behaviour and population genetics. The large number of publications related to chemical ecology is also the result of interest in these fields.

Research in synecology has concentrated on a better understanding of the structure and functioning of ecosystems. This phase, which began with the ideas of Lindeman (1942), has continued under the guidance of E. P. and H. T. Odum since 1955. It emphasises the energetics aspect of ecosystem function and tends to neglect species and individuals, being concerned mainly with biomass and the flow of energy and materials. A considerable amount of data has accumulated from this work, particularly with regard to the productivity of different ecosystems. This research has continued within the framework of the International Biological Programme and then in the project 'Man and Biosphere'. Finally, reference must be made to the increasing part played by mathematics in many aspects of ecology. The use of models is becoming increasingly frequent together with recourse to systems analysis so that tentative predictions may be made about the future development of ecosystems.

Ecology has thus become an extensive area within biological science. It lacks precise limits, overlapping with ethology (science of behaviour), with genetics, with biogeography and many others. The scope of applied ecology is also broad and only a few selected aspects can be included here. Although the science of ecology has now become widely known, at least in name, through its practical applications, it is essential that the importance of fundamental investigations is not overlooked. These have tended to be neglected recently but without them progress is not possible. This Introduction to Ecology is concerned with the fundamentals of ecology.

Chapter 1

ENVIRONMENTAL FACTORS

I THE CONCEPT OF ENVIRONMENTAL FACTOR

Every organism is exposed, in the habitat where it occurs, to the simultaneous action of a wide range of climatic, edaphic, chemical and biotic agents. *Environmental factors* are those components of the environment capable of acting directly on living organisms during at least one stage of their life cycle. This definition excludes components such as altitude and depth. In fact, altitude acts as a result of temperature, exposure to sunlight and atmospheric pressure, and not directly. In a similar way, depth affects aquatic animals through increased pressure and by the reduction of light.

Environmental factors act on living organisms in various ways:

(i) by eliminating some species from regions where climatic and physico-chemical conditions are unsuitable for them and, in this way, determining their geographical distribution;

(ii) by modifying the rate of fecundity and mortality of various species, by acting on stages in development and also by causing dispersal, and so influencing population density;

(iii) by encouraging the development of adaptive features, of quantitative metabolic changes and also of qualitative changes, for example diapause, hibernation, aestivation and photoperiodic reactions.

II THE LAW OF THE MINIMUM, THE CONCEPTS OF LIMITING FACTORS AND ECOLOGICAL VALENCY

The '*Law of the Minimum*' was introduced by Liebig (1840). The growth of plants is limited by any nutrient whose concentration falls below that minimum value necessary for synthesis to take place. For example, boron is an essential element, but it is always scarce in the soil. When it has been exhausted by crops, growth ceases even if other essential elements are present in excess.

Extension of the law of the minimum leads to the idea of *limiting factors*. An environmental factor plays the part of a limiting factor when it is absent or reduced below a critical minimum value, or when it exceeds a maximum tolerable value. In other words, it may be said that it controls the likelihood of success for an organism attempting to colonise a new habitat. Alternatively, under less severe conditions, while an organism can effectively live in a habitat, its overall metabolism may be affected by a limiting factor. Juday (1942) gives an example of two lakes in Wisconsin, relatively rich in calcium (21.2 and 22.4 mg/l respectively), which contain three and five times more plants and two and three times more animals (excepting fish) than two other lakes which are similar but poor in calcium (0.7 and 2.3 mg/l respectively).

The phosphate concentration in sea water acts as a limiting factor and regulates the abundance of plankton and, thus, productivity (fig. 1.1).

The idea of the limiting factor can be applied, not only to those nutrients essential for the survival of living organisms as in Liebig's conception, but also to all environmental factors, and equally for both lower and upper limits. Thus each living organism shows *limits of tolerance* with respect to different environmental factors, and the ecological optimum for each organism lies within these limits (fig. 1.2). This idea was expressed by Shelford (1911) in his 'Law of Tolerance'. He found that the reproductive period in tiger beetles was that when their requirements were the most exacting. For oviposition and survival of young larvae, it was necessary that eggs were laid in a well-drained, sandy soil containing little humus, and where temperature and moisture content lay between fairly narrow limits. In addition, eggs were generally laid in the shelter of small stones where light was reduced. The female beetle searched actively for a suitable site and, if this was not available, no eggs were laid, or young larvae did not survive.

By analogy to chemical valency, the concept of ecological valency may be introduced. The *ecological valency* (relative degree of tolerance) of a species is the ability of a species to live in different habitats which are characterised by a wide range of environmental factors. A species with low valency can only tolerate limited variation in environmental factors, and is called *stenoecious*. A species capable of tolerating a wide range of habitats and conditions is called *euryecious*. When these ideas are applied to factors such as temperature, salinity and nutrition, these species are stenothermal or eurythermal, stenohaline or euryhaline and stenophagic or euryphagic. In a habitat where one environmental factor shows wide variations, the fauna is generally poor in those species having low ecological valency. For example, in estuarine waters, which are characterised by wide variations in salinity, only euryhaline species occur in large numbers. Stenohaline species originating from either the sea or fresh water are unable to survive.

The ecological valency of a species may vary with the stage in its development. This has been shown above for the tiger beetles studied by Shelford. In the marine environment, the prosobranch snail *Littorina neritoides* spends its adult life at high water mark and survives long periods of exposure each day, but the larva is planktonic and leads a strictly marine life. Another good example is that of the Syncarida, crustacea living in underground waters at low constant temperatures. In culture, the species *Parabathynella fagei* can live at temperatures

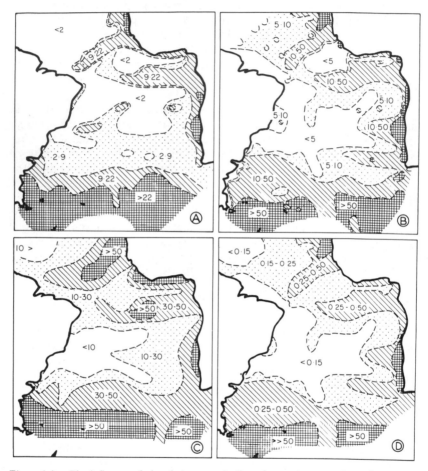

Figure 1.1   The influence of phosphate concentration of sea water on numbers of planktonic animals and productivity in the Atlantic Ocean.

A: phosphate concentration (mg/m³), mean for upper 50 metres.

B: abundance of plankton, in thousands of individuals per litre of sea water from the upper zone.

C: number of planktonic metazoa in 4 litres of sea water from the upper zone.

D: gross primary productivity in summer (grammes of carbon/m² /day).

Note the close correlation between all four factors (Steemann Nielsen and Aabye Jensen, 1957).

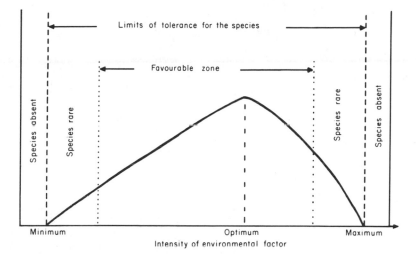

Figure 1.2   Limits of tolerance of a species in relation to the intensity of a particular environmental factor.

ranging from 6 to 32°C, and species of the genus *Bathynella* are also completely eurythermal. Under natural conditions, however, syncarids appear to be stenothermal in cold water, since, when the temperature exceeds 13°C, the single egg laid by the female degenerates and the species is unable to reproduce. Syncarids are thus eliminated from waters where the temperature exceeds 13° at the time of reproduction, and are confined to underground waters (Delamare Deboutteville, 1960).

Ecological valency can be used as a direct indication of the potential of a species to increase its geographical distribution. It is often observed that *eurytopic* species, i,e. those which are widely distributed, are also those with high ecological valency. Conversely, *stenotopic* species, i.e. those with a limited distribution, are often stenoecious. When, however, a species is restricted to a specialised type of habitat, which is not large but is widely distributed, the animal may be eurytopic yet at the same time stenoecious. The phyllopod crustacean *Artemia salina*, which occurs only in hyperhaline waters (habitats generally restricted in size but with world wide distribution), belongs to this group. A ubiquitous species will be eurytopic and euryecious, the house-fly being an example.

The idea of ecological valency cannot alone explain the distribution of living organisms. Allowance must be made for other causes such as geological history (i.e. paleogeography), the possibilities of active or passive movements, the capacity for increase, etc. Furthermore, the limits of tolerance for the same species may vary within its geographical distribution. This may result in physiological adaptations, the limits of

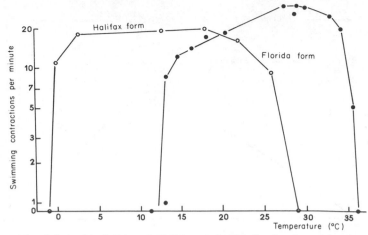

Figure 1.3   Relationship between temperature and rate of swimming contractions in the jelly-fish, *Aurelia aurita* (Bullock, 1955).

tolerance often being determined by a genetic mechanism in local forms, known as ecological races or *ecotypes*. If the genetic basis for this adaptive mechanism is ignored, then the term *physiological race* is used. Many examples have been described. The jelly-fish *Aurelia aurita* was studied by Mayer (*in* Bullock, 1955). Individuals collected in an area off Halifax, Nova Scotia, in water at a temperature of 14°C, showed a maximum rate of swimming contractions at this temperature. Other individuals collected in the vicinity of Florida have a maximum rate at about 29°C which is the water temperature in this region (fig. 1.3). Another example is that of the polychaete *Nereis diversicolor*. Some forms live in lagoons on the shores of Romania which have been isolated from the Black Sea for at least sixty years, and they have become acclimatised to a salinity of 62‰. If these are transplanted to the Black Sea, they die. Similarly, if Black Sea individuals are put into water from the lagoons, they also die. These have been described as physiological races, as no morphological difference exists between individuals from the two localities.

Tolerance to dilution varies in the crustacean *Palaemon squilla* with geographical location, and several physiological races can be detected. The concentration of salt that can be tolerated is related to the area of origin, as the following figures show (salinity decreases by regular intervals from Naples to the lake of Mangalia, Romania):

Naples	29 to 40 g/l
Bosphorus	26 to 36 g/l
Varna	20 to 31 g/l
Agigea	14 to 25 g/l
Sulina	8 to 19 g/l
Odessa	2 to 13 g/l
Mangalia (fresh water lake)	0 to 7 g/l

Living organisms are not always found, in nature, living under those optimum conditions that have been determined in the laboratory. For example, the American shad lays its eggs in fresh, clear water at a temperature of 12°C although its eggs develop best at a temperature of 17°C, in cloudy brackish water with a salinity of 7.5‰. As a result, there is a high mortality under natural conditions, which does not, however, prevent this species from surviving. Labeyrie (1960), while investigating the distribution of eggs between identical hosts, showed that, for the ichneumonid *Diadromus* sp., a parasite of the pear moth *Acrolepia assectella*, there is a modification in searching behaviour after the discovery of the first host. Eggs are not laid at random, but there is a tendency to concentrate eggs in the first host chosen. The distribution of *Diadromus* eggs is determined by two partly opposing influences: the tendency to return to the first host, and the choice between hosts with different characteristics. In other words, females do not lay their eggs in the most favourable conditions for the survival of their progeny, but they tend to concentrate them in the same host. Numerous other examples cited by Labeyrie show that this tendency is fairly general.

These examples show that care must be taken not to fall into the ultimate error of believing that all things are perfect in nature and that, in all instances, living organisms are found living under those environmental conditions that are most suitable for them. Bodenheimer (1938) observed that 'it is not correct to believe that each animal is always guided by its sense organs in the search for optimum conditions'.

1 INTERACTION BETWEEN DIFFERENT FACTORS

It is also necessary to point out that interactions between two or more environmental factors may modify the limits of tolerance with respect to those factors. Went (1957) has shown that certain tropical orchids develop better in full sunlight than in shade, when they are kept at a slightly lower temperature. Under natural conditions they develop only in shade, as they cannot tolerate the over-heating caused by direct sunlight.

Another example of the way that interaction between environmental factors modifies the limits of tolerance is found in the grasshopper *Podisma pedestris*. This typically boreo-alpine species is widely distributed in the mountains of Europe, the extreme north of Europe and Siberia. It is stenothermal, but its stenothermy, which varies with humidity, is far more pronounced in a humid climate than in a dry climate, as figure 1.4 shows. As a result, *Podisma pedestris* is more widely distributed in the drier regions of the southern Alps than in the more humid northern Alps (Dreux, 1962).

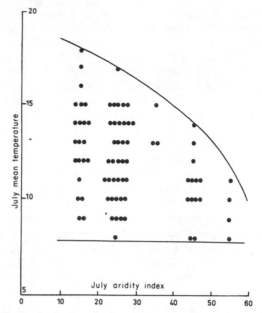

Figure 1.4   Distribution in the French Alps of the grasshopper *Podisma pedestris* in relation to the July mean temperature and the aridity index, *i*.

$$\left(i = \frac{12p}{t + 10} \text{, where } p \text{ is the rainfall for the month in mm, and } t \text{ is the mean temperature.} \right).$$

Each point represents a sample containing *Podisma pedestris* (Dreux, 1962).

2   THE SEARCH FOR LIMITING FACTORS

The ecologist is not content to set up a long list of possible environmental factors. It is necessary, on the contrary, to discover and analyse those factors that act directly on individuals, populations and communities.

There are some simple rules that make it possible, in some instances, to identify limiting factors. If a species has narrow limits of tolerance for an environmental factor which shows wide variation in the habitat studied, that factor is probably a limiting factor. On the other hand, if a species being studied has wide limits of tolerance for a factor that is relatively stable, the latter cannot be a limiting factor. Thus oxygen is unlikely to be a limiting factor in a terrestrial habitat (except for endogenous species and for those that live at high altitudes), but in an aquatic habitat it becomes a limiting factor as some waters contain little oxygen. The ecologist studying an aquatic environment ought, therefore, to measure oxygen tension and its effect on the various species present, while the terrestrial ecologist can nearly always ignore it.

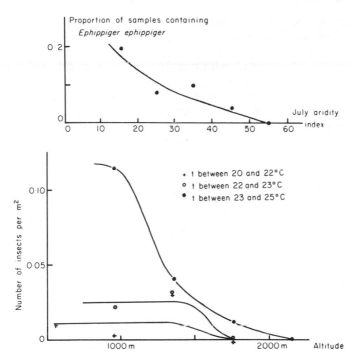

Figure 1.5   Habitat choice by the grasshopper *Ephippiger ephippiger.*
Above: relationship between frequency of this species and July aridity index.
Below: density of insect in relation to altitude for three different climatic regions (mean July temperatures corrected to sea level) (Dreux, 1962).

One practical method of finding out about limiting factors is to study species at the limits of their area of distribution. Dreux, for example, showed that the grasshopper, *Ephippiger ephippiger,* common in France, is found mainly in areas with a low index of aridity, and that it is absent when the July index of aridity is greater than fifty. It is therefore a xerophilous species. Furthermore, its disappearance at altitude shows that it is not able to live at a mean July temperature less than 13°C (fig. 1.5).

Another example is taken from the work of Kalela (1949). In Finland, this author noticed that there had been a slight increase in mean temperature since 1880 and at the same time an extension northwards of the area of distribution of some species of birds such as the lapwing *Vanellus vanellus.* This suggests that temperature is a limiting factor in the distribution of the lapwing, as other environmental factors did not change appreciably over the period studied.

15

Knowledge of limiting factors is of great practical importance in applied ecology, especially in the control of crop pests. Appropriate cultivations make it possible to modify the environmental characteristics and so to interfere with the life cycle of pest species. In the U.S.A., for example, the wireworm *Limonius* is limited by a clearly-defined level of soil moisture, the larval instars being the most sensitive. This insect can be controlled either by flooding irrigated fields, or by drying out non-irrigated fields using crops with a high affinity for water, e.g. maize or lucerne, and these measures result in the destruction of the larvae.

III   CLASSIFICATION OF ENVIRONMENTAL FACTORS

1   BIOTIC AND ABIOTIC FACTORS

It is traditional in ecology to distinguish between *abiotic factors* and *biotic factors*. The former include climatic factors, soil factors and water chemistry, and the latter include, among others, predation, competition and parasites. This classification is, however, rather arbitrary and it has no advantage other than its simplicity. It is sometimes difficult to decide to which category a factor belongs. For example, temperature, considered an abiotic factor, is often modified by the presence of living organisms. Michal (1931) has shown that, in a culture of *Tenebrio molitor,* the animals tend to collect in a group, which maintains a higher temperature nearer to the optimum for development, when the ambient temperature becomes too low. At an air temperature of 17°C the temperature in a group of larvae can reach 27°C. Reaumur (1734–42), through his work on honey bees, appears to be the first to have shown the effects of living organisms on the temperature of their habitat. When the ambient temperature fell to about 13°C, the bees became agitated and raised the temperature to 25–30°C. The microclimate of the hive is therefore much more stable than that of the external environment.

Leaves exposed to the sun at an air temperature of 24°C have a temperature which may be up to 9°C higher, while those in the shade may be 4°C lower than the ambient temperature. There is similar variation in the relative humidity of the air under these conditions. At a distance of 0.5 mm from a leaf, the relative humidity may be 93%, while it falls to only 52% at a distance of 16 mm from the surface. Mining larvae living deep in the parenchyma of leaves are exposed to a particular microclimate which is largely determined by biotic environmental factors.

A detailed study of changes in microclimate produced by populations of *Tribolium castaneum* and *T. confusum* was made by

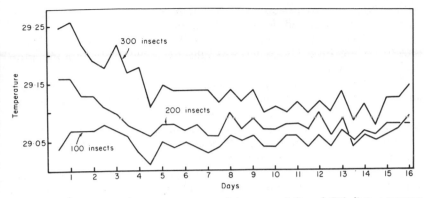

Figure 1.6   The increase in temperature caused by a population of *Tribolium castaneum* (Pimentel, 1958).

Pimentel (1958). The insects were raised in flour contained in Petri dishes kept in a water bath in darkness at a temperature of $29 \pm 0.5°C$ and a relative humidity of 70 ± 5 per cent. Each dish contained 8 grammes of flour and between 100 and 300 insects (their biomass being about 0.7 grammes for 300 individuals). Pimentel found a temperature increase of about $0.4°C$ after twenty-four hours under these conditions. This was followed by a gradual fall in temperature. The relative humidity increased by 11% over a period of a fortnight. This increase in temperature and relative humidity is related to population density and shows that the microclimate is determined by both abiotic and biotic factors, in this case the biotic factors being the insects themselves (fig 1.6). In silos where large quantities of grain are stored, the presence of grain-eating insects can raise the temperature by $25°C$ above ambient (Pimentel, *op cit.*)

The total elimination of life from a particular habitat may result in modifications of climate. At Copperhill, Tennessee, a luxuriant forest was killed by toxic fumes from factories processing copper ore, and the area was converted to a desert. Temperatures are now higher and rainfall lower in the denuded area as compared with neighbouring forests (Odum, 1971).

2   DENSITY-DEPENDENT AND DENSITY-INDEPENDENT FACTORS

Smith (1935), after re-examining previous work, especially that of Howard and Fiske (1911), and Thompson (1929), adopted a classification which used density-independent and density-dependent factors. The former act on a population by causing the death of a constant proportion of individuals, irrespective of the number present, while the

17

latter kill a proportion which increases as density increases. Allee (1941) subdivided density-dependent factors into those which are directly dependent and increase mortality as density increases, and those that are inversely dependent and lower mortality as population density increases.

According to these classifications, density-independent factors are mainly climatic, a cold spell killing a percentage of individuals in a population which is not related to density. Density-dependent factors are mainly biotic factors, the effect of competition, predation and parasitism being related to population density. This distinction between density-dependent and density-independent factors has been criticised by many authors. The controversy still continues and will be discussed later, in the section on population dynamics.

3. MONCHADSKII'S CLASSIFICATION

The views of Monchadskii (1958, 1961) are discussed at length here because of their originality. According to this Russian ecologist, a rational classification of environmental factors should consider details of the way living organisms exposed to these factors react, and, as far as possible, should consider the degree of adaptation, which gradually improves over a period of time. This recalls the ideas of Prenant, who stated that 'the basic idea behind ecology is that of adaptation, that is to say, of the relationship between the organism and its environment'.

1.1 Primary periodic factors

According to Monchadskii, organisms adapt, in the first instance, to those environmental factors that show a regular periodicity as a direct consequence of the earth's regular rotation about its axis and its revolution around the sun, or of the succession of the moon's phases. These factors are primary periodic factors. The regular cycles that bring about variations in these factors existed well before the appearance of life on earth and this explains why adaptations of living organisms to primary periodic factors may be so longstanding and so well established in their genetic make-up.

Temperature, light and *tides* are primary periodic factors. Reactions to seasonal variations of light are shown, for example, by all photo-periodic responses. The major terrestrial climatic zones are the result of primary periodic factors, and they also tend to play a large part in determining the geographical distribution of species. According to Monchadskii, variation in primary periodic factors can only control numbers of individuals by limiting their areas of distribution. Their action, if any, within these areas is never an important one. In general, adaptive responses to primary periodic factors are similar for all animal groups and do not show any specificity. For example, the mathematical laws describing the effect of temperature on vital processes are similar for groups as widely different as insects and vertebrates.

Primary periodic factors thus play an important role except in some specialised habitats like the abyssal region and in caves, where they show little or no variation. Allowance must be made for these factors, especially in experimental studies in ecology. Results obtained for animals kept under constant environmental conditions, as far as temperature and light are concerned, would be different from those obtained for animals exposed to conditions to which they are accustomed. Fecundity and longevity in females of *Diadromus varicolor* (Hymenoptera) show significant variations between individuals reared in a stable environment and those reared under cyclical variations in temperature and humidity. Pertunnen (1960) showed the existence of seasonal variations in phototropism in the scolytid *Blastophagus piniperda*.

The deep-rooted adaptations of organisms with regard to primary periodic factors can explain the harmful effects of constant temperatures. Shelford (1929) and Park (1930), among others, have shown the importance of temperature fluctuations to the growth and vitality of animals. 'The stimulating action of temperature variations, at least in temperate regions, may be regarded as a fundamental ecological principle, a principle that is all the more important since there is nearly always a tendency to carry out laboratory studies at constant temperatures' (Vibert and Lagler, 1961).

1.2   Secondary periodic factors

Variations of secondary periodic factors are the consequence of variations in primary periodic factors. The periodicity of the secondary periodic factor becomes more regular as it becomes more closely related to the primary periodic factor. Atmospheric humidity, for example, is a secondary factor that is closely related to temperature. In tropical regions or in monsoon climates, the rainfall follows a daily or seasonal periodicity. Another secondary periodic factor is the availability of plant food whose periodicity is related to the plant growth cycle. For predators and parasites, the adaptation is made in response to seasonal changes in the biology and physiology of the prey or host. In the aquatic environment, oxygen tension, dissolved salts, turbidity, horizontal and vertical water movements, variations in level and current speed are most frequent secondary periodic factors. Finally, intraspecific biotic effects are also secondary periodic factors, since interactions between individuals are related to annual cycles.

Compared with primary periodic factors, secondary factors are not so long established. Living organisms have had less time to adapt themselves, and their adaptations are less precise, but at the same time, more diverse in different animal groups. Relative humidity, for example, becomes an environmental factor for living organisms when they become adapted to terrestrial life. The corresponding adaptations are less well marked than adaptation to temperature, which is a primary periodic factor. The limits of tolerance to humidity changes are often narrower than corresponding limits for temperature, and at the same time the adaptive responses are more varied. Feeding adaptations also show considerable variation and are often restricted, as, for example, in oligophagous and monophagous animals.

As a general rule, secondary periodic factors affect the abundance of animals within the area of distribution, but have little influence on the extension of that area.

1.3 Non-periodic factors

These are factors which do not normally occur in the habitat of an organism, and which appear unexpectedly. Living organisms generally, therefore, have little time to adapt themselves to these factors because of the random nature of the latter.

Some climatic factors, e.g. wind, storms and fires, belong to this category. It is also necessary to include in this group the various types of human activity and the action of predators, parasites and pathogens, i.e. biotic factors with the exception of intraspecific interactions. The effect of a host on a parasite appears as a secondary periodic factor since the habitat formed by the host is the normal habitat of the parasite. On the other hand, the parasite (or pathogen) is not indispensable to the host. Thus it is non-periodic and generally produces no adaptation, except in special cases (acquired immunity for example) which are relatively rare in comparison with the large number of parasitic and pathogenic organisms.

This frequent lack of adaptation to non-periodic factors provides a foundation for control methods against animal pests, using pesticides or biological control. It is only when insecticidal treatments are repeated on many generations of insects that resistant races appear, and by then the insecticidal treatment has become a secondary periodic factor.

Non-periodic factors act mainly by regulating the abundance of animals within a particular area. They do not generally affect either the area of distribution or the life cycle of a species.

This classification of environmental factors by Monchadskii is allied to another classification of habitats which distinguishes: (i) *stable habitats*, where environmental factors are relatively constant, for example the abyssal region, caves and the habitats of parasites of homeotherms; (ii) *periodic habitats*, where conditions vary daily, seasonally or tidally, as is the case in the intertidal zone and in most temperate regions; (iii) *irregular habitats*, characteristic of zones subjected to unpredictable sporadic changes, such as volcanic regions and areas subject to burning

TABLE 1.1 Summary of the different classifications of environmental factors, showing how the different groupings are related to one another

CLASSIFICATION USED IN THIS BOOK	MONCHADSKII'S ENVIRONMENTAL FACTORS		
A. Climatic Factors		Abiotic factors	Density-independent factors
Temperature *Light*	Primary periodic factors		
Relative humidity *Rainfall*	Secondary periodic factors		
Other factors			
B. Physical Non-climatic Factors *Aquatic environmental factors*	Secondary periodic factors or non-periodic factors		
Edaphic factors	Non-periodic factors	Biotic factors	Density-dependent factors
C. Food as a Factor	In general, secondary periodic factors		
D. Biotic Factors *Intraspecific interactions*			
Interspecific interactions	Non-periodic factors		

or tornados; (iv) *sequential habitats*, where changes in environment and organisms occur, following a succession which is non-cyclic in origin. Sedimentation in deltas, and soil formation are examples. The periodic factors of Monchadskii correspond to the periodic habitats and non-periodic factors to the irregular and sequential habitats.

The work of Monchadskii presents new and interesting viewpoints. Its use, however, in a text-book would lead, with our present knowledge, to some repetition. In the following discussion of the effects of environmental factors on animals, the plan which is summarised above in table 1.1 is followed. This table compares some classifications of environmental factors.

Chapter 2

ABIOTIC FACTORS: CLIMATIC FACTORS

I INTRODUCTION

The characteristics of the climate to which a living organism is exposed depend on the size of the individual. This can be illustrated by the insects and large mammals found together on the African savanna. The environmental conditions for a giraffe living two metres above the ground will be very different from those of an ant living amongst the grass. The following types of climate can be distinguished: macroclimate, mesoclimate and microclimate.

The *macroclimate* (regional climate of Martonne, Grossklima of Geiger) depends on geographical and orographical location. An example would be the climate of the Paris region. Local modifications of macroclimate are *mesoclimates* (local climate of Martonne, Kleinklima of Geiger). The climate of a wood or of an embankment provide examples of mesoclimates.

Macroclimates and mesoclimates are described from measurements made with instruments installed under standard conditions, i.e. in a screen situated two metres above level ground which is covered by short grass and in an unobstructed position in order to avoid any disturbance of the climate.

The *microclimate* (ecoclimate of Uvarov) corresponds to the climate in the immediate vicinity of an organism, and its study shows how much more important it is to the organism than either macroclimate or mesoclimate. Measurements of microclimate can only be made with specially designed instruments which are often complex and difficult to use. Some aspects of microclimates are poorly understood and their role is not clear owing to the lack of suitable measuring instruments.

Macroclimates, mesoclimates and microclimates are discussed in this chapter. The distribution of the major types of vegetation over the surface of the globe is determined by macroclimate, while the distribution of xylophagous buprestid beetle larvae under the bark of a tree trunk is determined by the microclimate found underneath that bark.

II MAJOR CLIMATIC FACTORS ON A GLOBAL SCALE

1 RADIATION RECEIVED BY THE EARTH

With the exception of small quantities of energy supplied by the earth's internal heat, virtually all energy received by the earth is of solar origin. The wave length of the solar radiation received by the earth varies from less than one thousandth of a nanometre (1 nm = 10^{-9} m) to several thousand metres. Different types of radiation are able to penetrate the atmosphere to varying extents. Under normal conditions, the earth's surface receives only visible radiation of between 390 and 770 nm (about fifty per cent of the earth's energy requirements), a small amount of ultra-violet radiation (the atmospheric ozone layer at an altitude of about 25 000 metres absorbs radiation of a wave length less than 295 nm) and some infra-red radiation (to about 2 400 nm), as well as radio waves with wavelength greater than 10^{-4} m. From an ecological point of view, only the effects of radiation in the infra-red, visible and ultra-violet regions have been studied, and the effect of very short wave radiation in the upper atmosphere (cosmic rays) is still not clear although it may have mutagenic effects.

The amount of solar energy reaching the atmosphere appears to be almost constant, between 8.30 and 8.38 joules (J) per cm^2 per minute. This is equivalent to 2.1×10^{18} MJ per year for the entire earth's surface, and this quantity is known as the *solar constant*. This constant probably varies and, in recent years, an increase of the order of two per cent· has been observed in the sun's intensity. Maximum intensity coincides with an outburst of solar activity which occurs every eleventh year. It is possible to explain climatic changes during the Quaternary in terms of cyclical variations of this type, although of longer duration.

A proportion of the energy received is reflected back into the universe by cloud and is lost to the earth. Another fraction, in the infra-red region (about twenty per cent), is absorbed by water vapour and serves to warm the air. A large part of the ultra-violet radiation is

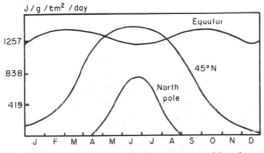

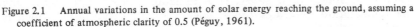

Figure 2.1   Annual variations in the amount of solar energy reaching the ground, assuming a coefficient of atmospheric clarity of 0.5 (Péguy, 1961).

23

absorbed by the ozone layer. The remaining solar radiation reaches the ground, either in the form of direct light or diffused light. This diffusion is brought about both by gas molecules in the atmosphere, causing the blue colouration of the sky, and by solid particles in suspension, causing a whitish or greyish colouration of the sky especially over large towns. The quantity of energy reaching the ground varies with time of day, angle of incidence of the sun's rays and clarity of the atmosphere. The graphs (fig. 2.1) show the quantity of energy reaching the ground at different times of the year and different latitudes, assuming a coefficient of atmospheric clarity of 0.5. The solar radiation received by a unit horizontal surface at mid-day at Potsdam in joules/cm^2/minute is as follows:

	FINE WEATHER	MODERATE CLOUD
June	4.65	2.39
December	1.02	0.25

The quantity of energy received at different latitudes is shown in table 2.1.

TABLE 2.1   Quantities of solar energy (kJ/cm^2) received at different latitudes

	DURING THE 4 SUMMER MONTHS	DURING THE YEAR
Arctic region (80° North)	57.0	70.4
Boreal region (60° North)	128.2	182.7
Cold temperate region (48–52° North)	152.9	230.2
Warm temperate region (39–45° North)	171.8	343.6

The atmosphere at high altitudes is less dense and water vapour is less abundant. As a result the amount of solar radiation reaching the ground increases with altitude, as the following table shows:

Thorenc (Alpes Maritimes)	1 200 m	6.79 joules/cm^2/minute
Davos (Switzerland)	1 600 m	6.66 joules/cm^2/minute
Tacubaya (Mexico)	2 300 m	6.96 joules/cm^2/minute
Tlamacas (Mexico)	3 900 m	7.08 joules/cm^2/minute
Popocatepetl (Mexico)	5 300 m	7.17 joules/cm^2/minute
Griesheim (Germany), balloon	7 500 m	7.21 joules/cm^2/minute
Omaha (U.S.A.), from a rocket	22 000 m	7.46 joules/cm^2/minute

Data for the Neouvieille massif of the Pyrenees are shown in table 2.2.

More solar radiation is absorbed by water than by air, as is demonstrated in table 2.3.

Figure 2.2 shows light penetration in several natural water bodies for yellow light of wave length 500–600 nm.

The radiation received at ground level comprises heat radiation (infra-red radiation together with a small part of the visible spectrum), radiations showing chemical activity (ultra-violet rays) and visible radiation. These sub-divisions are not clearly separated and are simply made to facilitate their study. Heat radiation, for example, by raising the temperature, increases the efficiency of photosynthesis, which is itself controlled by visible radiation.

TABLE 2.2   Solar radiation received at two places in the Neouvieille massif of the Pyrenees.

	SOULANE* AT FABIAN (1100 m)		AUMAR PLATEAU (2200 m)	
Radiation received on a unit horizontal surface during a period of 10−15 hours in fine weather (joules/cm²/minute)	mean	6.29	mean	6.91−7.08
	maximum	6.66	maximum	7.21
Index of atmospheric clarity	mean	40−50%	mean	55−65%

*soulane is a local term for a south facing slope

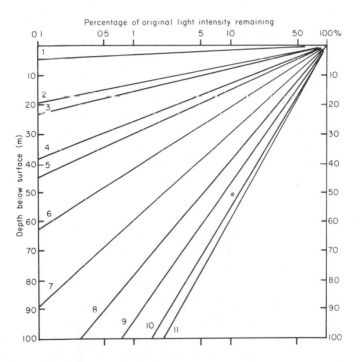

Figure 2.2   Light penetration in several natural water bodies for yellow light of wavelength 500 to 600 nm (Clarke et al., 1939).

1. Midge lake, Wisconsin.
2. Gunflint lake, Minnesota.
3. Woods Hole Bay, Massachusetts.
4. Baltic Sea.
5. Straits of Dover.
6. Gulf of Maine.
7. Sea off Vancouver.
8. Crater lake, Oregon.
9. Gulf Stream.
10. Caribbean Sea.
11. Distilled water.

Table 2.3   Absorption of solar radiation by air and water

	ABSORPTION IN THE VISIBLE REGION AT 550 nm (% OF ORIGINAL VALUE REMAINING)	ABSORPTION IN THE INFRA-RED REGION at 800 nm (% OF ORIGINAL VALUE REMAINING)
Radiation entering the atmosphere	100	100
At Mount Whitney (California) (altitude 4420 metres)	93	97
At sea level	75	88
Two metres below the surface of the water	71	2
30 metres below the surface of the water (very clear oceanic water)	17	–

2   TEMPERATURE

Some idea of temperatures in different parts of the world can be obtained from the maps in Figures 2.3–2.5, which show annual, January and July isotherms. The temperatures are shown at sea-level and correction factors of 0.53°C for the annual mean, 0.40°C for January and 0.61°C for July must be subtracted from these values for every 100 metres increase in altitude. If temperatures were not expressed as sea-level equivalents. the isotherms would simply represent, to a large extent, the relief of the various regions. It can be seen that annual isotherms run almost parallel to the equator, although they show some modification due to the continental land masses. The northern hemisphere tends to be warmer than the southern hemisphere, and the thermal equator is almost entirely situated in the northern hemisphere. A mean annual temperature of over 30°C in the Sahara and central Africa makes this the warmest continent.

In the tropics, the diurnal change in temperature is greater than the annual variation (i.e. the difference between the means for the warmest and coldest months). This phenomenon has important biological consequences. In temperate regions, however, there is a distinct annual temperature cycle. January is the coldest month and July the warmest in the northern hemisphere, and the converse is true for the southern hemisphere. These extremes of temperature also have important biological consequences. In the southern hemisphere, which is largely covered by sea, the isotherms for January and July are almost regular. In the northern hemisphere, however, they are considerably modified by the large continental masses. In January, for example, the 0°C isotherm runs almost north-south in Europe from 46 to 71 degrees of latitude north, and, along the 60th parallel, the mean temperature

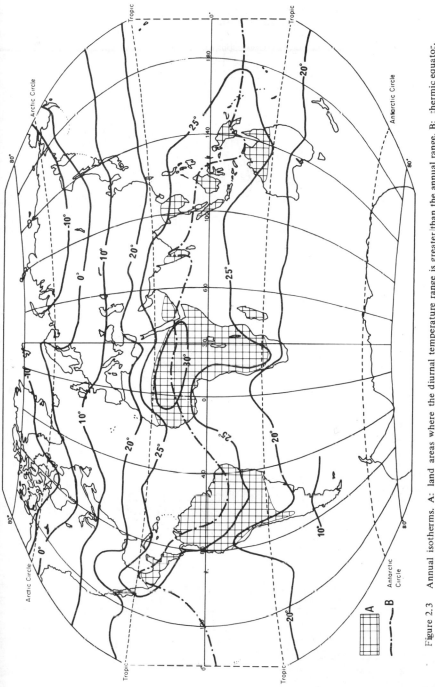

Figure 2.3   Annual isotherms.   A: land areas where the diurnal temperature range is greater than the annual range.   B: thermic equator.

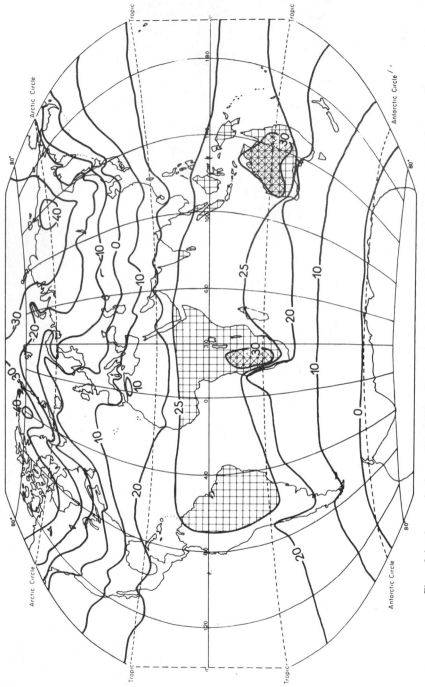

Figure 2.4   January isotherms. The shading corresponds to land areas with temperatures greater than 25°C and 30°C.

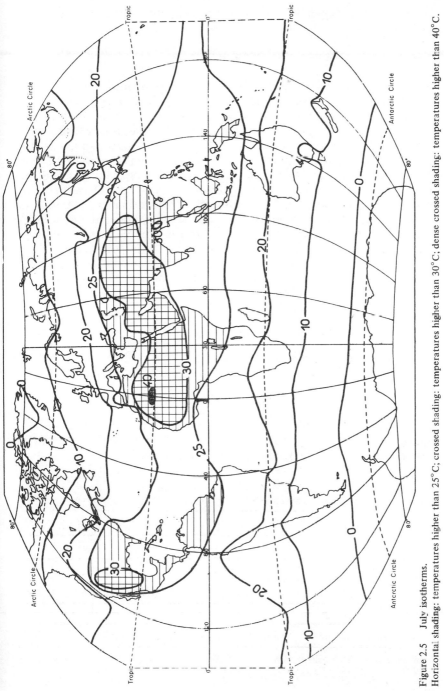

Figure 2.5   July isotherms.
Horizontal shading: temperatures higher than 25°C; crossed shading: temperatures higher than 30°C; dense crossed shading: temperatures higher than 40°C.

varies from +5°C in the south of Norway to −38°C in Siberia, a variation of 43°C.

3 LIGHT

From an ecological point of view, the duration, intensity and wave length of light are important. The last two vary widely with local conditions and will be discussed in the section on mesoclimates and microclimates, only daylength being dealt with under this section.

The angle of inclination of the ecliptic to the earth's polar axis is about 66°33', and this is responsible for the unequal lengths of days and nights. There are two regions, between the polar circles and the poles, where the year is divided into four periods. There are two periods, with the solstices (21st June and 21st December) at their mid points, during which the sun is either permanently above the horizon in summer or permanently below the horizon in winter. During the other two periods, the sun rises and sets each day as in other parts of the earth. The lengths of continuous day or night increase with latitude, and some values for the northern polar day and polar night are shown in the following table (note that the length of day is greater than that of night as a result of atmospheric refraction):

	POLAR DAY	POLAR NIGHT
Latitude 70°	70 days	55 days
Latitude 75°	107 days	93 days
Latitude 80°	137 days	123 days
Latitude 85°	163 days	150 days

Between the two polar circles, daylength in winter increases when moving towards the equator and in summer when moving away from it. Some examples of daylength are given in table 2.4. The tropics are

TABLE 2.4   Daylength (in hours) at several latitudes in the northern hemisphere

LATITUDES	DATES 15 Jan	15 Feb	15 March	15 April	15 May	15 June
60°N	6.3	8.9	11.6	14.5	17.0	18.8
40°N	9.5	10.6	11.9	13.2	14.3	15.0
20°N	11.0	11.4	12.0	12.5	13.0	13.3

LATITUDES	DATES 15 July	15 Aug	15 Sept	15 Oct	15 Nov	15 Dec
60°N	18.0	15.6	12.9	10.2	7.5	5.8
40°N	14.7	13.7	12.4	11.1	9.9	9.3
20°N	13.2	12.8	12.2	11.6	11.2	10.9

parallels of latitude approximately 23°27′ north and south of the equator. They represent the most northerly and southerly positions on the earth's surface where the sun passes overhead at noon. This event takes place during the June solstice in the northern hemisphere and during the December solstice in the southern hemisphere. At the equinoxes (21st March and 21st September) all parts of the earth have equal lengths of day and night.

The angle of incidence of the solar radiation on the surface of the earth is important biologically since the heating effect is proportional to the size of this angle, absorption by the atmosphere being inversely proportional to the angle of incidence.

4 RELATIVE HUMIDITY

The saturation vapour pressure (F) is the maximum partial pressure that water vapour can attain in air, and is proportional to temperature. At any given temperature, the actual vapour pressure (f) is less than, or equal to, the saturation vapour pressure at the same temperature. Relative humidity (e) is the ratio, expressed as a percentage, of the actual water vapour pressure (f) to the saturation vapour pressure (F) at the same temperature

$$e = 100 \frac{f}{F}$$

This definition can be used for negative temperatures, when F becomes the saturation vapour pressure relative to frozen water. The map (Figure 2.6) shows the distribution of relative humidity for January.

The saturation deficit (Δf) is the difference

$$\Delta f = F - f$$

Relative humidity is measured with the aid of a hygrometer (e.g. the Assmann psychrometer) consisting of two similar thermometers, one of which has its bulb covered by a thin film of water and shows a temperature (t') which is less than the air temperature (t). The difference $t - t'$ becomes larger as the air becomes drier, and the following relationship then applies

$$f = F - Kh(t - t')$$

where h is atmospheric pressure and K is a constant equal to 0.000 79 for $t' > 0°C$ and equal to 0.000 69 for $t' < 0°C$. Relative humidity is found from tables as a function of t and $t - t'$.

Another piece of apparatus frequently used to measure relative humidity, the hair hygrometer, is less precise and less reliable.

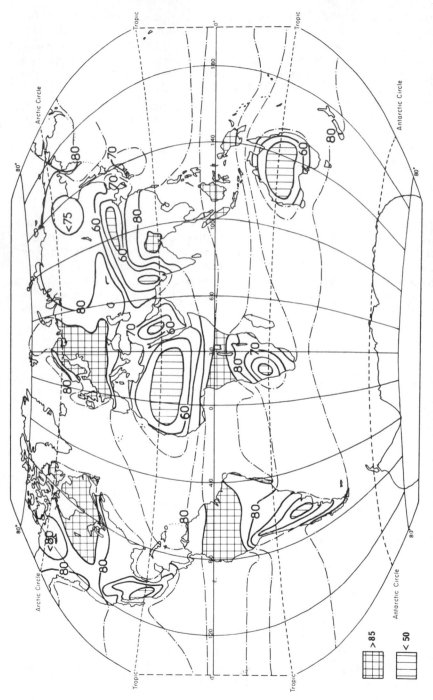

Figure 2.6   World distribution of relative humidity for January.

32

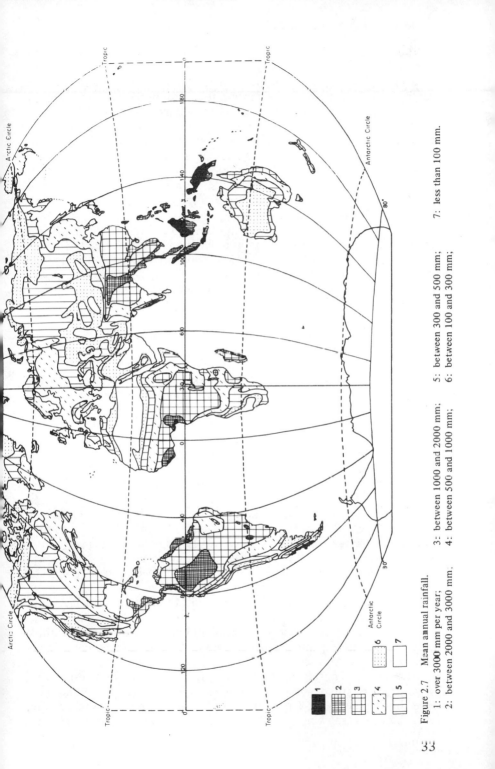

Figure 2.7 Mean annual rainfall.

1: over 3000 mm per year;   3: between 1000 and 2000 mm;   5: between 300 and 500 mm;   7: less than 100 mm.
2: between 2000 and 3000 mm;   4: between 500 and 1000 mm;   6: between 100 and 300 mm;

33

5 RAINFALL

The map (figure 2.7) shows the annual rainfall over the surface of the earth. The tropical zones are the wettest, Indonesia, the Amazon basin and parts of Africa receiving more than two metres of rainfall per year. Very dry regions also occur in the tropics, especially in the Sahara and the north of Chile where the coastal desert at Arica has only received 1.8 mm of water in ten years. In temperate zones rainfall is generally less abundant except in mountainous areas like the Alps, Pyrenees, Scandinavia, Himalayas and Andes. In Asia between the Caspian Sea and eastern China, and in the extreme north of America and of Asia the annual rainfall is less than 250 mm.

It can be seen by comparing figures 2.6 and 2.7 that low rainfall areas correspond to dry regions with a relative humidity of less than fifty per cent. The desert regions, arid, semi-arid and very arid (fig. 2.8), are also dry regions. They are located either in the vicinity of the cold marine currents at the coast, or else at the centre of the large continental land masses (central Asia, Sahara).

6 THE COMBINED ACTION OF SEVERAL FACTORS: CLIMATIC INDICES AND CLIMOGRAPHS

The ecological classification of climates is based largely on the two most important and well known factors: temperature and moisture. A number of climatic indices have been suggested to explain the distribution of vegetation. Some of the more important indices are the following:

(i) Martonne's aridity index (I), which is given by

$$I = \frac{P}{T + 10}$$

where T is the mean annual temperature in °C and P the annual rainfall in mm. Where it is necessary to calculate the index of aridity for a single month, the formula $I = 12\,p/(t + 10)$ is used, p being the rainfall for the month concerned and t the mean temperature for that month. This index becomes smaller as the climate becomes more arid (table 2.5).

The index for July, which corresponds to the season of maximum plant growth and animal activity in the northern hemisphere, has a markedly higher value than the annual index. The July index of aridity can be used to explain the distribution of some insects like *Podisma pedestris* in the Alps (cf. p. 13).

(ii) Gaussen considered that drought conditions were established when the monthly rainfall, P (in mm), was less than twice the mean monthly temperature, T (in degrees Celsius). Graphs can be constructed for a locality showing months on the abscissa and temperature and

34

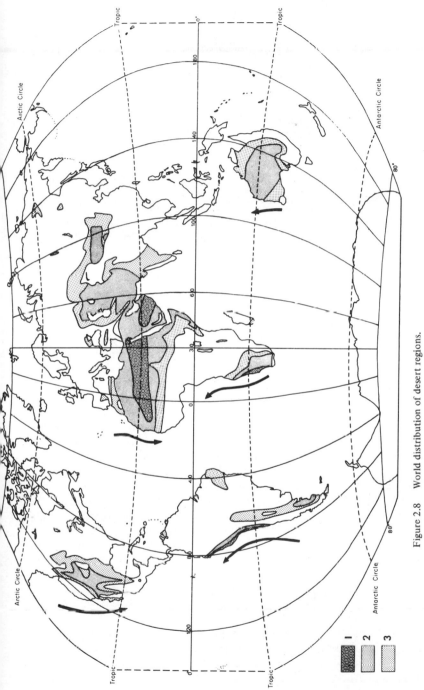

Figure 2.8 World distribution of desert regions.
very arid (1); arid (2); semi-arid (3) regions. The arrows indicate the cold marine currents.

35

TABLE 2.5 Martonne's aridity index
for some locations in France and North
Africa

	P (mm)	T ($^{\circ}$C)	I
Biarritz	1182	14	49
Brest	820	12	38
Paris	560	10	28
Marseille	540	13.5	23
Oran	428	18	15.3
Tamanrasset	20	21	0.7

rainfall on the ordinate, the scale for rainfall being double the temperature scale (fig. 2.9). This type of graph is a pluviothermic diagram, and the dry season is shown stippled. When a series of dry Mediterranean localities lying almost on the same meridian are examined, it can be seen that the length of the dry period increases progressively from north to south:

Valence (Drôme), slightly north of the Mediterranean region	0 dry months
Montélimar, first olive trees appear	1 dry month
Marseille, average Mediterranean climate	3 dry months
Batna, semi-arid Mediterranean climate	4 dry months
El Kantara, steppe climate	7 dry months
Touggourt, desert climate	12 dry months

The *xerothermic index* has been introduced to compare localities having similar numbers of dry months. It allows for the occasional showers that arrive unexpectedly during the dry months, for periods of dew and mist, and for days when the relative humidity exceeds 40%. The xerothermic index is given by the annual number of dry days; for Valence this is seven, and for El Kantara 206. The limits are clearly 0 and 365 and in the north and west of France this index is zero. It varies between 0 and 40 in the non-Mediterranean south of France, 60 and 100 in the Mediterranean region and on the coast of North Africa, 100 and 300 on the Hauts Plateaux region of Morocco, and exceeds 300 in the Sahara. In some tropical areas it may be low (Abidjan: 12; Tahiti: 17) but, where there is a dry season, it can be high (Dakar: 190).

(iii) Emberger suggested a more complicated formula which would allow for annual variations in temperature. His *pluviothermic quotient* (Q) is given by the formula:

$$Q = \frac{(M + m)(M - m)}{100\ P}$$

where P is the annual rainfall in mm, M the mean maximum of the warmest month and m the mean minimum of the coldest month. The different types of Mediterranean climate can be classified using Emberger's method. These areas have well-marked hot, dry seasons and

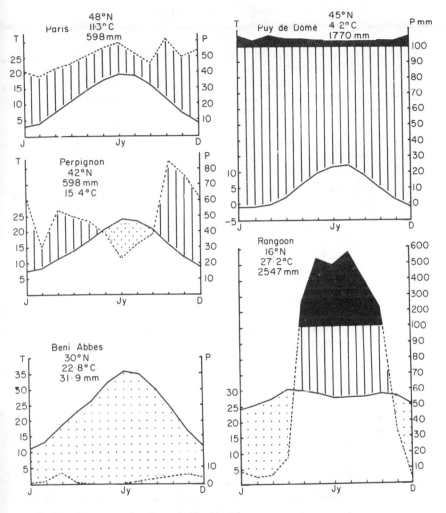

Figure 2.9   Pluviothermic diagram for five localities.
   Left: increase in the length of the dry season along the same meridian (no dry season in Paris, brief in Perpignon and throughout the year at Beni Abbes).
   Right: in a cold, moist alpine climate at Puy de Dome and a tropical, monsoon climate (Rangoon) which consists of alternating wet and dry seasons.
   The numbers by the name of each station show mean annual temperature and rainfall.
   Dotted areas: dry season;
   hatched areas: wet season;
   black: rainfall over 100 mm,
   continuous line: temperature graph;
   broken line: rainfall graph.

37

rainfall is restricted to the cooler part of the year so that the ratio

$$\frac{\text{mean maximum temperature of warmest month}}{\text{summer rainfall}}$$

is less than or equal to 7. A Mediterranean type of climate is also found in other parts of the world, including the Cape territory, the edge of southern Australia, central Chile, and California. In all these areas, groups of plants occur that are similar from the ecological point of view. The distribution of some insects can also be explained by similarities in climate. The culicine flies *Theobaldia longeareolata* and *Culex theileri* occur together both in the Mediterranean region proper and in the biogeographic subregion of the Cape. In France, the limits of the Mediterranean region correspond closely with the limits of the evergreen oak (fig. 2.10).

The classification of world climates proposed by Emberger is based on the rhythms of climatic factors, including rainfall, temperature, humidity and light, and this classification is particularly useful from the ecological point of view since it depends on the variability of environmental factors, the importance of which was emphasised by Monchadskii. Emberger's classification is as follows:

Desert climates: where rainfall is erratic and does not occur every year. Desert climates are found in equatorial and temperate regions.
Non-desert climates: with a regular annual rainfall cycle.
Tropical climates: these are equatorial and sub-equatorial climates and tend to be isothermal. Some include a dry season while others may show some sign of seasonal temperature changes.
Temperate climates: these show daily and seasonal changes in illumination. They include oceanic climates without a dry season, continental climates with a winter dry season, and Mediterranean climates with a summer dry season. Others showing seasonal changes in illumination, include polar and sub-polar climates.

Each of these types of climate can be defined in terms of humidity, altitude and by the mean temperature of the coldest month.

(iv) Thornthwaite introduced a series of climatic indices which are more complex and which, although more realistic, are laborious to calculate and require data, like evapo-transpiration, that are more difficult to obtain.

The *rainfall index* (*i*) of Thornthwaite is calculated from

$$i = 0.1645 \left(\frac{P}{T + 12.2} \right)^{10/9}$$

Thornthwaite also proposed an *index of humidity* $I_h = 100 \, s/n$ and an *index of aridity* $I_a = 100 \, d/n$ in which n is the water requirement of the vegetation, s is the water surplus during the wet period (the water

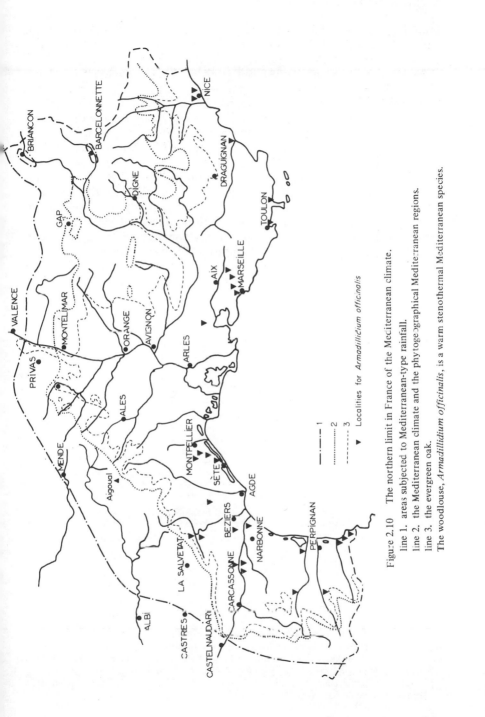

Figure 2.10   The northern limit in France of the Mediterranean climate.

line 1.   areas subjected to Mediterranean-type rainfall.
line 2.   the Mediterranean climate and the phytogeographical Mediterranean regions.
line 3.   the evergreen oak.
The woodlouse, *Armadillidium officinalis*, is a warm stenothermal Mediterranean species.

39

retaining capacity of the soil being arbitrarily set at 100 mm per year), and d is the water deficit during the dry period. These two indices can be related to the type of vegetation. In the U.S.A., for example, along the 41st parallel, dense forest is found in areas where there is a large excess of water during the wet period and only a very small deficit during the dry period. Prairie grasslands occur in areas having only a small deficit during the dry period (tall-grass prairie) or with only a small excess of water during the wet period (short-grass prairie).

Another index of humidity I_h was suggested by Thornthwaite using the formula $I_h = (100s - 60d)/ET_p$ where s and d have been defined above and where the potential evapo-transpiration ET_p is the total amount of water lost to the atmosphere. ET_p is the quantity of water lost in unit time from a given area of ground plus the vegetation covering that ground when provided with an excess of water. ET_p can be measured directly using an evapo-transpirometer (fig. 2.11), or it can be calculated from one of several empirical formulae (Geiger, 1957).

Semi-arid climates have an index between -20 and -40, arid climates less than -40 and hyper-arid climates have an index less than

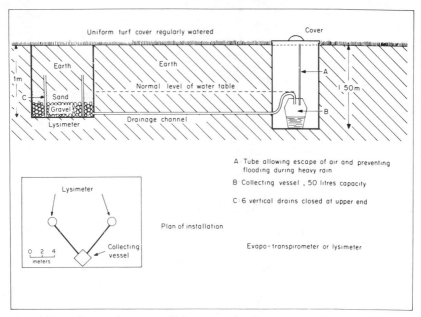

Figure 2.11   Diagram of apparatus (lysimeter) used to measure potential evapo-transpiration. Regular watering of the measuring tank or lysimeter and the surrounding soil with a known quantity of water keeps this soil in the region of field capacity. The weekly rainfall (P) is measured with a rain gauge, the quantity of water (A) applied to the lysimeter is known, together with the amount of water (C) draining into the measuring tank (an air-tight container of 50 litres capacity) from the lysimeter.

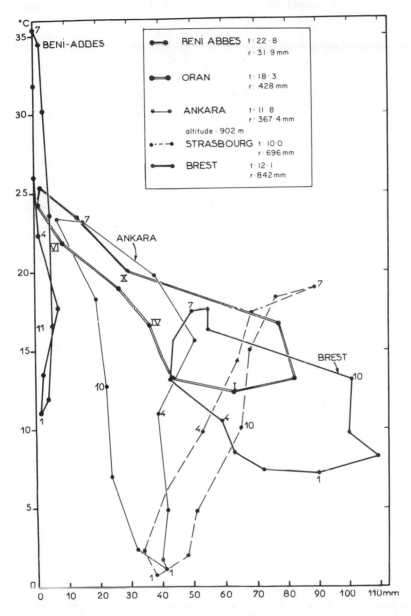

Figure 2.12   Climographs for some characteristic stations. Beni Abbes: desert climate; Oran: semi-arid mediterranean climate; Ankara: cold continental climate; Strasbourg: semi-continental climate; Brest: west maritime climate. The figures (arabic or roman) on the curves relate to months of the year.

−57, with no seasonal rainfall cycle and often more than a year without rain. Some authors describe all arid zones as deserts while others regard only hyper-arid zones as deserts (these are the 'true' deserts of Emberger) (fig. 2.8).

(v) The climograph is a well known method of describing a regional climate. The climates of different localities can be easily compared from their climographs. Temperature is usually plotted on the ordinate and rainfall on the abscissa, although occasionally humidity is substituted for rainfall and evaporation for temperature. Figures 2.12 and 2.13 show climographs for some localities with characteristic climates.

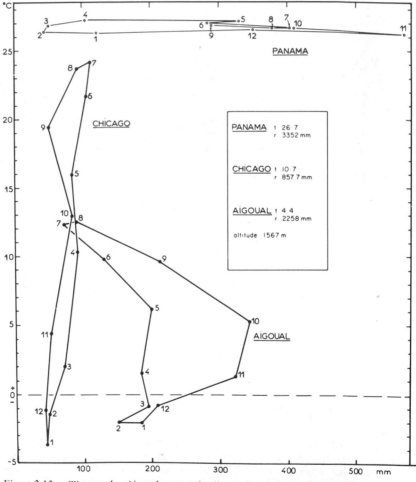

Figure 2.13 Climographs. Aigoual: mountain climate; Panama: wet-tropical climate; Chicago: continental climate with wide temperature variation.

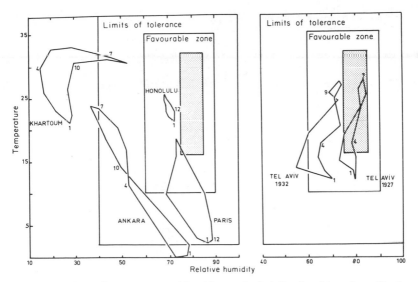

Figure 2.14   Effect of temperature and humidity on the fruit fly, *Ceratitis capitata*. The three rectangles mark out an optimum zone (dots), a favourable zone and a limiting zone for the development of the insect. The climographs superimposed show that conditions at Tel Aviv were favourable throughout 1927; in Paris, six months are too cold although they do not exceed the limits of tolerance; at Ankara the climate is either too cold or too dry for three months; at Khartoum the climate is too dry for nine months; at Honolulu the climate is favourable and, if the fly were accidentally introduced, it could become established (Bodenheimer, 1938).

Climographs can be used to predict the possible spread of a species. Bodenheimer (1938), as a result of his studies on the ecology of the Mediterranean fruit fly *Ceratitis capitata,* was able to predict regions that this pest species might invade, from a comparison of the optimum environmental conditions, range and limits of tolerance for this species with climographs for several localities (fig. 2.14).

III   THE MESOCLIMATES AND MICROCLIMATES OF SELECTED HABITATS

It has already been shown how regional or macroclimates are very different from mesoclimates or microclimates. Some typical terrestrial and aquatic mesoclimates and microclimates are described below.

1   THE MONTANE MESOCLIMATE

Much research has been carried out on the climate of mountainous areas in various parts of the world, especially in western Europe. Two basic types of montane climate have been distinguished: the *xerothere*, having a dry period with a mean monthly rainfall (in mm) less than

43

Table 2.6   Mean values for atmospheric pressure and the partial pressure of oxygen at different altitudes

ALTITUDE	ATMOSPHERIC PRESSURE IN mm OF MERCURY	PARTIAL PRESSURE OF OXYGEN EXPRESSED AS A % OF THAT AT SEA LEVEL
0 m	760	100
1 000 m	674	89
2 000 m	595	78
3 000 m	520	68
4 000 m	468	61
5 000 m	398	52
6 000 m	346	46
7 000 m	280	37
10 000 m	180	24

twice the temperature; and the *hygrothere*, which has no dry period. The former type occurs in the tropics and in North Africa (the Atlas of Morocco), and the latter in the mountains of temperate Europe. Only the hygrothere climate is discussed here.

Atmospheric pressure is the one factor that regulates all others at high altitudes. Table 2.6 shows mean values for atmospheric pressure at different altitudes.

Although the proportions of gases in the air (4/5 nitrogen, 1/5 oxygen) show little change, there is a significant reduction in the partial pressure of oxygen with altitude. Man and homeothermic animals are particularly susceptible to this reduction and, as a result, do not live at such high altitudes as insects and plants (fig. 2.15).

The rarefied atmosphere at high altitudes affects other climatic factors, especially solar radiation, temperature and relative humidity of the air. It has already been shown (p. 24) that the intensity of solar radiation increases slowly with altitude, while the intensity of ultra-violet and infra-red radiation increases rapidly. Ultra-violet radiation is four times greater at the altitude of Briançon than at sea level (cf. the data on page 25). Alpine regions tend to be exposed for longer periods of insolation than those at sea level. A comparison of the degree of insolation on the Pic du Midi de Bigorre (2860 m) with Toulouse during the period July to September showed a mean of 658.5 hours for Toulouse and of 720 hours for the Pic du Midi, ie. nine per cent higher for the latter. These factors, especially the amount of infra-red radiation, explain the unusually high temperatures that are often recorded on the surface of the soil in mountainous areas during periods of bright sunshine.

Temperatures of 42–50°C have been recorded in the shade of vegetation in a meadow at an altitude of 2000 m at 14.00 hours, while temperatures of 60–80°C were recorded at the surface of dark rocks sheltered from the wind.

As a result of solar radiation, the superficial layers of the soil become

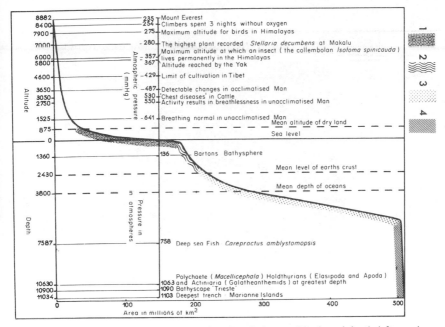

Figure 2.15   Distribution of living organisms in relation to altitude and depth (after various authors).
1: edge of continents; 2: continental shelf; 3: abyssal plain; 4: deep trenches.

rather warmer than the surrounding air. The result is that the mean annual temperature of the soil in mountainous areas is often higher than the air temperature and, at an altitude of 1 800m, this difference may be 2–3°C. The air temperature decreases by about 0.5°C with every one hundred metres (0.56°C in the Alps and 0.44°C in the High Atlas of Morocco). The mean temperature at the summit of the Pic du Midi de Bigorre is +3.0°C in June, +6.2°C in July, +6.5°C in August, +3.1°C in September and −1.0°C in October. An increase in altitude of 1 000 metres is equivalent, as far as temperature is concerned, to a movement of 1000 km northwards. Taking into account the fact that the snow does not melt until almost the end of June or even July at 2500 metres, it can be seen that the period of plant growth and animal activity is reduced to three months. At high altitudes above a certain limit it is no longer possible to distinguish between the seasons. In addition, the difference between the mean temperature of the warmest month and that of the coldest month decreases with altitude, being 19.4°C at 460 m, 17.1°C at 880 m, 14.5°C at 1800 m, 13.8°C at 2500 m and only 2.0°C at 7700 m.

The diurnal temperature variation is much greater in mountainous areas. There is a considerable fall in temperature at night, and, in the

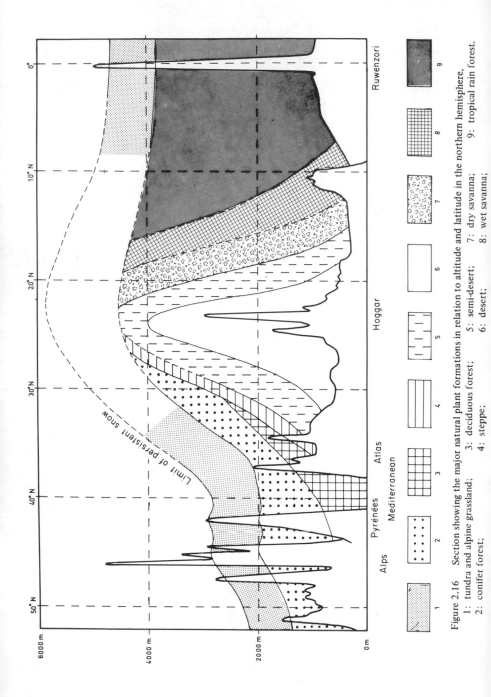

Figure 2.16 Section showing the major natural plant formations in relation to altitude and latitude in the northern hemisphere.
1: tundra and alpine grassland; 3: deciduous forest; 5: semi-desert; 7: dry savanna; 9: tropical rain forest.
2: conifer forest; 4: steppe; 6: desert; 8: wet savanna;

Pic du Midi de Bigorre for example, there are often frosts in July and August. These wide variations are of great importance to living organisms, and Lascombes has shown that the characteristic form of alpine plants is largely determined by the low temperatures experienced at night.

Rainfall increases up to altitudes of about 3500 or 4000 metres (cf. the climatic diagram for Puy de Dome, fig. 2.9). A comparison between Toulouse and the Pic du Midi de Bigorre shows that the summer rainfall is 67% greater for the latter (192 mm compared with 115 mm, based on a four year mean). However, there are often alternate wet and dry periods, a very wet period being followed by a period of drought. This explains the xerophilous adaptations of alpine plants and the fact that many insects shelter under stones during dry periods. At 2000 m, the atmospheric humidity is only one half, and at 4000 m only a quarter, of the corresponding value at sea level.

Exposure plays an important part on mountains, and there are often marked differences in both temperature and vegetation between sunny and shaded slopes. In France the warmer exposed slopes are orientated towards the south and south-west, and the colder slopes towards the north-east and north. There may be differences in extreme values of between three and four degrees at the same altitude. The tree line is

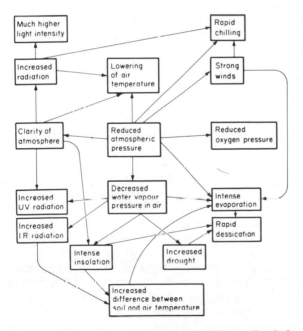

Figure 2.17   Diagram showing relationships between the different climatic factors in mountains.

47

from 100 to 200 m higher and the lower limit of permanent snow is between 150 and 500 metres higher on slopes exposed to the sun.

Snow is an important environmental factor in mountain regions, a snow cover preventing the soil from freezing. At Davos (Switzerland) the minimum winter temperature is about $-0.6°C$ underneath one metre of snow, while it may be $-33.7°C$ at the surface. There is a difference in altitude of about 800 m between the lower limit of snow and the tree line. The permanent snow line varies with latitude. In the Alps and Pyrenees it is situated at about 2500 m, while in tropical mountain areas it rises to about 5000 m (fig. 2.16).

Finally, another climatic factor of considerable importance in mountains is that of *wind*. The relationships between different components of mountain climates are summarised in figure 2.17.

2   FOREST MICROCLIMATES

The change in light intensity as it penetrates the tree canopy is probably the most important feature of the forest microclimate. In conifer stands, light intensity is considerably reduced but there is little qualitative change. In hardwood stands a marked selectivity in light absorption produces a greenish-yellow colouration when leaves are on the trees, and the variation in light intensity varies considerably depending on whether leaves are present or not (fig. 2.18). In temperate forests, illumination at ground level may be as low as two per cent of the corresponding value over bare ground. In tropical forests it varies between one tenth of a per cent and one per cent according to the circumstances.

The mean annual temperature is lower, and annual rainfall higher, in forests than in neighbouring non-wooded areas. In the forest of Fontainebleau, for example, a mean temperature of $8.9°C$, $1.5°C$ less than that of the surrounding countryside was recorded together with a rainfall of 696 mm, 17% higher than the mean for the Paris region as a whole.

Studies of temperature changes in a forest show a distinction between diurnal and nocturnal types of variation. During the day the crowns of trees bearing leaves are the warmest zones and, where crowns are in contact with one another, they form a false ground level and there is a temperature inversion. The zone of maximum temperature slowly changes its position during the day. At sunrise, the warmest part is the boundary of the canopy, towards mid-day it is in the centre of the crowns, and in the evening it moves again towards the top of the crown. The reason for this change is that the sun's rays are only able to penetrate through the canopy into the forest in the middle of the day. The maximum temperature in the canopy is reached later when the sun's rays are more oblique and can be more easily absorbed by the crowns. At night the temperature is practically the same at all levels,

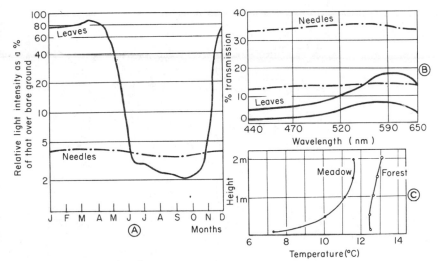

Figure 2.18   Forest microclimates.

A: Annual variations in relative light intensity in conifer and deciduous stands.

B: Selective transmission of light by leaves and conifer needles. The graphs relate to stands which have been heavily thinned and moderately thinned.

C: Distribution of temperature in forest and a neighbouring meadow, up to two metres above the ground, at 18.30 hours in Alsace on the 11th April, 1934. The temperature difference between forest and meadow varies from 5.2°C at 0.10 m to 1.4°C at 2 m.

although it may be slightly higher at ground level and in the layer of air up to two metres above the ground. Temperature is higher at night in forest than over bare ground, where inversion is always well marked (fig. 2.18). When the trees shed their leaves, the phenomenon of temperature inversion disappears almost completely and maximum temperatures are recorded at ground level.

The carbon dioxide content of the air is always slightly higher in woodland than over open ground. Relative humidity is also higher, especially at night. The rainfall is only partly responsible for this higher humidity, since the tree canopy holds back a large part of the rainwater (two-thirds of rain showers and at least a fifth of storm water in coniferous forests; much more in the case of deciduous forests). The higher humidity is largely due to a reduction in evaporation, partly caused by a reduction in wind speed.

Similar observations have been made for tropical forests, for example, in the forest of Banco on the Ivory Coast, readings were taken from a tower forty-six metres high which carried recording apparatus at different levels. The Banco forest consists of primary tropical rain forest, the most characteristic trees including *Turraenthus africana* (Meliaceae, 45 m high), *Guarea thompsoni* (Meliaceae, 37 m), *Combretodendron africanum* (Lecythidaceae, 43 m), *Homalium aylmeri* (Samydaceae, 32 m) and *Piptadenia africana* (Leguminosae, 43 m). A profile

49

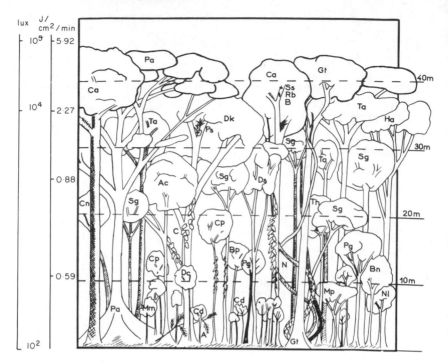

Figure 2.19   Botanical diagram of tropical forest at Banco in the Ivory Coast, showing a
section of vegetation 30 m long and 20 m wide. Note the diversity of tree species. On the
left hand side are shown illumination (lux) and energy received by plants (J/cm²/min) in
relation to the height above ground level. A: *Ancistrophyllum* sp (climbing palm) Bn: *Baphia
nitida*;   Ca: *Combretodendron africanum*;   B: *Bulbophyllum* and epiphytic orchids;
Dk: *Dacryodes kleineana*; Cn: *Cola nitida*; Cp: *Carapa procera*; Gt: *Guarea thompsoni*;
Ha: *Homalium aylmeri*; Nl: *Napoleona leonensis*; Sg: *Strombosia glaucescens* var. *lucida*;
Rb: *Rhipsalis baccifera* (epiphyte); Ta: *Trichoscypha arborea*; Th: *Trichilia heudelotii*;
Mp: *Microdesmis puberula*; Ss: *Solenangis scandens* (epiphyte); Dc: *Drypetes chevalieri;*
Ds: *Diospyros sanza-minika;* Pa: *Piptadenia africanum;* N: *Neuropeltis* (climber); Mn: *Mono-
dora myristica;* Cd: *Conopharyngia durissima;* Pg: *Parinari glabra;* Bp: *Baphnia pubescens*
(after Cachan, 1963).

of the forest is shown in figure 2.19. The most striking feature of
tropical rain forest is the extent of vertical change in climatic
factors and, from this point of view, tropical rain forest can be
compared with the sea. The vegetation shows a characteristic zonation. The tops of the highest trees reach forty metres, those of the
smaller trees between twenty and thirty metres, and others reach up
to ten metres. Moving down from the crowns of the trees to the ground
on a sunny day gives the impression of descending a well; the
temperature falls from 32°C to 27°C, relative humidity increases from
thirty per cent to eighty per cent, and illumination falls from 100 000

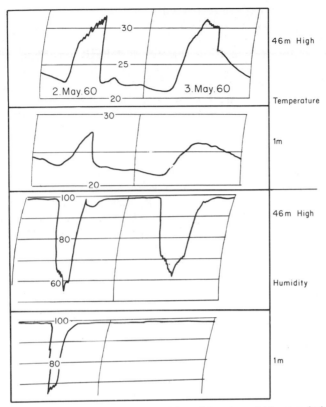

Figure 2.20   Temperature and humidity at heights of 1 metre and 46 metres during two days in the dry season in the forest at Banco in the Ivory Coast (Cachan, 1963).

lux to between 100 and 200 lux. The wind speed, which may attain a velocity of seven metres per second at the tops of the trees, falls rapidly. The vertical distribution of insects is related to this stratification in climatic factors. There is a temperature inversion similar to that found in temperate forests, with a maximum being reached in the forest canopy. Soil temperatures rarely exceed 25°C, even at the warmest part of the year; and, at this temperature, the losses and gains of nitrogen in a tropical soil are in equilibrium. This is the reason that these forests are able to grow on poor soils that produce only poor crops when the forest is cleared. Temperature variations are reduced and a daily temperature variation of 10°C at the top of the canopy, forty-six metres above the ground, is reduced to only 5.1°C at one metre above the ground (fig. 2.20).

The microclimates in each stratum of the tropical rain forest change during the course of the year. A particular microclimate is not characteristic of a particular level, but shows vertical movements with

51

seasons. Animals that are restricted to particular micro-climates (especially some insects) show corresponding vertical movements in the course of a year.

It can be said in conclusion that forest animals experience a climate that is more moderate than that of the open ground, with lower mean temperatures, higher relative humidities, restricted air movements and reduced illumination which includes a higher proportion of red and infra-red light and less yellow-green light. There are also considerable variations in temperature and humidity dependent on the height above the ground.

Some specific habitats have very different microclimates from those described here for the forest. These include forest glades, tree trunks and tree holes. The glades or clearings receive more sunshine than the surrounding undergrowth. As a result, their temperatures may show greater variations than those of unforested areas, and frosts become more frequent in winter as the size of clearing increases. Drought may also be important, since much of the rainfall is trapped by the trees located around the edge of the clearings.

Tree trunks and fallen trees have temperatures that vary widely with the degree of exposure. The temperature under the bark depends on the type of bark (colour, thickness and structure) and on the exposure of the tree. In the case of readings taken from a fallen tree in a clearing, exposed directly to solar radiation, the temperature under the bark on the upper side of the trunk reached a maximum of 36.9°C with a daily mean of 23.1°C, while on the lower side of the trunk the maximum was only 24.2°C with a daily mean of 20.1°C. The air temperature reached a maximum of 26.7°C with a daily mean of 19.3°C (fig. 2.21). The relative humidity also varied considerably with environmental conditions. Where the bark was thick, covered with moss and located in the shade or in moderate sunlight, the relative humidity exceeded 80%, and often reached 100%. Where, however, the bark was thin, peeling and exposed to sunlight, the relative humidity was similar to that of the surrounding air. In the interior of the trunk, temperature variations were dampened and similar to those of the surrounding air (Savely, 1939). The relative humidity in the galleries of wood-boring insects may be either very high (more than 80%) or relatively low (of the order of 50%), according to the physical condition of the wood. These conditions very largely determine the composition and distribution of the fauna of dry wood.

Tree holes filled with litter and other organic material are habitats where changes in temperature and relative humidity are reduced and where humidity is usually high. The organic material serves as an insulator, with the result that temperatures are higher at night and lower in the day in comparison with those of the ambient air (Dajoz, 1966).

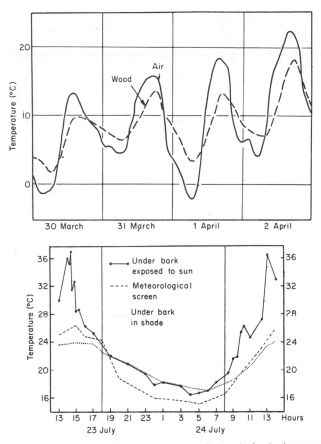

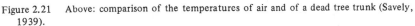

Figure 2.21   Above: comparison of the temperatures of air and of a dead tree trunk (Savely, 1939).
Below: comparison of temperatures in a meteorological screen, under bark exposed to the sun and under bark on the lower, shaded side of the trunk of a dead, felled beech tree (Dajoz, 1966).

3   THE SOIL MICROCLIMATE

One important feature of the soil microclimate is the considerable diurnal change in temperature that takes place at the surface, and this variation becomes reduced with increased soil depth. During the day there is a considerable increase in temperature in the upper layers of the soil. The temperature at a distance of 0.1 mm from a warm body of surface temperature 87.5°C is only 77.4°C, while at 1 mm distance it falls to only 56.8°C, i.e. a fall in temperature of 30°C. Similarly, there may be a temperature difference of more than six degrees between one mm and one cm above the ground. The microclimates experienced by two insects on sand dunes, a small ant and species of *Pimelia* (Coleoptera, Tenebrionidae), are very different because *Pimelia* is lifted off the ground by its legs. The presence of a layer of very warm air close to the surface of the soil explains the behaviour of the Sarcophaginae, thermophilous flies which, in summer, are often found resting on the ground with their legs spread out as if they were trying to get as close as possible to the warm surface. Reptiles, like lizards and vipers, behave in a similar way, flattening themselves against the soil with the sides of their bodies spread out and their backs exposed to the sun in order to obtain the maximum benefit from the heat in the soil which reaches them by conduction (St-Girons *et al*, 1956).

The phenomenon of temperature inversion occurs shortly before sunset, when the air becomes colder nearer the ground. A soil that is a good conductor warms up more quickly during the day and cools more quickly at night, while these variations are less well marked for soils that are poor conductors. Diurnal temperature variations decrease rapidly with increasing soil depth, and are hardly perceptible at a depth of 40 cm (fig. 2.22). A similar reduction is shown in annual temperature variations: at 7.5 m depth the temperature difference between summer and winter is less than 1.5°C.

Humidity variations are also more marked near the ground. In

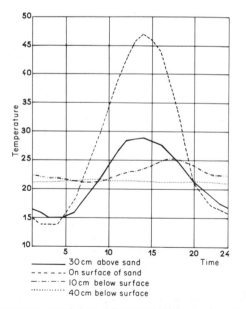

Figure 2.22   Temperatures recorded at different levels above and below the surface of a sand dune in Finland on 26 July 1927 (Krogerus, *in* Balogh, 1958).

general, the air is more humid during the daytime, and drier at night when dew begins to form.

Wind speed is considerably reduced at ground level. This reduction is more marked when there are small irregularities in the surface of the ground, and wind may be of considerable ecological importance. At Terre Adélie (Sapin Jaloustre, 1960), for example, where violent winds or blizzards, carrying ice particles, blow on the average for 300 days in the year at speeds in the region of 70 km/h, there is virtually no temperature gradient near the ground as a result of the air movement, and wind speed is considerably reduced (fig. 2.23). Sapin Jaloustre defined the cooling potential of a blizzard as the amount of heat lost

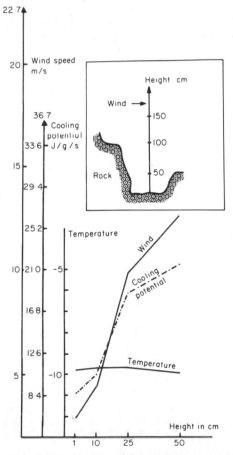

Figure 2.23   Temperature, cooling potential (J/g/s) and wind speed (m/s) at different heights above the nest of an Adelie penguin. At the height of the meteorological tower, the wind speed was 22.7 m/s and the cooling potential 36.7 J/g/s. The rectangle shows the location of the nest sheltered from the wind by a rock (Sapin Jaloustre, 1960).

per unit time from a given body at known temperature as a result of the action of the wind. This is a useful concept when studying temperature regulation by a homeothermic animal in a very severe climate where the mean annual temperature is about $-12°C$. For example, an Adelie penguin surrounded by snow at a temperature of $-20°C$ will be subjected to a cooling potential of about 2.5 J/g/s, although it could reach 79.6 J/g/s, i.e. thirty-two times more, in a blizzard blowing at 40 m/s under the most severe conditions. The location of the nest of this penguin is shown in figure 2.23. This diagram shows that the wind speed 10 cm above the ground is only 4.4 m/s and the cooling potential about 10.5 J/g/s, while the general climatic conditions may be a wind speed of 22.7 m/s and a cooling potential of 36.0 J/g/s, the temperature in both instances being about $-10°C$.

A study has been made of microclimates in the High Atlas of Morocco. At an altitude of 2100 m, the diurnal variation in temperature was small. The temperatures at scree surfaces showed wider fluctuations than those of boulders. The temperature in the shade of boulders remained fairly constant and similar to the air temperature. The relative humidity of the air in the shade did not fall below sixty per cent, although it was almost zero at scree surfaces. The air temperature above the ground fell rapidly, being $52°C$ at the scree surface and $31°C$ at 25 cm above the scree, when the air temperature was $28°C$. A knowledge of these variations is important as it may explain the behaviour of animals. Overheated rocks, for example, are deserted by reptiles when the temperature reaches a critical value.

The temperature and humidity beneath stones is very different from that at the soil surface, and this provides a suitable environment for many animals. In the Himalayas, a temperature of $+10°C$ and a relative humidity of 98% were recorded under a stone at a time when the temperature of the air above was $-1.5°C$ and relative humidity 40%; the upper surface of the stone exposed to the sun was at a temperature of $+30°C$.

Edney (1953), during his work on the behaviour of the isopod crustacean *Ligia oceanica*, recorded the temperature and humidity of the habitat of this animal. When the air temperature was $20°C$ and relative humidity 70%, the temperature in the spaces between the shingle at the foot of a cliff was $30°C$ and the relative humidity was 98%. Readings in a rock crevice were $27°C$ and 70%, compared with $34°C$ for the vertical face of the rock and $38°C$ on the upper surface of boulders. The body temperature of *Ligia* specimens living in the shelter of the shingle was $30°C$. *Ligia oceanica* is a species which is normally active at night and shelters during the day. If the temperature increases, however, as in this example, *Ligia* leaves the shelter of the shingle and moves on to exposed rock surfaces in full sunlight. The reason for this movement is that this animal, which is poorly adapted to terrestrial life, has a very permeable cuticle. It loses considerable

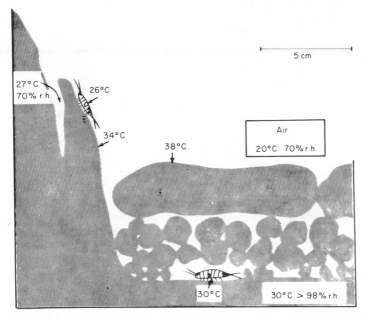

Figure 2.24   When the temperature increases, the crustacean *Ligia oceanica* leaves the shelter of the shingle and moves on to the rock in full sunlight. Its body temperature is reduced by evaporation (Edney, 1953).

quantities of water through evaporation when the relative humidity of the air is low, and this is the case on rocks exposed to the sun. This high rate of evaporation lowered the body temperature to about 26°C in this example, although, under the shingle, it would have remained at 30°C because the air was saturated and evaporation almost nil (fig. 2.24).

The microclimates in caves and mammal burrows are similar to those of soil. Mammal burrows have a relatively constant temperature and high humidity. These conditions have been studied in burrows of the gerbil, a rodent of desert regions in North Africa and the Middle East. The microclimate in burrows of *Psammomys obesus*, another desert rodent, was described by Petter (1961). This diurnal rodent retires to the deeper parts of its burrow, where the stable temperature is vital for its survival, especially during cold periods. In winter it emerges from the burrow only during the warmest part of the day when the shade temperature is about 20°C. In summer its activity outside the burrow is limited to sunrise and sunset, when the shade temperature is less than 35°C (fig. 2.25).

In caves, temperature fluctuations are always considerably lower than those recorded outside the caves. The annual variation is only 0.7°C in the Moulis cave (Ariège), and about 2.2°C in the Corveissiat cave (Ain). The temperature of caves is usually similar to that of the

57

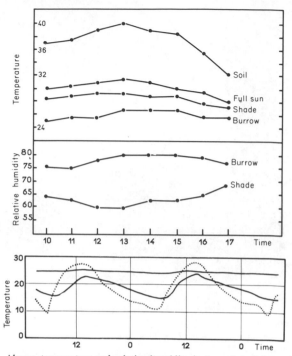

Figure 2.25   Above: temperature and relative humidity in the soil, in full sun, in shade and in the burrow of the gerbil (Kirmiz, 1962). Below: temperature in burrows of the rodent *Psammomys obesus* at depths of 10 cm (lower line) and 1 metre (upper line), and shade temperature in the vicinity of the burrow (dotted line). Measurements made at Beni Abbes on 18 March 1953 (Petter, 1961).

mean external average temperature. The relative humidity of the air is consistently high, about 95–100% (Vandel, 1965), in caves occupied by true cave-dwelling animals (troglodytes). There are also dry caves, in the Sahara for example, where the relative humidity is about 38% and where troglodyte animals are not found (Pierre, 1958).

4   MICROCLIMATES OF ARABLE LAND AND NATURAL GRASSLANDS

There are many publications dealing with the microclimates of arable land. Crops studied have included wheat, rye, potatoes, maize, beet and lucerne, and also grassland.

4.1   Light

Light undergoes changes as it passes through the vegetation. Significantly more visible light (0.4 to 0.7 μm) is absorbed than near infra-red

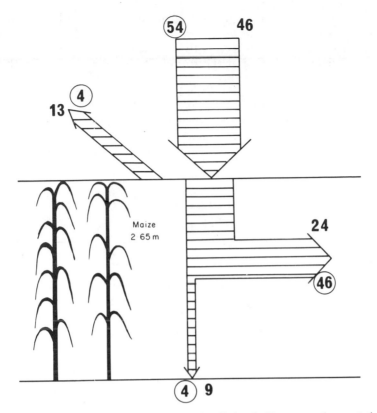

Figure 2.26   The diagram shows how the total radiation incident upon the vegetation is divided into reflected, transmitted and absorbed components, the proportions of total radiation being shown as percentages. The figures enclosed in circles relate to the visible radiation, while the other figures relate to the near infra-red. This particular example is for a maize crop 2.65 m high with a leaf area index $F = 4.3$ (The leaf area index F is the total leaf area per unit area of ground).

(0.7 to 3.0 μm) and, of this visible light, the red and blue show a greater degree of absorption than the green (fig. 2.26). Because much of the blue light is absorbed by the vegetation, the spectral distribution of the light reaching the ground resembles that of direct sunlight rather than of diffused cloud light.

This selective absorption of light by the vegetation results in a marked reduction in available energy. In *Dactylis glomerata* grassland where the incident energy is about 4.53 J/cm^2/minute above the grass, the available energy is reduced to only 0.80 J/cm^2/minute at ground level.

The selective absorption affects the distribution of those insects that live in the shelter of the plant cover. The following example may be used to illustrate this. In the Landes region, where maize and climbing beans are grown together, the level of infestation of the bean crop by the seed beetle *Acanthoscelides obtectus* is always higher at the edges of the field; it also increases with distance from the ground, being highest in pods developing at the top of the bean plants level with the flowers of the maize plants that support the beans. There are only slight variations in temperature and humidity over the field and it appears that light intensity determines the abundance of the beetle. If a clearing is made at the centre of a large field by removing the maize, and if a non-climbing variety of bean is used, then the beetle damage becomes considerably greater than that under maize. The following data have been given:

STATION	NUMBER OF SEED BEETLES PER 100 BEAN SEEDS	RELATIVE AMOUNT OF DAMAGE
Field edges	7.0	50%
Among maize	1.3	9%
Clearings	5.7	41%

If the maize becomes etiolated, allowing more light to pass through, then the damage tends to become more evenly distributed over the field. These observations show clearly the importance of light in the distribution and abundance of the seed beetles.

4.2 Temperature

Even a thin cover of vegetation is sufficient to reduce significantly the temperature gradient which exists above and below the surface of exposed soil, since the plants form a barrier to the radiation. In summer, when the temperature is 25°C at the surface of a sandy soil, a layer of moss will reduce the temperature to 23°C for the same soil. Under the same conditions the temperature falls to only 12°C when the ground is covered with short, compact turf. If the plant cover is destroyed by cultivation, the microclimate becomes more severe, with wider temperature fluctuation, and this explains why up to a half of the plants in a potato crop that has been weeded may be damaged by frost while no plants are damaged in an unweeded crop. The changes in the microclimate of weeded fields are also responsible for decreases of 25–50% in numbers of soil-dwelling wireworms in ploughed-out grassland. These larvae are very susceptible to heat and drought, and cultivation is harmful to them. For similar reasons chafer larvae hibernate only two or three centimetres below the soil surface in grassland, but at depths up to fifty centimetres in bare soil.

The temperature in the plant cover depends on the type of vegetation, its height, density and arrangement of the leaves. Temperature differences between the ground and the tops of the grasses in dense *Brachypodium* grassland are less well marked than sparse *Bromus*

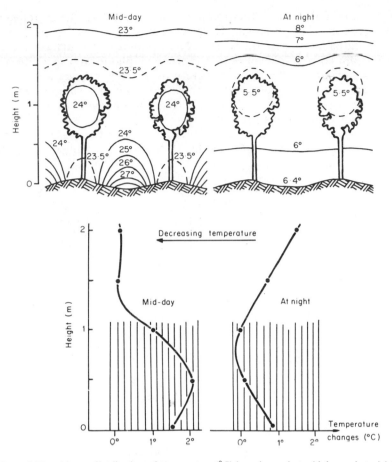

Figure 2.27   Above: distribution of temperature (°C) in a vineyard at mid-day and at night, in Germany on 17 September 1933. Below: temperature changes in a field of winter rye near Munich in May (Geiger, 1957).

grassland. As a general rule, the temperature at the centre of the vegetation is greater than that of the air during the daytime and less at night (fig. 2.27). Geiger has shown that in vineyards the foliage together with that ground between the vines which received direct sunlight is the warmest region during the day. At night the foliage is the coldest region, and this is of considerable importance in the utilisation of dew by the plants. The dew condenses on the coldest surfaces, and the leaves receive moisture while the soil and stems remain dry. The vertical distribution of temperature in a lucerne field before and after cutting, and over bare ground has been investigated. The plant cover forms a transition zone where temperature fluctuations are reduced, temperature within the vegetation remaining lower than the air temperature, even in cloudy weather. Heaps of drying lucerne in fields form a specialised habitat where the temperature depends very much on degree of exposure (figure 2.28). The temperatures of plants are usually found to be between the soil and air temperatures. A temperature of 56°C has been recorded at the surface of a houseleek (*Sempervivum*) when the air temperature was 31°C, and temperatures in the region of 50°C have been recorded by various authors.

Waterhouse (1950) found from measurements made on agricultural grassland that, in long grass (50 cm high), the warmest zone occurred in

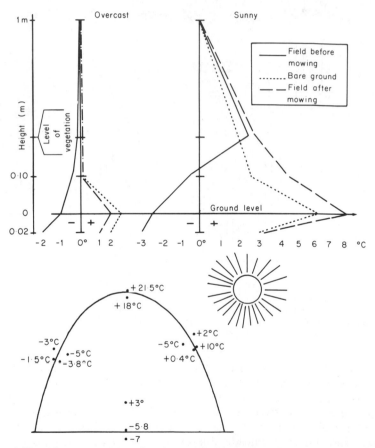

Figure 2.28   Above: temperatures in a lucerne field before and after mowing and over bare ground, in overcast and sunny weather. The graphs represent the differences in temperature in relation to air temperature measured one metre above the ground. Below: temperatures at different points on the surface and in the interior of a lucerne rick with regard to its orientation towards the sun. The figures correspond to differences in temperature from the air temperature.

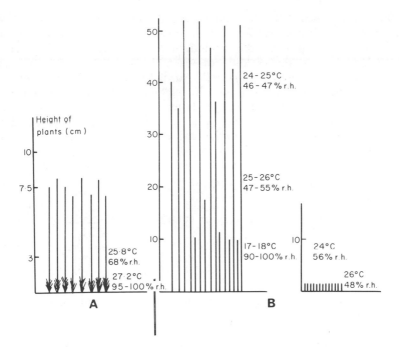

Figure 2.29   Distribution of temperature and relative humidity in three different types of plant formation.
A: in Normandy (Ricou, 1967);
B: in England (Waterhouse, 1950).

summer at a height of about 20 cm, the highest part of the vegetation being at a lower temperature but always greater than that of the ambient air (figure 2.29). In autumn, temperatures tended to equalise within the plant layer. Temperature fluctuations were also reduced in moving down from the top of the plants to ground level.

4.3   Relative humidity

Relative humidity is highest nearer the ground. In pastures in Normandy the relative humidity was found to be 10−20% higher than that of the air above the grass, rarely falling below 50% and usually varying between 70% and 85% (Ricou, 1967).

The effect of the wind is considerably reduced and, near the ground, its speed is even more reduced as the vegetation becomes higher and more dense. There is a greater reduction in wind speed in a wheat crop than in a field of beet or over short turf.

There may be significant changes in the carbon dioxide content of the air, with a maximum at night and minimum in the daytime.

Rainfall is trapped by the vegetation and, in a field of wheat, 60% of the water from heavy rain and 90% from fine rain was retained by the plants and never reached the ground. In an oat crop, 45—75% of the water from a rain shower did not reach the soil. The percentage of water trapped by plants is very similar to that retained by tree cover.

In conclusion, the microclimates of arable ground and grassland are characterised by wide variations in temperature which depend on the height above the ground and time of day, by a high humidity, reduction of wind speed and by a spectral distribution of light which is rich in infra-red and poor in red and blue radiation.

5 THE AQUATIC ENVIRONMENT

Two important climatic factors in aquatic environments are temperature and light.

5.1 Light

It has already been shown (p. 25) that the absorption of solar radiation by water is very rapid (figure 2.2). For visible light, absorption in clear water becomes greater as wavelength increases. For this reason, blue light penetrates to the greatest depths, reaching 100 metres or more. When the water contains suspended materials or dissolved substances, the latter absorb an additional part of the radiation, and this explains the emerald green colour of the water in some clear lakes. When the transparency of water is measured for the middle region of the spectrum (between 500 and 600 nm), it is found that the depth at which light intensity falls to one per cent of its value at the surface can vary from two to thirty metres depending on the water body. This value of one per cent corresponds to the limit below which green plants are generally unable to grow. Ultra-violet radiation is absorbed rapidly, and has little effect beyond a depth of a few metres. Some incident light is reflected at the surface of the water, and the proportion increases with the angle of incidence. For this reason a greater proportion of light enters the water as the sun rises higher in the sky.

In the marine environment it is possible to distinguish three light zones (Pères and Devèze, 1963). The *euphotic zone* extends from the surface to a mean depth of about fifty metres (range: twenty to 120 metres). There is suffcient light here to allow normal photosynthesis. The *dysphotic zone* extends to a mean depth of 500 metres (limits: 200 to 600 metres). The human eye can still detect slight illumination, but autotrophic plants can only survive there a short time and are unable to grow with the exception of some deep-sea coccolithophores which possess sensitive red pigments, allowing them to make use of blue light which penetrates the furthest. The *aphotic zone* of complete darkness, lit only by flashes from occasional luminescent marine animals, extends from 500 metres down to the greatest depths.

In lakes and ponds, Vibert and Lagler (1961) distinguish three zones. These are the *littoral zone*, where light penetrates easily and which is frequently occupied by rooted vegetation; the *limnetic zone*, which is an open-water zone free from rooted plants and bounded by the photosynthetic compensation level, i.e. the level above which the rate of photosynthesis is greater than that of respiration and so phytoplankton is able to develop. The limnetic zone is absent from shallow ponds. The third zone, the *profundal zone*, lies below the compensation level. It is only found in deeper lakes.

5.2 Temperature

Temperature variations in flowing water tend to follow those of the air, but with lower amplitude. The temperatures of spring waters usually vary only slightly. In Florida, for example, the water from Silver Springs is maintained at a temperature of between 22.2°C and 23.3°C throughout the year. The waters of streams with open banks tend to be warmer than those of streams whose banks are shaded by trees or cliffs. This is ecologically important since the composition of aquatic faunas is partly determined by temperature.

In lakes and ponds of sufficient depth in temperate regions, a temperature regime is set up as the result of a well-known physical phenomenon; the fact that the maximum density of water occurs at a temperature of 4°C. In winter the surface of a lake may be completely frozen and the temperature of the surface water about 0°C, while that of the deeper water is about 4°C because this water is the most dense. Temperature thus increases with depth. In spring the warming up causes ice to melt and, as the water reaches 4°C, it flows to the bottom of the lake causing complete circulation, since water from the depths moves upwards and there is an equalisation of temperature. In summer the surface water, whose temperature is more than 4°C, becomes warmer and less dense, and it floats above the colder denser water which warms up only slowly. Temperature is now found to decrease with depth. In summer three zones can be distinguished in a lake. There is a superficial or *epilimnion* which is agitated by the wind, rich in dissolved oxygen and phytoplankton, with good illumination, and where temperature gradually decreases with depth. There is a transition zone, or *thermocline*, where temperature decreases rapidly by at least one degree per metre; and there is a deeper zone, or *hypolimnion*, extending to the bottom, which is poorly oxygenated, with little or no illumination, poor in phytoplankton and where temperature varies only slightly throughout the year. In autumn the surface water cools to a temperature of 4°C and moves downwards to the bottom, causing a second mixing of the water within the year, an equalisation of temperature and

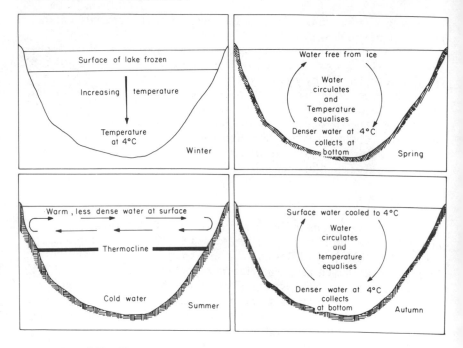

Figure 2.30   Water movements and thermal stratification in a lake during a year.

redistribution of oxygen and dissolved nutrients (figures 2.30 and 2.31).

When the temperature at the bottom of a lake is similar to that at the surface, water continues to circulate and no thermocline is established. If the temperature at the bottom remains at about 4°C, there is little circulation and, as a result, the surface water is rich in oxygen but lacks nutrients while the deeper layers are rich in nutrient salts but lack oxygen. In this last example productivity will obviously be very low. Lakes showing a complete seasonal mixing of water are called *holomictic*, while those with no seasonal mixing are *meromictic*.

The marine environment also shows temperature stratification of water with increasing depth. The following layers can be distinguished in the oceans:

(i) A superficial layer where the waters are agitated and mixed by the wind and which can be regarded, by comparison with the atmosphere, as a marine *troposphere* or *thermosphere*. Temperatures here show daily variations down to depths of about fifty metres, and seasonal variations beyond fifty metres. This thermosphere is about 400 metres in depth.

(ii) An intermediate layer extending down to about 1500 metres in

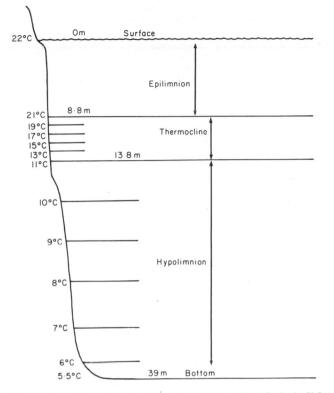

Figure 2.31   Thermal stratification during the summer in a lake in the U.S.A.

depth and corresponding to a permanent thermocline. Temperature is reduced in this layer to between 1 and 3°C according to the ocean.

(iii) A deep layer with a uniform temperature of 1–3°C except in polar regions where the temperature is near freezing point. At great depths the water pressure is responsible for a gradual adiabatic increase in temperature in the vicinity of the bottom. The temperatures recorded at the bottom of the oceans are, however, less than 4°C, in contrast to those recorded in lakes, because the density of saline water increases as temperature decreases instead of showing a maximum at 4°C as in fresh water.

Table 2.7 shows temperatures recorded at different depths in several seas and oceans.

The mean temperature of sea water at the surface varies widely according to the region, from −2°C in polar water to +27°C in equatorial waters. In enclosed seas the temperature may be even higher, reaching 34°C in the Red Sea. Underwater sills are responsible for significant changes in the distribution of temperature. The sill of

Table 2.7   Temperatures recorded at various depths in seas and oceans

DEPTH (m)	PACIFIC OCEAN (PHILIPPINE TRENCH)	SOUTH ATLANTIC OCEAN	ARCTIC OCEAN	MEDITERRANEAN SEA
0	28.80	25.72	−1.23	24.00
100	25.90	14.55	2.15	15.55
200	15.15	12.44	2.70	15.16
1000	4.45	4.02	−0.20	13.70
2000	2.25	3.35	−0.85	13.70
3000	1.64	2.65	−0.82	13.70
4000	1.58	2.03		13.70
5000	1.72	0.72		
8000	2.15			
10000	2.48			

Gibraltar, at a depth of 300 metres, allows only the warmer superficial waters of the Atlantic Ocean to enter the Mediterranean. The temperature of this water is 13.7°C, and this is the water temperature in the Mediterranean below a depth of 1000 metres. The sill known as the Wyville Thomson ridge, which extends between the Shetland Islands and the Faroes, isolates two water masses at different temperatures. On the south side of the ridge the waters are warm (6°C at 1100 m), and on the north side they are colder (−0.4°C at 1100 m). The faunas on either side of the sill are very different, with only forty-eight species common to both sides out of the 385 species recorded (Pères and Devèze, 1963).

IV   THE ECOLOGICAL ROLE OF CLIMATIC FACTORS

1   THE EFFECT OF TEMPERATURE

1.1.   Limits of tolerance. Stenotherms and eurytherms

As a rule, living organisms can only survive at temperatures between 0 and 50°C, these temperatures being compatible with normal metabolic activity. There are, however, a few remarkable exceptions which it is interesting to note.   Living bacteria have been found in hot springs at temperatures of nearly 90°C. Blue-green algae of the genera *Phormidium* and *Oscillaria* have been recorded from localities where the temperature is over 85°C. The prosobranch mollusc *Hydrobia aponensis* lives in a spring which emerges at 46°C near Padova in Italy. Thecamoebae (Protozoa) have been recorded in springs at 58°C. The crustacean *Thermosbaena mirabilis* was discovered in springs where the temperature varied between 45 and 48°C near Gabes in Tunisia. Another crustacean *Thermobathynella adami* was found in water at 55°C at

Katanga. Among the insects, the ant *Cataglyphis bombycina* is active at the surface of the sand in the Sahara when the temperature is over 50°C. The following figures have been given for aquatic insects, and show the maximum temperatures reached in warm water springs where they were living:

Culex pipiens	40°C (Europe)
Chironomus sp.	51°C (N. America)
Stratiomyidae	50°C (N. America)
Ephydra sp.	49.1°C (Himalayas)
Tabanidae	42°C (N America)
Cloeon dipterum	45°C (Europe)
Ceratopogonidae	52°C (Java)

The food of these phytophagous insects is supplied by green plants in waters reaching between 60 and 63°C, and by other kinds of plants in waters with temperatures reaching between 68 and 71°C. The number of species of aquatic Coleoptera living in warm water springs decreases with temperature in the following way:

Temperature	32°C	36°C	39°C	41°C	43°C	45°C	46°C
Number of species	51	34	24	11	6	3	2

Amongst the vertebrates, some fish and frog tadpoles live in warm waters (40°C) in the Yellowstone Park. The teleost *Cyprinodon macularis* from California lives in springs at temperatures of 52°C.

Tolerance of high temperatures depends on the temperature to which an animal is normally exposed in its habitat. The bristletail *Thermobia domestica*, which is found in warm places in dwellings, for example near baker's ovens, can live indefinitely at 42°C, while the related species *Lepisma saccharina*, of cooler habitats, is killed by temperatures over 36°C (Sweetman, 1938).

Living organisms can survive very low temperatures in a resting state; myriapods have been exposed to −50°C, rotifers and tardigrades to −192°C, and nematodes to −272°C without harm. Insects in diapause are able to tolerate temperatures as low as −80°C. Some living organisms have become adapted to temperatures in the region of 0°C or slightly below, even in an active state. Algae of the genus *Chlamydomonas* are able to develop on the snow surface, often giving it a reddish tint, and they die if the temperature rises above + 4°C. The marine bivalve *Yoldia arctica* lives in polar waters, where the temperature is permanently in the region of 0°C. The silphid beetle *Isereus xambeui* occurs at all stages of development in an ice grotto of the Trou du Glaz in the Isère, where the temperature is 0°C in winter and only 1 or 2°C in summer. Another silphid *Pholeuon glaciale* completes its life cycle at temperatures between +0.1°C and +0.8°C in an ice grotto in Romania (Vandel, 1964). Some collembola belonging to the genera

Hypogastrura and *Proisotoma* were found to be active at night on the surface of a glacier in the Himalayas at an altitude of 5000 m and a temperature of $-10°C$. The oribatid mite *Maudheimia wilsoni* occurs on the Antarctic continent at latitudes greater than $72°C$ at an altitude of 1250 m. It lives under stones because the soil, which is frozen throughout the year, cannot provide it with shelter. The temperature minima are $-30°C$ in summer and between -60 and $-65°C$ in winter. This species can develop at temperatures permanently less than $0°C$.

Among the vertebrates, the teleost *Boreogadus saida* leads an active life in waters at $-2°C$, and the fish *Dallia pectoralis* from Alaska can survive for forty minutes at $-20°C$. Some homeothermic vertebrates such as penguins are able to colonise areas of the Antarctic continent where the mean annual temperature is less than $-12°C$. However, they possess elaborate ecological and physiological mechanisms which enable them to maintain a constant high internal temperature under very harsh conditions. The emperor penguin *Aptenodytes forsteri* breeds during the Antarctic winter while the temperature is below $-18°C$, sometimes even falling to $-62°C$. The reindeer, *Rangifer tarandus*, survives at temperatures between $-50°C$ and $-60°C$. Some woody plants and insects can withstand temperatures of $-30°C$ and survive the formation of ice crystals within their tissues. They only die when all their tissue fluids are frozen.

For any particular species, it is possible to define the following: (i) a lower lethal temperature, i.e. the temperature at which death occurs as a result of the cold; (ii) an upper lethal temperature, i.e. the point where heat causes death; (iii) a lower threshold temperature, which is the lowest temperature at which an organism can lead an active life; (iv) an upper threshold temperature, which is the highest temperature at which an organism can remain active; (v) a cold torpor temperature; (vi) a warm torpor temperature; (vii) an optimum or preferred temperature, which is the one that an animal tends to seek. The latter is usually nearer to the upper lethal temperature than the lower lethal temperature.

Stenothermal species can only tolerate narrow temperature variations, while *eurythermal* species are active over a wide temperature range. *Megathermal* species (warm stenotherms or *thermophilous stenotherms*) are adapted to high temperatures while *microthermal* species (cold or *psychrophilous stenotherms*) are adapted to low temperatures.

The temperatures at which animals are able to live permanently are often very different from the limiting values that have been described for them. For example, the flatworm *Crenobia alpina* is unable to live in springs where the temperature is greater than $10°C$, although its upper lethal temperature is about $30°C$. Conversely, animals living in habitats with relatively constant temperatures may, however, be eurythermal, e.g. the cavernicolous beetle *Speonomus diecki*, whose lethal temperatures are $-5°C$ and $+25°C$, and whose limits of activity are $0°C$ and $+20°C$ (Glacon *in* Vandel, 1965).

1.2 Some examples of stenothermal and eurythermal species

a Warm stenothermal species

The hot spring crustacean *Thermosbaena mirabilis* dies at temperatures below +30°C and lives in waters at temperatures of 45 to 48°C. The marine copepod *Copilia mirabilis* requires temperatures of between 23 and 29°C. Many other marine invertebrates are warm stenotherms, including most appendicularians, many oligotrich protozoa, and the siphonophores. A number of woodlice are warm stenotherms, searching for fairly high temperatures. The distribution of the Mediterranean species *Armadillidium officinalis* is restricted in France to a region noted for its high mean temperature and small seasonal variations, and these conditions only occur near the sea. The 'isopod Mediterranean zone' is a much narrower zone than the Mediterranean zone described by botanists and based on plants (figure 2.10). Insect ectoparasites of mammals and birds are also warm stenotherms.

b Eurythermal species

The aquatic snail *Hydrobia aponensis* can survive at temperatures from −1°C up to +60°C. The butterfly *Vanessa cardui* and the blowfly *Lucilia sericata* are broadly eurythermal. Some eurythermal species are found among oribatid mites living on bare rocks or in the lichens growing on these rocks in mountainous areas. They can survive temperatures as high as +60°C during the day and as low as 0°C at night (Trave, 1963). The flatworm *Planaria gonocephala* survives temperatures from +0.5°C to +24°C, while the oyster can withstand variations from −2°C to +20°C. Amongst the vertebrates, the toad *Bufo bufo* is widely distributed from latitude 65°N to North Africa, and reaches an altitude of 2200 m in the Alps. The puma, *Felis concolor,* is distributed in America from Canada to Patagonia. The tiger in Asia is found from the tropical jungles of India as far up as Irkutsk in Siberia, and in the Himalayas up to an altitude of 4000 m. (In the case of the tiger, there are local races showing structural modifications in their fur, and this enables them to withstand low temperatures.) These three vertebrates are eurytherms who encounter wide variations in temperature over their extensive areas of distribution.

c Cold stenothermal species

From a study of the snow fauna of the Himalayas, it has been possible to distinguish between those species that were active below 0°C and down to −10°C (collembola and the tipulid fly *Chiona* sp.), those species active between 0 and +5°C (numerous collembola and flies), and those species active at temperatures between +5 and +10°C during periods of sunshine in the daytime. Species in the first two categories were very sensitive to temperature increases, being killed in a few minutes by the warmth of a human hand

Dreux (1962) showed that several species of Orthoptera are steno-thermal, being adapted to fairly low temperatures, and in this way explained the local distribution of these species at high altitudes in France, as well as on low ground in Scandinavia and Siberia (boreo-alpine species). *Aeropus sibiricus* requires a mean July temperature of between 7 and 14°C, while *Aeropedellus variegatus* needs a mean July temperature of between 6 and 8°C and this explains its restricted distribution to a few stations in the Alps at altitudes over 2500 metres. Many marine amphipods are cold stenotherms, being restricted to polar seas. A large number of the species found at great depths are also cold stenotherms. surviving only at temperatures in the vicinity of 0°C.

1.3 The relationship between activity and temperature. Temperature preference

The preferred temperature (or temperature preferendum) varies widely according to the species and to the stage in development. The human louse has a preferred temperature of between 24 and 32°C, and the house fly of about 42°C. Temperature preference may be modified by the physiological state of the animal (hunger, sexual activity, etc.) and by environmental factors (light, humidity or the temperature at which the animal has been kept previously). If the ant *Formica rufa* is kept at an initial temperature of 3 to 5°C, the preferendum is between 23 and 29°C, but this increases to between 32 and 52°C for individuals previously kept at temperatures between 27 and 29°C (Herter *in* Uvarov, 1931). Individuals of the beetle *Tribolium confusum* kept at an initial temperature of 15 to 18°C show a preference for temperatures between 24.7 and 26.5°C, but, after being kept for a month at 25°C, their preferendum is altered by 9 to 11°C and only returns to normal after several hours. Temperature preference varies according to the stage in the life cycle. For the locust *Schistocerca gregaria*, Boden-heimer (1930) found the following preferenda:

adult (during oviposition)	29.4°C
first instar nymph	30.1°C
second instar nymph	28.8°C
third instar nymph	31.6°C
fourth instar nymph	37.1°C
fifth instar nymph	36.7°C
young adults	39.3°C

The preferred temperature is of considerable importance. It may explain certain features about the distribution of the animals in their habitats, as well as their behaviour.

From studies on the activity of *Schistocerca gregaria* in relation to temperature, Bodenheimer was able to distinguish the following stages: (i) onset of cold torpor; (ii) slight movements of the antennae and the

TABLE 2.8   Temperatures (°C) corresponding to the various phases in activity in two insects: the locust *Schistocerca gregaria* and the beetle *Calathus fuscipes* (Bodenheimer, 1938)

	PHASES OF ACTIVITY							
SPECIES	i	ii	iii	iv	v	vi	vii	viii
Schistocerca gregaria								
Male adult (March)	4.9	10.5	18.0	20.1	32.6	41.8	48.4	50.5
Fifth instar nymph	5.5	7.6	20.7	26.0	36.9	45.3	49.2	51.0
Calathus fuscipes								
Imago	0.5	3.5	20.2	31.3	35.0	35.8	–	39.7

tips of the legs; (iii) intermittent walking movements; (iv) normal activity; (v) exaggerated activity; (vi) strong excitation; (vii) onset of heat torpor; (viii) heat death. Similar activity stages have been recorded for other insects, and some examples are shown in table 2.8.

The following temperatures corresponded with phases of activity in the caterpillar of *Lymantria monacha:*

lower lethal temperature	−10.5°C
cold torpor	−10.5 to −0.5°C
reduced activity	−0.5 to 8°C
normal activity	8 to 33°C
maximal activity	33 to 38°C
hyperactivity	38 to 43.5°C
heat torpor	43.5 to 45.5°C

and the following values have been given for several species of scolytid beetles:

lower lethal temperature	−15 to −10°C
cold torpor	−10 to +5°C
onset of activity	5 to 9°C
normal activity − no swarming	10 to 15°C
swarming activity	16 to 18°C
maximum activity	18 to 29°C
hyperactivity	30 to 40°C
heat torpor	40 to 49°C
upper lethal temperature	50 to 51°C

It has already been shown that the limits of tolerance for temperature can vary within the same species where there are geographical races (e.g. in *Aurelia aurita*, cf. p. 12), or else at different stages of development (e.g. in *Bathynella*, cf. p. 10 and in *Schistocerca gregaria*, cf. table 2.8), or even as the result of interaction with other environmental factors (e.g. in *Podisma alpestris*, cf. p. 13). The limits of temperature tolerance may also vary with environmental conditions, and there is a mechanism for physiological adaptation which will be examined later.

In homeotherms the body temperature is relatively constant, with a few rare exceptions (e.g. some hibernating animals). Among the poikilotherms there are three categories. *Cyclotherms* have a body temperature that more or less follows that of their environment. When this temperature exceeds 30°C or is below 10°C, the body temperature of cyclotherms is either slightly reduced or slightly increased by regulating mechanisms which may be either physical (evaporation) or chemical (raising of the metabolic rate by increased muscular contraction, for example). *Chemotherms* are able to increase their temperature by intense muscular activity; this is a characteristic, for example, of the hawk moths, which vibrate their wings on the spot for a short period before flying. In the spurge hawk moth *Deilephila euphorbiae*, the threshold temperature for flight is 34.5°C. Many insects (Orthoptera, buprestids) and some reptiles (lizards and snakes) are *heliotherms*, warming themselves in the sun by orientating themselves in such a way that they obtain the maximum benefit from the sun's rays.

1.4 Effect of temperature on the geographical distribution of living organisms

The limits of geographical distribution are often determined by temperature acting as a limiting factor. It should be noted that it is usually extreme temperatures rather than mean temperatures that have this limiting role.

a Minimum temperature as a limiting factor

In France the maximum altitude reached by many species in mountainous areas depends on the minimum temperatures to which the area is exposed. Many species are restricted to the Mediterranean region because temperatures fall too low outside this area. The praying mantis is generally considered to be a relic of a tropical fauna that has been able to survive in a few isolated places in the Paris area, in northern France, in Belgium and throughout the Mediterranean region. The phasmid *Clonopsis gallicus*, which is common in the South of France, also occurs at a few selected stations as far north as the forest of Fontainebleau and in Brittany. The large grasshoppers with reduced wing belonging to the genus *Ephippiger* are restricted to the Mediterranean region, except for the less exacting species *E. ephippiger* which occurs throughout France and even reaches Belgium and Holland. Of the 215 species of Orthoptera belonging to the French fauna, only sixty are found in the Paris area. Another group of thermophilous insects with a mainly southern distribution is the family Buprestidae (Coleoptera). The French fauna includes 156 species, of which 121 are found in the departement of Var, 104 in Pyrenees Orientales, sixty-one in the valley of the Seine, thirty-four in the departement of Normandie and only six in Finistère. The woodlouse *Chaetophiloscia elongata* has a southern distribution, the limit of which corresponds with a mean summer temperature of 18 to 19°C, this temperature being the minimum for reproduction in this species.

The migratory locust *Locusta migratoria* is not found in eastern Europe north of a line corresponding to the June isotherm of 20°C. In tropical Africa the tse-tse fly is restricted to those areas where the mean annual temperature is higher than 20°C. The elk is found at latitudes more than ten degrees further north in Scandinavia than in Siberia, although the latter country has a higher mean temperature, because the winters there are more severe (cf. figure 2.4 showing January isotherms). The northern limit for snakes occurs at latitude sixty-seven degrees in Europe, sixty degrees in Asia and fifty-two degrees in America because of temperature differences between these three continents. In the southern hemisphere their southern limit is latitude forty-four degrees in South America and in Queensland. Reptiles of the higher northern latitudes (*Vipera berus, Natrix natrix, Lacerta vivipara, Anguis fragilis*) are all viviparous, since oviparous species are unable to reproduce in colder regions because their eggs do not develop. It is for this reason that most reptiles living at altitudes greater than 1200 m in the Alps and Pyrenees are viviparous, although below 1200 m oviparous species predominate (Hock, 1964). Examples of plants include the wild madder *Rubia peregrina*, whose northern limit in Europe corresponds with the +4.5°C January isotherm, and the beech which is limited in the north by the −2°C isotherm.

Differences in the ecology of species of North American frog determined their geographical distribution. The first species, *Rana sylvatica*, which lives in forests, breeds at the end of March, laying its eggs in water at 10°C. These eggs are able to develop if the temperature does not fall below 2.5°C, and larval development takes sixty days. This is the most northerly species, occurring in Alaska and Labrador and reaching latitude 67°N. The second species, *Rana palustris*, lays its eggs in small pools and streams on the prairies at the end of April when the temperature is about 15°C. Larval development lasts about ninety days, and its northern limit is reached in southern Canada around latitude 55°N. The third species, *Rana clamitans*, requires water at a mean temperature of at least 25° C for egg-laying, and this explains why it only just reaches the northern U.S.A. at latitude 50°N.

In the sea, reef building corals are only able to survive in waters at temperatures greater than 20°C and are, therefore, restricted to tropical regions. Temperature variations in the sea provide efficient barriers to dispersal because many animals are stenothermal (cf. the example of the Wyville Thomson ridge given on page 68).

b   Maximum temperature as a limiting factor
A temperature that is too high may limit the range of a species. The southern limit of the distribution of the large white butterfly, *Pieris*

brassicae, which is found in Europe and north-west Africa, occurs in Palestine, where summer temperatures are too high for it to survive and where it could only reproduce in winter. Species from temperate climates are found at higher altitudes as they approach the equator, and this explains the part played by the Andean mountain range as an 'equatorial bridge'. Some North American species have been able to invade the Neotropical region using the high plateau of the Andes. Examples include some Lepidoptera and the 'stiff-tail' duck *Oxyura jamaicensis*, which is widespread in North and Central America and also occurs in the Andes of South America. These examples show the importance of the Andes as a dispersal route for the North American fauna (Dorst, 1967). In the sea some species react in the opposite way. They occur at the surface in polar regions and move into deeper water near the equator in order to find the lower temperatures necessary for survival. The holothurian *Elpidia glacialis* is one such example. This phenomenon is known to oceanographers as 'tropical submersion'.

The glacial periods made it possible for boreal species, which were unable to survive high temperatures, to migrate further south, in some instances as far as the Mediterranean. After the glaciers retreated, these species were able to survive on higher ground over a wider area than

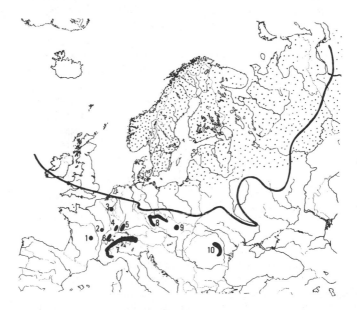

Figure 2.32   Distribution map for the butterfly *Colias palaeno*, a boreo-alpine species. The dots correspond to the distribution of this species in northern Europe. The black corresponds to mountain localities. 1: Mont-Dore; 2: Morvan; 3: Baraque Michel; 4: Hautes Vosges; 5: Black Forest; 6: Jura; 7: Alps; 8: Bohemia; 9: Tatra; 10: Carpathians. The black line corresponds to the southern limit of the quaternary ice cap.

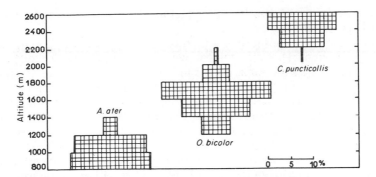

Figure 2.33   Distribution in altitude of carabid beetles in the Haute Vesubie valley (Alpes Maritimes). *Abax ater*: species inhabiting low ground and extending up to 1,400 m. *Oreophilus bicolor*: species of the montane and subalpine zones. *Cyrtonotus puncticollis*: alpine species, unable to survive at the higher temperatures of lower altitudes. The width of the bands in the diagram indicates the frequency of each species (as a %) at each altitude (Amiet, 1967).

their normal distribution. They are known as boreo-alpines, and a good example is that of the butterfly *Colias palaeno*. In the south on low ground the distribution of this species is restricted by the January isotherm of -1 to $-2°C$ and, outside its normal range, it is restricted to the larger mountain massifs (figure 2.32). Other examples include the blue hare *Lepus timidus*, the ptarmigan *Lagopus mutus*, and various plants such as the dwarf birch *Betula nana*, least willow *Salix herbacea* and *Dryas octopetala*. The glaciers also forced out of southern Europe those species requiring higher temperatures like the palms, magnolias and laurels, and these have been prevented from returning by the Mediterranean which acts as a barrier. In North America, where the mountains run from north to south and there is no sea barrier, there has been some re-colonisation, and this explains why the fauna and flora contain more thermophilous species than in Europe. Finally, in mountainous areas above a certain altitude, new species have evolved as the result of different temperature requirements. There are many examples among the insects (figure 2.33). These species, which have evolved *in situ* as a result of geographical isolation, do not occur on low ground in northern regions and are not, therefore, boreo-alpine species.

1.5   Action of temperature on the distribution of species within a habitat

Living organisms can sometimes avoid unfavourable conditions by searching for more favourable microclimates within the habitat. The ideal solution would be migration to avoid unsuitable environmental conditions. Migration by animals, especially birds, is partly controlled

77

by temperature, but also by other factors like daylength which may be equally important.

The degree of exposure has a marked effect on vegetation, especially on high ground. This effect is shown, for example, by the different altitudes at which the same organism occurs on exposed and sheltered slopes of mountains. The tree-line is higher on the south than on the north side. Exposure also has a marked effect on the distribution of animals. Dreux (1962) gives some examples for grasshoppers. On the southern side of Chavière (2800 m) in the Vanoise massif, isolated individuals of *Melanoplus frigidus* are found at 2700 m. At about 2600 m they also occur with *Aeropus sibiricus*, while *Stauroderus scalaris* is found up to an altitude of 2000 m. *Arcyptera fusca* occurs up to 2300 m, together with *Chorthippus longicornis*. These species are not found so high up on the northern slopes and are usually less abundant.

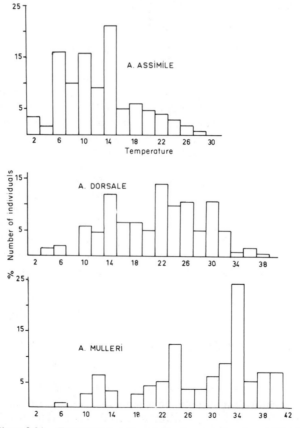

Figure 2.34   Temperature preferenda for three species of carabid beetle.

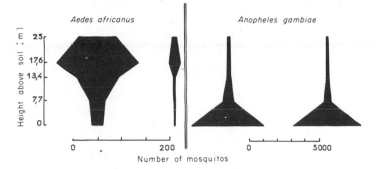

Figure 2.35   Vertical distribution of two species of mosquito in tropical forest in Uganda. For each species, the diagram on the left represents the wet season while that on the right is the dry season. The latter reduces the difference in distribution between the different levels by making the microclimate more uniform (Haddow, 1947).

The temperature preferenda for many woodland species are fairly low (together with a high humidity and low light intensity). Species characteristic of open ground and fields prefer higher temperatures together with lower humidities and stronger light. The distribution of three related species as a function of temperature clearly illustrates this point (figure 2.34). The carabid beetle *Agonum assimile*, a species of woodland and hedgerows, shows a preference for lower temperatures (optimum 14°C), while *A. dorsale*, which occurs in fields as well as woodland, prefers intermediate temperatures, and *A. mulleri*, a field species, favours higher temperatures (optimum 34°C).

The distribution of xylophagous beetle larvae under the bark of dead trees is related to their temperature tolerance. The thermophilous larvae of the buprestid *Chrysobothris affinis* occur on the exposed surface of fallen trees where temperatures are highest. They are uncommon on the sides, and absent from the sheltered under-surface of the trunk.

The insect fauna of nettles varies according to whether the plants are in the sun or shade. Aphids and psyllids are most common on plants in direct sunlight, while collembola and mites are dominant in the shade.

The vertical distribution of mosquitos in tropical forests is related to temperature variations. The lowest strata have high humidities and fairly constant, relatively low temperatures, while the upper strata show wide fluctuations in temperature and a climate more similar to that outside the forest. The most common culicid mosquito is *Aedes africanus*, which is restricted to the forest canopy, while *Anopheles gambiae* occurs mainly near the ground (figure 2.35).

1.6   Effect of low temperatures on population density

Low temperatures often have a very marked effect on animal populations, which may be greatly reduced or even eliminated from those areas near the northern limits of their geographical range. The effect of the unusually cold spells in February 1956 and in the winter of 1962/63 on the fauna of the Camargue has been described by several authors (e.g. Blondel, 1964). The vertebrates most affected were reptiles and some birds, including wild duck, rails and coots, some waders and, among the passerines, the warblers. The flamingo population was almost destroyed. Terrestrial invertebrate numbers were severely reduced, and also numbers of crustacea in those shallow waters which were completely frozen. Cold also has an indirect effect by reducing the available food when the layer of ice and snow is too thick. The role of low temperature as a limiting factor for population density will be discussed in chapter 8.

1.7   The effect of temperature on biological processes

a   Development at constant temperatures

In poikilotherms the speed of development and number of generations per year both depend on temperature. The threshold of development, or *developmental zero*, is the temperature (K) below which the speed of development is zero. The *effective temperature* $(T - K)$ is the difference between developmental zero and the temperature at which an animal is reared (T). It has been shown that at constant temperatures the equation

$$S = (T - K)D$$

can be used to describe the relationship between temperature and speed of development. In this equation S is a constant and D the duration of development (developmental period). The curve for this equation (figure 2.36) forms part of an equilateral hyperbola. The reciprocal of the developmental period (i.e. rate of development) plotted against temperature (figure 2.36) intercepts the temperature axis at a point K, the developmental zero. This reciprocal represents the mean percentage development that takes place per unit time. It can be seen that the rate of development is a linear function of temperature. For example, at seventy per cent relative humidity, the fruit fly completes its development in twenty days at 26°C, and in 41.7 days at 19.5°C, the developmental zero being 13.5°C. Using the measurements made at two different temperatures, it is possible to draw a line and calculate graphically the duration of development at other temperatures. S is the thermal constant or $\Sigma_i(T_i - K_i)D_i$ temperature necessary for the development of an animal, and is sometimes expressed in units of day-degrees. In the case of an insect, if n_1 is the development period of the egg, n_2 that of the larva, and n_3 that of the pupa, then the sum of the effective temperature is

$$S = (T - K)(n_1 + n_2 + n_3)$$

assuming that the developmental zero (K) is the same for each stage.

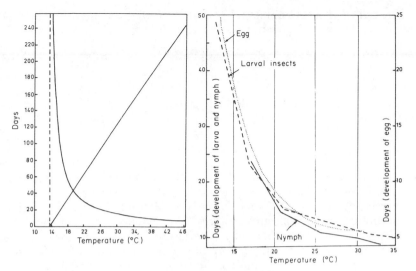

Figure 2.36   Left: descending curve shows the length of time required to complete the development of *Ceratitis capitata* plotted against temperature; the ascending curve shows the reciprocal of this time (i.e. speed of development) plotted against temperature.
Right: time taken to complete development for various stages in the life cycle of the Colorado beetle plotted against temperature (after Bodenheimer, 1926).

In the example of the fruit fly given on the preceding page

$$S = (26 - 13.5) \times 20 = 250 \quad \text{or} \quad S = (19.5 - 13.5) \times 41.7 = 250.2$$

In the case of the Colorado beetle, the developmental zero is 12°C. At a constant temperature of 25°C, larval development takes from fourteen to fifteen days, while at 30°C it only lasts 5.5 days. Development does not take place above 33°C. The sum of effective temperatures is between 330 and 335 day-degrees. This figure has been used in eastern Europe to predict the developmental period of the Colorado beetle and the number of generations per year in terms of the mean temperature recorded; from this information can be calculated the number of control measures necessary to protect potato crops against this pest. The first treatment, against young larvae, is made when the sum of effective temperature reaches 150 day-degrees, and the second is applied against larvae of the second generation when the sum of effective temperatures reaches 475 day-degrees.

The exponential curve is used in place of a hyperbola to describe the relationship between developmental period and temperature. According to Davidson (1944) the reciprocal curve derived from the curve of development is a logistic curve with the equation

$$\frac{1}{y} = \frac{K}{1 + e^{a-bx}}, \quad \text{or} \quad \frac{100}{y} = \frac{K}{1 + e^{a-bx}}$$

if the mean percentage of development in unit time instead of the reciprocal of the developmental period is plotted on the ordinate. The curve for the developmental period can then be calculated from the equation

$$y = \frac{1 + e^{a - bx}}{K}.$$

Davidson obtained the following equation for the development of eggs of *Drosophila melanogaster:*

$$y = \frac{1 - e^{4.451 - 0.207x}}{0.07} \qquad \text{(figure 2.37)}$$

In practice the multiplicity of formulae put forward to describe the relationship betweeen rate of development and temperature may be explained by the fact that each formula is valid only for certain species. In addition, theoretical curves often differ widely from experimental results at extreme temperatures. Variations in permeability of the integument with temperature, resulting in the loss of water by evaporation, are important and complicate the experimental data.

Life span increases as temperature decreases down to a limiting value which represents the critical low temperature. For example, adult *Drosophila* fed with a one per cent solution of cane sugar live for 6.2 days at 34°C and 12.3 days at 19°C. The total length of life for this animal (from eclosion to death of the adult) varies between 21.1 days at 30°C and 123.9 days at 15°C. Similar examples can be found among the vertebrates. The Atlantic salmon lives only for about six or seven years at the southern limit of its geographical distribution, but reaches an age of twelve years in Norway. In the rat and the brook trout, rapid growth is not compatible with longevity.

One of the results of an increased lifespan is the overall increase in size that frequently occurs in animals living at low temperatures. This is

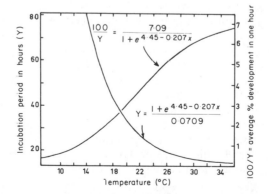

Figure 2.37   Number of hours required for the development of the egg of *Drosophila melanogaster*, and the reciprocal curve showing rate of development (Davidson, 1944).

especially true for many marine abyssal species, which tend to be larger than related species living in warmer, shallow waters. Radiolarians of the family Challengeridae living at depths of between 50 and 400 metres have mean measurements of between 0.11 and 0.16 mm; those living between 400 and 1500 metres measure between 0.22 and 0.28 mm; and those at 1500 to 5000 metres measure between 0.33 and 0.58 mm. The crustacean *Storthyngura fragilis* measures 12 mm when collected from water at 1.8°C, and 30 mm in water at 0.3°C. Birstein, however, criticises this hypothesis of increased size at lower temperatures. He believes that increased hydrostatic pressure, and not decrease of temperature, is responsible for the increased size of abyssal animals. Increased size is shown, however, by species near the surface in polar waters; the largest jelly-fish known, *Cyanea arctica*, measures two metres in diameter and lives in the cold waters of the Arctic.

The study of development at constant temperatures cannot be related to that under natural conditions. It has been found for some insects that growth rates at low temperatures are between five and eight per cent more rapid at fluctuating temperatures than at constant temperatures. Development over the middle of the temperature range takes place at the same rate for both constant and fluctuating temperatures, whilst at higher temperatures growth is retarded by fluctuation in temperature.

b   Variations in the number of generations per year.
   Estivation and hibernation
The rate of development in poikilotherms is dependent on ambient temperature, and species of tropical regions will, generally, grow more rapidly and have a larger number of generations per year than related species from temperate regions. For example, the fruit fly *Ceratitis capitata* has eleven or twelve generations in Honolulu, nine in Cairo, five in Algiers and only two in Paris (cf. figure 2.14). *Homodynamic* insects are species in which the number of generations depends on the length of the favourable season; *heterodynamic* insects, on the other hand, are those with a fixed number of generations. In the latter, development is spontaneously arrested, even when environmental conditions are still favourable. This break or *diapause* is essentially under the control of internal factors, but is also partly dependent on environmental factors, including temperature and daylength (cf. p. 107).

The state of *dormancy* (or *quiescence*) is immediately and directly brought about by unfavourable environmental conditions, and takes the form of either *estivation* or *hibernation*. Estivation is a temporary arrest of development which takes place when temperature is too high or, more frequently, when humidity is too low, the two factors often being related. Some dragonfly larvae are able to survive for several months in dry sand in Australia. Mosquito eggs laid in dry places remain dormant until the rains come and fill the depressions where they have been deposited. Hibernation takes place when temperature falls sufficiently low for all development to stop. The onset of hibernation can take place at relatively high temperatures for some tropical species (13°C for the cotton weevil).

Dormancy may occur at any stage in development. The animal usually finds a site where the microclimate is more favourable than the general climate. For example, coccinellids tend to aggregate in crevices, while some fish and amphibians group together in mud.

The relatively few hibernating homeotherms have one unique property. They are able to adapt themselves for a limited length of time to become poikilothermic, with the advantages that this offers. Their metabolic rate falls during hibernation, sometimes to as little as 1/30 or even 1/100 of the active basal metabolic rate. The body temperature of hibernating homeotherms varies with ambient temperature although, when this approaches 0°C, their metabolic rate shows a slight increase. Hibernating mammals include the dormouse, hamster, marmot and various other rodents, and also bats. In birds, hibernation is recorded for the swift *Apus apus*, which can reduce its body temperature from 38 or 39°C to about 25°C when it is unable to feed. The nightjar *Phalaenoptilus nutalli* hibernates in California, its rectal temperature oscillating between 18.0 and 19.8°C when the ambient temperature is between 17.5 and 24.1°C.

The number of generations per year is determined for poikilotherms, such as insects, by the onset of dormancy. According to Bodenheimer

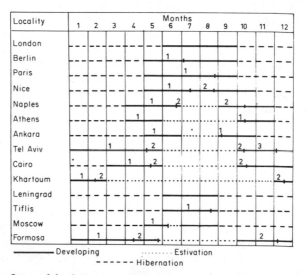

Figure 2.38   Stages of development, estivation and hibernation for the coccinellid *Coccinella septempunctata* in different regions. Figures show successive generations (Bodenheimer, 1938).

the coccinellid *Coccinella septempunctata* has one generation per year in London, Paris and Berlin, two generations in Nice and Naples and three generations in Tel-Aviv. The onset of dormancy occurs at different parts of the year according to geographical location and climate (figure 2.38). For scolytids, development cannot take place when the mean annual temperature falls below 1.5°C. When the mean temperature is between 2 and 6°C, the favourable period is too short for the insects to become established and aggregation is not possible. There is a single generation per year when the temperature is over 6°C, and when the temperature reaches 12°C there are two generations.

c   Other effects of temperature

Temperature affects the quantity of food eaten. A fourth instar Colorado beetle larva will eat 638 mm^2 of potato leaf at 36°C, but only 215 mm^2 at 16°C. There is a temperature threshold between 10 and 13°C below which the adult beetle will not feed, and the maximum amount of food is consumed at 25°C. There is also a daily cycle of feeding which is reversed above the optimum temperature. The insect feeds more during the day than at night below 25°C, and more at night than in the day above 25°C.

Temperature has an effect on fecundity which, in the Colorado beetle, reaches a maximum at 25°C, longevity being greatest at this temperature. The mean number of eggs laid per female per day increases, however, up to 34.9°C because oviposition becomes more rapid to compensate for the shorter life span.

The number of eggs laid by the lucerne weevil *Phytonomus posticus* is closely related to mean temperature. If, however, the daily number of eggs laid is increased by higher temperatures, the total number of eggs produced remains the same.

Song in the male, usually associated with sexual activity, often shows a clear relationship with temperature. Song frequency of crickets shows an almost linear relationship with temperature, while grasshoppers sing only when it is fairly warm. Australian desert frogs are able to breed at any time of the year and are able to take advantage of infrequent rain showers. The song of the male, however, is dependent on temperature and only occurs at a set time for each species.

1.8   Adaptations to extreme temperatures

a   Morphological adaptation

Several 'ecological rules' have been put forward to describe the relationship between environmental factors and morphological adapt-ations of animals. The rules of Bergmann and Allen relate to temperature.

Bergmann's rule states that, within any particular taxonomic group of homeotherms, the smaller-sized species generally occur in warmer regions, provided other environmental factors are similar. This rule can be explained by the fact that it is essential for homeotherms to maintain a constant body temperature. Heat loss occurs mainly from the surface of the body. The volume and weight of the body increase as the cube of linear dimension, while body surface area only increases as the square of linear dimension. Thus, the larger an animal becomes, the smaller its surface area/volume ratio; heat loss by convection is therefore, relatively, reduced. A sphere is the geometrical shape that, for a given volume, has the smallest surface area. It is, therefore, an advantage for an animal to develop the largest possible body, and to approach as nearly as possible to a spherical form.

Bergmann's rule is illustrated among the birds by penguins. The largest species, *Aptenodytes forsteri*, which has a height of 120 cm and weighs 34 kg, lives in the heart of the Antarctic continent and is rarely found north of latitude 61°S. The smallest species, *Spheniscus mendiculus*, measuring only 50 cm high, lives in the Galapagos islands near the equator. Between these two extremes there are a number of species like *Aptenodytes patagonica*, 90 cm to 1 m high and weighing 15 to 17 kg, from the Macquarie Islands (latitude 55°S); or *Pygoscelis adeliae*, living near latitude 66°S and measuring between 70 and 75 cm and with a weight of 6 kg; and *Spheniscus demersus*, which measures 55 cm, weighs between 5 and 6 kg and lives at latitude 34°30'S.

In the northern hemisphere a giant race of the puffin *Fratercula arctica naumanni*, with a wing length of 175 to 195 mm, occurs on Spitsbergen and in northern Greenland. Other races of *F. arctica* found on the coasts of Norway, south Greenland and Iceland have wing spans of only 155 to 177 mm, while those from Heligoland and islands off the British Isles and Normandy measure only 155 to 166 mm. There is also a dwarf race which overwinters in Majorca which has a wing length of only 135 to 145 mm. The largest humming bird, *Patagona gigas*, lives in the temperate zone and at high altitudes in the Andes.

A number of criticisms have been made of Bergmann's rule. It has been pointed out that races of the puma, raccoon and otter show an increase in size from north to south in contradiction to this rule. Scholander (1955) considered that the differences in size observed (usually less than 5%) were too small to have any significant effect on thermoregulation. He suggested that the insulating properties of the skin and fur play a much more important part in thermoregulation than body surface area.

Allen's rule is a corollary of Bergmann's rule. This states that there is a significant reduction in the surface area of body appendages in mammals from cold regions. The ears and tail are reduced, the neck and limbs are shorter and the body generally more stocky. The Old World foxes provide a good illustration. The fennec (*Megalotis zerba*) from hot deserts has very long ears, the European fox (*Vulpes vulpes*) has shorter ears, while the arctic fox (*Alopex lagopus*) has very short ears and a shortened muzzle.

The fur of mammals from colder regions is thicker than those from warm regions. This thickness increases with the size of animal, but there is a limiting thickness to the fur beyond which movement would be impeded. Smaller species are able to compensate for their reduced thermal insulation by hibernating in burrows beneath the snow where temperatures are more favourable than in the open air. Examples include the squirrel *Spermophilus undulatus*, the lemming *Dicrostonyx* sp. and the weasel *Mustela rixosa*. Mammals from tropical regions show no correlation between size, thickness of fur and its insulating properties. Two exceptions were described by Scholander. These are the sloths *Bradypus griseus* and *Cholaepus hoffmanni*, whose metabolic rate would be too low to maintain their body temperature during the fairly cold nights to which they are exposed at the tops of tropical forest trees where they live permanently. The same is true for the douroucouli monkey, *Aotus trivirgatus*, which is also exposed to cold in the same habitat. These three species have a fur with better insulating properties than that of other species that have been studied from tropical forests (figure 2.39).

Other morphological adaptations include seasonal changes of a type known as *cyclomorphosis* or *allotropy* which have been recorded for some cladoceran crustaceans. In *Daphnia cucullata* the crest on the head capsule is high and pointed in summer and low and rounded in winter. The same is true for another cladoceran, *Bosmina coregoni*, which also develops antennules that are twice as long in summer as in winter (Wesenberg Lund, 1910; Coker, 1939). These changes of form help flotation in water when it becomes less dense at higher summer temperatures.

Several morphological changes designed to protect the animal against excessive temperatures have been described for desert insects. There is a cavity underneath the elytra in some beetles (especially in tenebrionids) which affords some measure of thermal insulation. The body is insulated from the ground by long legs, and these are extended to their maximum as the ground becomes too hot. There are several types of pigmentation. A whitish incrustation, which reflects the sun's rays, covers the body of the tenebrionids *Zophosis complanata*, *Mesostema angustata* and *Sternodes caspicus*. Some tenebrionids of the genus *Pimelia* are covered with white pubescence. Other beetles of the genera *Anthia* (Carabidae) and *Rhytirrhinus* (Curculionidae) also develop a whitish coloration. The metallic colouring of buprestids and of chrysid wasps may also act as a protection against overheating by reflecting radiant heat.

In cooler regions, dull, matt colouration is more common, especially in polar regions and alpine habitats. If the two forms of the grasshopper

87

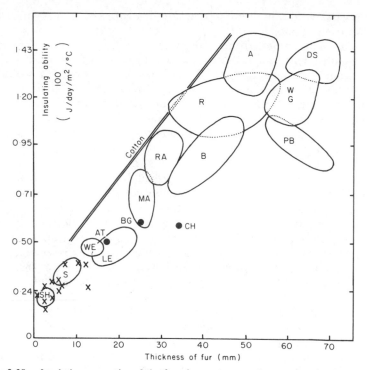

Figure 2.39   Insulating properties of the fur of several mammals plotted against its thickness. The insulating ability is shown as the ratio 100/joules lost per day per m², for a temperature difference of one degree across the fur. The lines marking out small areas represent mammals from arctic and temperate regions. The crosses and black dots represent tropical mammals. The double line represents the insulating ability of cotton. A: arctic fox; WG: wolf and grizzly bear; PB: polar bear; R: reindeer; DS: Dall mountain sheep; B: beaver; RA: rabbit; MA: marten; LE: lemming; WE: weasel; S: squirrel; SH: shrew (summer); CH: sloth *Choloepus hoffmanni*; BG: sloth *Bradypus griseus*; AT: nocturnal monkey *Aotus trivirgatus* (after Scholander, 1950).

Calliptamus sp. are exposed to sunlight, the dull form reaches a body temperature four to five degrees higher than the bright form.

b   Physiological adaptations

The physiological adaptation of organisms to unfavourable environmental conditions is known as *acclimation*. Some authors use the word *acclimatisation* for this kind of adaptation. The use of acclimation is often restricted to experimental investigations. This type of adaptation to temperature is achieved in homeotherms by changes in metabolic rate. Some tropical mammals begin to show a higher metabolic rate when the ambient temperature is about 25°C. At 10°C heat production is tripled, and at 0°C they die. At the other end of the scale, small arctic mammals only increase their metabolic rate when the ambient temperature is −30°C and larger mammals like the musk ox withstand temperatures as low as −40°C before their metabolism is increased. Adaptation to high temperature includes a reduction in metabolic rate and an increase in heat loss as a result of peripheral vasodilation and evaporation from the skin.

In poikilotherms, adaptation to high temperatures is also achieved by the evaporation of water (cf. the example of *Ligia oceanica*, see p. 56).

Animals respond to low temperatures by progressive acclimatisation. This process involves a gradual reduction, during the cold season, of the temperature at which the body fluids of an organism begin to freeze and thus bring about death. In aquatic insects, living in an environment where temperature variation is small, there is only a slight lowering of the lethal temperature in winter. It is a little more marked in the scarabid beetle *Popillia japonica*, which hibernates in the soil; and even more so in *Dendroides canadensis*, a beetle hibernating under bark and exposed to very low temperatures (figure 2.40). This progressive increase in cold resistance is also found in plants. It explains why mild

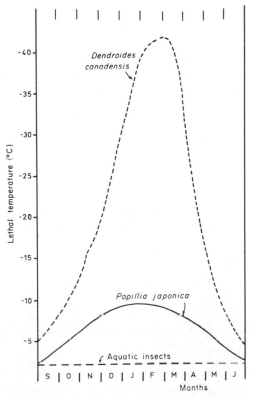

Figure 2.40   Cold tolerance in insects (Payne, 1927).

TABLE 2.9   The relationship between acclimation temperature and temperature at which cold torpor occurs in three insect species (Mellanby, 1939)

SPECIES	ACCLIMATION TEMPERATURE		
	14 to 17°C	30°C	36°C
Blatta orientalis	2.0°C	7.5°C	9.5°C
Calliphora erythrocephala, larva	1.0°C	5.4°C	–
Cimex lectularius	4.5°C	7.0°C	7.5°C

winters or alternating periods of freezing and thawing destroy a much greater number of insects than severe winters which allow time for cold adaptation to take place.

The temperatures tolerated are related to acclimation temperatures, and this has been demonstrated in three species of insect by measuring the relationship between the temperature at which cold torpor occurs and the acclimation temperature (table 2.9).

A similar process of acclimation takes place in fish (figure 2.41). In the cat fish, *Ameiurus nebulosus*, the upper lethal temperature, which in experiments results in the death of fifty per cent of the animals within twelve hours, is 35.8°C during August. In October it is 31°C, and during the winter 29°C (Brett, 1944).

Resistance to lower temperatures may be due to a progressive dehydration which increases the osmotic potential of the tissue fluids, thus lowering their freezing point. This has been shown to exist in Arctic species of chironomid flies, the larvae of which hibernate in frozen pools. At the same time the metabolic rate for insects gradually

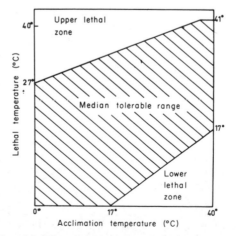

Figure 2.41   Variation of lethal temperature with acclimation temperature for goldfish (Fry, Brett and Clausen, 1942).

falls, between 0 and $-5°C$ the temperature quotient (Q_{10}) for respiration changing from a mean value of between two and four to a mean value of between twenty and fifty. The energy reserves that were sufficient for ten days are sufficient for 1000 days at $-23°C$, and this explains how insects can survive in frozen soils (Downes, 1965).

Other cold resistant insects contain significant quantities of glycerol produced from glycogen. The glycerol protects the tissues and acts as an 'antifreeze'. It can lower the freezing point of the tissue fluids to $-20°C$. The concentration of glycerol may be as much as twenty-five per cent of the live weight in the Canadian parasitic wasp *Bracon cephi*, which can survive temperatures below $-20°C$. A similar phenomenon is seen in the Labrador cod, *Gadus ogac*, where the freezing point of the blood plasma, which is normally $-0.8°C$ in summer, falls to $-1.6°C$ in winter as a result of the production of trimethylamine which acts in a similar way to glycerol in insects.

Another physiological adaptation to extreme temperatures is shown by a change in the timing of the active season in the year. Pierre (1958) described two distinct regions in the Sahara based on the activity of their insect faunas. In the northern warm temperate zone, the active season is the hot season and resting period is in the cold season. In the southern tropical zone of the Sahara, insects appear at the end of the hot season and during the cool season, and the resting period covers most of the hot season.

c   Behavioural adaptations
The behaviour of animals often allows them either to escape from, or seek out, the warmth of the sun's rays. The firebug, *Pyrrhocoris apterus*, gathers in large numbers at the feet of south facing walls in spring in order to obtain the maximum warmth. Migratory locusts climb up on the vegetation as the sun sets in order to remain in the warmest zone; and then, when the sun rises, they move down to the ground because this warms up more quickly. Many locusts orientate themselves during the day with their bodies parallel to the sun's rays in order to present the least possible surface area. In this way they maintain their body temperatures at between 2.5 and 4 degrees below that of the surrounding air (Chopard, 1938). The wood ant, *Formica rufa*, closes the openings of its large nest of pine needles at night, maintaining the temperature between $23°C$ and $29°C$, which is about ten degrees higher than the soil temperature. The termitaria of the 'compass' termite *Hamitermes meridionalis* in Australia are orientated with their long axis from north to south, and are thus less exposed to the sun's heat.

The research of Dorst (1967) on the humming birds inhabiting higher ground in the Andes provides a good example of the behaviour associated with low temperatures. These birds are found on the sides of steep slopes warmed by the sun which form 'oases of warmth in a cold desert'. *Oreotrochilus estella* not only nests on the sides of rocks, but also hunts for the insects and spiders that make up its diet in the gaps between the rocks where a flora adapted to this type of habitat has developed. Their nests are attached to the rocks by a sugary nectar which, during cold nights, conducts the heat stored in the rocks. These nests are also arranged facing east so that they are warmed by the sun's rays during the morning and yet sheltered at the hottest time of the day. Another feature of the nest of *O. estella*, and of many other birds in the Andes, is its large size. For a bird weighing between eight and nine grammes, the nest reaches a height of 100 mm and a width of 140 mm. The nests are constructed of plant material, hair and feathers. A furnarid, *Asthenes d'orbignyi*, about the size of a finch, constructs a nest of branches and twigs about fifty centimetres in diameter with an interior lined with a thick layer of plant debris and animal fur. The thermal insulation of these nests is very efficient, and their construction shows a remarkable adaptation to low temperatures. There are a number of other Andean birds that nest in holes sheltered from low temperatures and from unfavourable weather.

Many animals burrow into the ground to escape from high surface temperatures. According to Pierre (1958), out of 125 sand-dwelling insects in the Sahara, only three remain on the surface of the ground when the temperature of the sand exceeds 50°C. These are the ant *Cataglyphis bombycina* and two tenebrionids. The woodlouse, *Hemilepistes reaumuri*, burrows many centimetres into the soil by digging a hole five millimetres in diameter, when the temperature exceed 35°C in a dry soil or 45°C in a moist soil.

Changes in daily rhythm have been observed in animals from warm climates. Many desert species are nocturnal, and scorpions and insects are usually active only at night. In Palestine some species are active during the day in spring but only at night in the summer, and this includes many tenebrionids, Diptera and Hymenoptera. Reptiles show similar rhythms of activity related to changes in temperature (St.-Girons, 1956). The times of activity for the rodent *Psammomys obesus* are arranged so that it encounters the most suitable temperatures when it emerges from its burrow (cf. p. 57). The activity of the ant *Messor semirufus* outside its nest is dependent on temperature, and it rarely emerges in the early afternoon when temperatures reach 52°C (Bodenheimer and Klein, 1930).

2   THE EFFECT OF HUMIDITY

Water is an essential component of living protoplasm. As a general rule, the water content of protoplasm in active animals is between seventy and ninety per cent, although it is only fifty per cent in the larva of *Tenebrio molitor* and may reach as much as ninety-eight per cent in some jelly-fish. The water content of younger, rapidly growing tissues is

TABLE 2.10   The amount of water that can be lost by various animals without a fatal outcome (after Hall, 1922 and others)

SPECIES	% LOSS OF BODY WEIGHT	% LOSS OF WATER
Various species of terrestrial pulmonate molluscs	50 to 60	–
Crustacea: various woodlice	50	–
Annelids: *Lumbricus terrestris*	43	–
INSECTS:		
Tenebrio molitor, larva	53	100
Austroicetes cruciata, eggs in diapause	70	87
AMPHIBIA:		
Amblystoma punctatum	47	–
Rana pipiens	31	45
REPTILES:		
Chrysemys marginata	33	–
Scleroporus spinosus	48	–
MAMMALS:		
Peromyscus leucopus	31	–

higher than that of older tissues. The availability of an adequate supply of water and protection against water loss are important ecological problems for terrestrial animals (table 2.10).

2.1   The classification of living organisms according to their water requirements

Living organisms can be classified into several ecological groups according to their water requirements and, therefore, according to their distribution in various habitats. The following categories can be distinguished:

(i) *Hydrophile*, or aquatic organisms living permanently in water.

(ii) *Hygrophile*, or organisms restricted to habitats where the atmosphere is nearly or completely saturated. This group includes most adult amphibia, many terrestrial gastropods, earthworms and other soil animals, and most cave-dwelling animals, especially troglodytes.

(iii) *Mesophile* organisms, having less rigorous water requirements and surviving alternate wet and dry seasons. These include the majority of animals from temperate regions, together with most cultivated plants.

(iv) *Xerophile* species, living in dry habitats where there is a marked water deficit in the air as well as the soil. Xerophiles occur especially in deserts and coastal sand dunes and, among the animals, are mainly represented by insects and mammals. Cacti and lichens are the commoner xerophiles in the plant kingdom. Adaptations to drought are found in this category, and a good example is provided by the snail

Helix desertorum, which can survive for more than four years by estivating when the climate becomes too dry.

On the basis of ecological valency it is possible to distinguish between *stenohygric* species (i.e. xerophiles and hygrophiles), and *euryhygric* species (most mesophiles).

Some taxonomic groups show a range of species from xerophiles through to extreme hygrophiles, and each of these species provides a good indication of the environmental conditions in a particular habitat. Grasse (1929) established such a scale for Orthoptera, and this was completed by Dreux (1961). The more important species listed by these authors are as follows:

Hygrophiles	*Tettigonia cantans* *Pholidoptera griseoaptera* *Chrysochraon dispar* *Metrioptera roeselii*
Mesohygrophiles	*Decticus verrucivorus* *Omocestus viridulus*
Indifferent species	*Gryllus campestris* *Tettigonia viridissima* *Stenobothrus lineatus*
Mesoxerophiles	*Stauroderus scalaris*
Xerophiles	*Antaxius pedestris* *Ephippiger ephippiger* *Ephippiger bormansi* *Ephippiger terrestris minor* *Oecanthus pellucens* *Psophus stridulus* *Oedipoda coerulescens*

Maillet (1959) has drawn up a table (table 2.11) for Homoptera of the family Jassidae at Périgord.

TABLE 2.11   Water requirements of Homoptera of the family Jassidae

Extreme hygrophiles	*Jassargus sursumflexus* *Streptanus marginatus*	swampy grassland
	Cicadella viridis	wet grassland
	Macrosteles viridigriseus *Balclutha punctata* *Cicadula persimilis*	moist grassland
	Goldeus harpago *Goniagnathus brevis*	dry grassland
Extreme xerophiles	*Adarrus taurus*	very dry grassland

94

2.2   Water balance in living organisms

a   Water gain

Animals can absorb water in a number of ways:

(i) Absorption of water through the walls of the digestive tract in those species that drink.

(ii) The use of water contained in the food. Many desert animals never drink, and rely on this source of water (rodents like the kangaroo rat).

(iii) Absorption of water through the skin in Amphibia. In the frog *Rana pipiens*, it has been estimated that, at $20°C$, the amount of water absorbed through the skin daily and eliminated in the urine represents 31% of the total water content of the animal. Water can also be absorbed through the cuticle of some insects and mites. Water vapour can be absorbed from air which is not saturated by several species, including the Colorado beetle, the bed bug and *Tenebrio molitor*.

(iv) The use of metabolic water formed by the oxidation of fat. This method of obtaining water is used by a few rodents, and also by some insects living in very dry stored food products. The latter include the clothes moth *Tineola biselliella*, the caterpillar of which feeds on wool, the grease moth *Aglossa pinguinalis*, and the grain weevils *Sitophilus granarius* and *S. oryzae*. There is often a relationship between the water content of animals and that of their food. Robinson, 1928 (*in* Uvarov, 1931) has given the figures shown in table 2.12.

TABLE 2.12   The relationship between the water content of some animals and that of their food

SPECIES	FOOD	WATER CONTENT OF FOOD (%)	WATER CONTENT OF INSECT (%)
Sitophilus granarius	wheat grains	9 to 11	46 to 47
Sitophilus oryzae	wheat grains	15 to 16	48 to 50
Cyllene robiniae	dead wood	30 to 32	56 to 60
Leptinotarsa decemlineata	leaves of potato plants	70 to 74	62 to 66
Vanessa antiopa	willow leaves	70 to 73	77 to 79
Pieris rapae	cabbage leaves	88 to 89	83 to 84

b   Water losses

Water loss occurs through sweating, by evaporation through the integument, during respiration, via the urine and in the faeces.

2.3   Protection against dessication

This is achieved in a number of different ways. They can be grouped under three headings: reduction of water loss, the use of metabolic water and behavioural adaptations.

a   Reduction of water loss

Water loss can be reduced in a number of ways, both physiological and morphological. Some of the important ones are mentioned below.

(i) *An impermeable integument*   is essential for those animals that live in dry habitats. It is found amongst mammals, birds and reptiles and also in insects, where there are many species adapted to desert life.

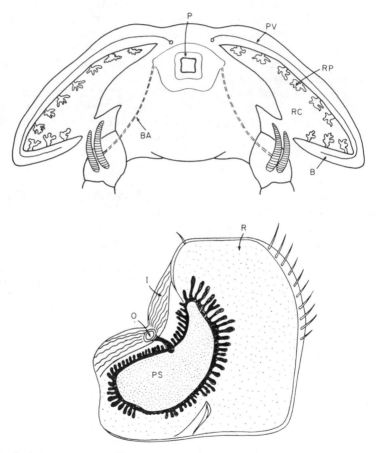

Figure 2.42   Above: Diagrammatic transverse section across the carapace of the coconut palm crab *Birgus latro*. PV: 'pulmonary' vessel running to heart; RP: vascular branched respiratory processes; BA: branchial artery; B: branchiostegite extended to enclose a large respiratory chamber RC; P: pericardium.

Below: exopodite of the first pleopod in the woodlouse *Porcellio scaber*, showing pseudotrachea. PS: pseudotracheal region; I: internal border of pleopod; R: ramus of pleopod; O: opening of trachea. Pleopod shown in transverse section.

Many crustacea are not well adapted to terrestrial life, and the integument remains permeable. This is true, for example, in the isopod *Lygia oceanica*, which is restricted to the littoral zone and which is able to benefit by the evaporation of water in order to regulate its body temperature (cf. p. 56).

(ii) *The development of internal respiratory organs*, which can take several forms. The gills are lost in terrestrial animals, and water loss from the respiratory organs is reduced in several ways. These include the tracheae in insects, myriapods and spiders, the 'lung' of pulmonate snails, and the lung of terrestrial vertebrates. They also include pseudotracheae located on the exopodite of the pleopod in the more terrestrial woodlice such as *Porcellio scaber* (figure 2.42). Some decapod crustacea, adapted to terrestrial life, are able to leave the water because the gills are enclosed by the carapace. In the coconut palm crab *Birgus latro*, which lives in burrows and climbs trees, the lining of the branchial cavity is richly vascularised and serves as a lung.

(iii) *The reduction of water lost by excretion* can be achieved by the production of a more concentrated urine, this method of water economy being used by many animals. Many vertebrates (especially reptiles), insects and terrestrial molluscs excrete solid uric acid instead of ammonia, which requires large quantities of water to reduce its concentration to a non-toxic level. In the lung fish *Protopterus aethiopicus*, nitrogenous waste is excreted in the form of ammonia while it is active in the water, but urea is produced during its terrestrial life when it is enclosed in a cocoon of mud and mucus. The reduction of sweat glands in rodents and desert antelopes, and the production of very dry faeces, are other methods of controlling water loss.

b . The use of metabolic water
Insects that feed almost entirely on dry food depend largely on metabolic water. The best known example is the beetle *Tenebrio molitor* (Govaerts and Leclercq, 1946). When the larva is fed on flour, it eats more in a dry atmosphere (at 0% relative humidity) than in a moist atmosphere (55% relative humidity). The growth rate of the larva is almost zero under dry conditions as most of the food taken is converted to metabolic water. According to Fraenkel and Blewett (1944), this is also true for the larvae of *Tribolium, Dermestes* and *Ephestia*. When adults of *Tenebrio molitor* are fed under optimum conditions of temperature and humidity, they show a gradual water loss as they become older. This dehydration is accompanied by an increase in the fat and dry matter content of the body. This conversion occurs at humidities between 0% and 86%, with an optimum situated between 35% and 75%. Adults that are then starved, survive for twenty days in a dry atmosphere and for twenty-four days in a moist atmosphere, this small difference showing how tolerant the species is of dessication. Bernard (1951) showed that some ant larvae in the Sahara make use of

metabolic water derived from fats. Their haemolymph also has a high protein content and this increases its osmotic potential, thus reducing the rate of water evaporation. The reduced size of the salivary glands in these ants may also result in some water economy. Adaptations to desert life are, however, most marked in mammals such as the camel and the kangaroo rat.

The camel may produce a little water by the oxidation of fats stored in the hump. It can also reduce its urinary excretion to only five litres per day. In winter, when it is feeding on green plants rich in water, it is able to remain for sixty days without drinking; while in summer it can survive for two weeks feeding only on dry plant material. The camel is able to survive for long periods without drinking for two reasons. Firstly, when the water content of the body is reduced, sweating is reduced, the body being able to withstand temperature increases of up to 6.2°C. This water economy is equivalent to about five litres per day for an animal weighing about 400 kg. Peripheral vasodilation takes place at night, and the camel loses excess heat by radiation, particularly near dawn when the air temperature is at a minimum. Secondly, the camel is able to withstand a water loss equivalent to more than thirty per cent of its body weight. Most other mammals cannot survive a loss of even twenty per cent. The camel, unlike other mammals, can quickly replace this loss when water becomes available (Schmidt-Nielsen *et al*, 1956.).

Despite these remarkable peculiarities, a camel working in full sun needs to drink every three days in order to carry out thermoregulation by sweating.

The kangaroo rat *Dipodomys merriami* is even better adapted than the camel. It is the only mammal which can produce adequate supplies of metabolic water. It excretes a very concentrated urine, and its faeces are very dry. It can withstand increases in body temperature and does not sweat as it is without sweat glands. The kangaroo rat is only active at night and spends the day hidden in its burrow. The figures in table 2.13 comparing three species of mammal show the extent to which water economy has been developed in *Dipodomys*.

TABLE 2.13   Comparison of water balance in the kangaroo rat, *Dipodomys merriami*, in the rat and in man (Schmidt-Nielson, 1964)

	DIPODOMYS	RAT	MAN
Water lost through evaporation in dry air, measured in mg water per ml of oxygen used for respiration	0.54	0.94	–
Concentration of urine (% urea)	23	15	6
Total electrolytes (as % NaCl equivalent)	8.5	3.4	2.09
Water content of faeces (as %)	45	68	–
Water lost in faeces per 100 g of barley eaten (grammes)	2.5	13.5	–

The kangaroo rat is also adapted to desert life through its ability to withstand some dehydration of its body tissues. In an individual feeding normally, the tissues contain 70.4% of water. In an individual fed for eight weeks on dry grain, the water content falls to 66.1%.

c Behavioural adaptations

Adaptation to life in burrows, where humidity is almost constant and usually high, is common in desert animals. Many are also active mainly at night, for example, rodents, and many insects of sand dunes and deserts, which remain buried in the soil at the base of plants during the daytime. Dispersive movements and migrations are also mechanisms that allow animals to escape from regions which become too dry.

2.4 Effect of humidity on animals

a Effects on longevity and rate of development

Two groups of animals can be distinguished. The first are water 'economisers' like the gazelle and some tenebrionid beetles from arid regions (for example the genus *Pimelia*). For these animals longevity does not change significantly when they are starved or exposed to different relative humidities. The second group are water 'wasters' like the tree frog *Hyla arborea* for which, under the same conditions, longevity increases with humidity and is greatly reduced at low relative humidities.

There is a linear relationship for many insects between rate of development and saturation deficit, for example in the flies *Musca domestica* and *Lucilia serricata*. In *Locusta migratoria migratorioides* the rate of development increases with relative humidity up to 70% and then decreases with further increases in humidity (figure 2.43).

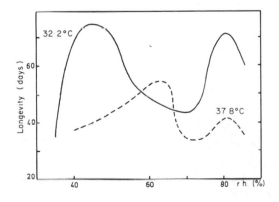

Figure 2.43   Effect of humidity on the longevity of adult *Locusta migratoria* for two different temperatures (Hamilton, 1950).

b Effect on fecundity

In eighty per cent of females of the noctuid moth *Panolis flammea* copulation is hindered and fewer eggs are laid at 100% relative humidity than at 90%. The rate of oviposition in the weevil *Sitophilus oryzae* is higher at a relative humidity of 70% than at 50% (figure 2.44). Individuals of *Locusta migratoria migratorioides* reach sexual maturity earlier at a relative humidity of 70%, and the fecundity of the females is higher at this humidity which seems, therefore, to represent an optimum value for many physiological processes in this species of *Locusta.*

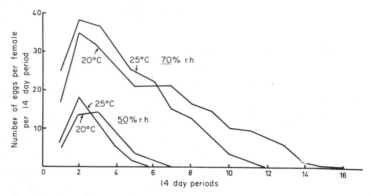

Figure 2.44   Effect of temperature and humidity on oviposition in the weevil *Sitophilus oryzae.*

c Effect on behaviour

The mosquito *Culex fatigans* stops feeding and biting if the relative humidity falls below forty per cent. The activity of the xylophagous beetle *Passalus cornutus,* which lives in small groups under the bark of dead trees, is reduced when humidity is increased. Activity is then increased if the humidity is lowered. If the tree trunks where these beetles live are cut up, they are exposed to lower humidities and their activity increases, causing them to escape and search for more suitable shelter.

Some species are very sensitive to changes in relative humidity. The crustacean *Ligia italica* is able to detect differences between compartments at 100% and 97% relative humidity and chooses the higher one. It is equally sensitive to humidity differences in the drier part of the humidity scale. The millipede *Orthomorpha gracilis* is also very sensitive to changes in relative humidity near saturation, but these reactions disappear at low humidities. These differences in behaviour between the two species are related to the conditions found in their habitats. *Ligia italica* is able to leave its refuge in broad daylight, should this become too dry, and run rapidly over the sunbaked rocks to find a new hiding place. This species must react to humidity differences in dry places as to stay there would prove fatal. *Orthomorpha gracilis*, on the other hand, is a tropical species which is never exposed to low humidities and which is not, therefore, adapted to avoid dry habitats, where it would quickly die.

Experiments have also shown considerable differences in reactions to humidity within the same species, depending on the physiological state of individuals. An insect like the beetle *Harpalus serripes* or *H. punctulatus*, which is more or less a xerophile, may show two different types of behaviour, depending on its physiological condition. Individuals reared under average humidity conditions (65%) and regularly fed tend to group together in the driest zone (25% to 40%), while those reared on dry sand for several months with limited food tend to collect in the most humid zone (90% to 100%) for the first three days.

d Effect on geographical distribution and on distribution within a habitat

In insects the most susceptible stages are usually the immature ones. Species of tse-tse fly which have a pupa with a relatively permeable cuticle deposit their larvae near water bodies, while those species having a pupa with a fairly impermeable cuticle deposit their larvae some distance from water. In the adults there is no difference, however, in resistance to drought, all species being strongly hygrophile.

Only the Isopoda (woodlice), among the Crustacea, have successfully colonised terrestrial habitats. The cuticle is still permeable in *Ligia oceanica*, which is a littoral species. It is less permeable in terrestrial species, such as *Oniscus asellus* and *Porcellio scaber*, but woodlice still require an atmosphere that is almost saturated, and the distribution of these terrestrial isopods depends largely upon the amount of water vapour present in the atmosphere. They are abundant at the coast, where the moisture content of the air is high; while, at high altitudes, where the air is dry, they are restricted to the underside of stones or under bark. Some species, in rainy weather, climb up on to bushes, and these include *Porcellio monticola* and *Philoscia muscorum*. A comparison of the fauna of a beech and spruce forest in the central Pyrenees (relative humidity 86% to 87%) with that of a pine forest at Cerdagne (relative humidity 57% to 62%) showed that the former was more diverse than the latter, which contained only a single species of woodlouse, *Trichoniscus pusillus*, which is widely distributed in western Europe.

The marked resistance of the chironomid larva *Polypedilum vanderplanki* to dessication allows it to colonise small water bodies exposed to full sunlight in Nigeria and Uganda. When these pools dry out, the larvae are able to survive for long periods. Their water content is less than four per cent at zero relative humidity, but, if they are exposed to

TABLE 2.14   Effect of dessication on the gastropods *Nucella lapillus* and *Littorina* spp. after seven days at 18°C

SPECIES	WATER LOST (AS % OF LIVE WEIGHT AT THE BEGINNING OF THE EXPERIMENT)	WATER LOST (AS % OF LIVE WEIGHT PER DAY)	% MORTALITY
Nucella lapillus	37.2	5.31	100
Littorina littorea	37.5	5.35	70
Littorina littoralis	56.5	8.35	80
Littorina saxatilis	39.7	5.60	8 to 17
Littorina neritoides	26.0	3.71	–

a saturated atmosphere at a temperature of 25°C, their water content increases to 33% within six hours (Hinton, 1960).

Locust migrations are partly determined by drought, which reduces the amount of food available to the nymphs. Mammals will move from one source of water to another without returning, and the progressive drying out of the Sahara during the Quaternary period has resulted in a permanent movement of larger savanna animals like the elephant and giraffe away to the south.

The role of water in the distribution of animals is clearly shown in the intertidal zone. Lewis studied the effect of dessication on *Nucella lapillus* and several species of *Littorina* (table 2.14).

These differences may, at least in part, be explained by differences in structure and behaviour. *Littorina littoralis* tends to open its operculum and to move about even in dry air, while *L. neritoides* carefully closes its operculum and has, in addition, a shell which is not permeable. Table 2.14 shows clearly how it is that *L. neritoides* is able to live higher up the shore than *Nucella lapillus*. It can be seen from Table 2.15 that, in terrestrial animals, uric acid excretion tends to replace the excretion of ammonia.

The vertical distribution of *Balanus* species up the shore is similarly limited by their resistance to dessication, and they do not occur much

TABLE 2.15   Uric acid content of the tissues of the gastropods *Littorina* spp. in relation to their distribution

SPECIES	DISTRIBUTION	URIC ACID IN THE TISSUES IN mg PER g OF LIVE WEIGHT
Littorina littorea	lower shore	0.21
Littorina littoralis	intertidal zone	0.37
Littorina saxatilis	intertidal zone, but higher than L. littoralis	1.05
Littorina neritoides	partly terrestrial, found up to 2 m above EHWS	1.83

above MHWN. tides. On a more exposed shore, species that are more susceptible to dessication may occur higher up the shore because the spray provides the necessary humid conditions for them. Algae show a clear zonation in the intertidal zone of rocky shores, each species occupying a level related to its exposure tolerance (cf. figure 14.2, page 336).

e Effect on population density
Relative humidity may affect the density of populations by reducing numbers of individuals when moisture conditions are unfavourable.

2.5 Simultaneous effect of temperature and humidity

Temperature and humidity, two basic environmental factors, often show some interaction in the way that they affect organisms. A number of examples have already been described (cf. p. 13). There are certain optimum conditions of temperature and humidity which result in the lowest mortality, the greatest longevity, the highest fecundity, most rapid development, etc. These effects have been mainly studied in insects.

Figure 2.45 shows the relationship between mortality rate, temperature and humidity for the pupa of *Carpocapsa pomonella* which is a

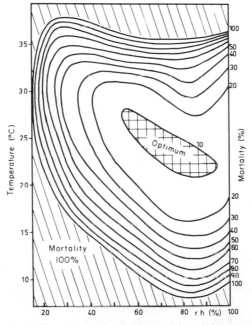

Figure 2.45   Combined effect of temperature and humidity on the rate of mortality for the pupa of *Carpocapsa pomonella* (Shelford, 1927).

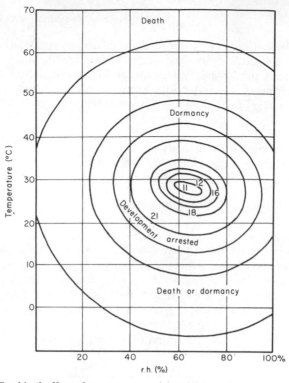

Figure 2.46 Combined effect of temperature and humidity on the life cycle of the cotton boll weevil, *Anthonomus grandis*. The developmental period (days) is shown in each zone.

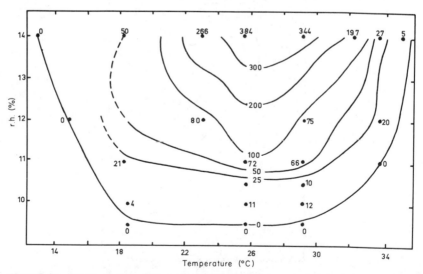

Figure 2.47 Number of eggs laid per female of *Sitophilus oryzae* in relation to temperature and humidity (Birch, 1945).

pest of apples. The figure shows that curves for equal mortality are concentric, and that the optimal zone is limited approximately by relative humidities of 55% and 95% and by temperature of 21°C and 28°C. Figure 2.46 gives similar data for the cotton boll weevil.

Fecundity is affected in a similar way by temperature and humidity (figure 2.47) in the weevil *Sitophilus oryzae*. Oviposition is highest at temperatures between 26°C and 28°C, and falls to zero when the relative humidity falls below 9.5%, irrespective of the temperature (Birch, 1945).

3   THE EFFECT OF LIGHT

Light is a fundamental environmental factor, and was classified by Monchadskii as a primary periodic factor. It also affects several physiological processes, especially photosynthesis, and light intensity has a considerable effect on productivity in ecosystems (cf. chapter 13).

The effect of light depends on its intensity, wave-length, polarisation, direction and duration. Its main ecological role lies in the synchronisation of biological rhythms of varying lengths, such as daily, lunar and seasonal rhythms. (The problem of light receptors belongs to the study of physiology and is not dealt with here. Similarly phototaxis and orientation belong to the study of behaviour.)

3.1   Biological rhythms

Three types of light-synchronised rhythms can be distinguished.

a   Seasonal rhythms induced by photoperiod

The rhythms induced by photoperiod are of two types. The first is synchronisation of the life-cycle within the seasons, so that the reproductive period coincides with the most favourable season. The second is responsible for the onset of diapause at a time when conditions are unsuitable for active life.

There are many examples of breeding seasons controlled by photoperiod in vertebrates, and a few are discussed here.

Among the fish, the brook trout *Salvelinus fontinalis* normally spawns in the autumn. If daylength is increased experimentally during the spring and then reduced in the summer in order to simulate the lighting conditions usually found in the autumn, the fish spawns in the summer (Hazard and Eddy, 1950).

Maturation of the gonads occurs as daylength increases in birds of temperate regions. The diagram in figure 2.48 shows how the timing of the breeding season changes when passing from the northern to the southern hemisphere. The extent of this change varies between twenty and thirty days for every 'ten degrees of latitude. There is also a reduction in the length of breeding season nearer the poles, although

105

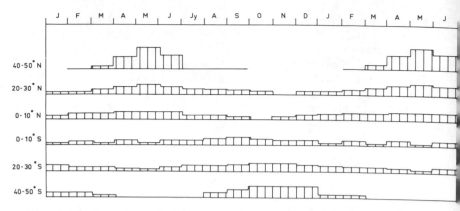

Figure 2.48   The timing of the breeding season and the relative intensity of egg laying for the common sparrow at different latitudes (Baker, 1938).

several orders of birds, such as the Anseriformes, do not follow this rule. In tropical regions, where the seasonal variation in daylength is not so marked, the relative importance of this factor is reduced, and it is the alternation between dry and wet seasons that determines the breeding period for many animals.

There is little variation in breeding season for mammals and birds within any particular region. Some mammals mate during the period of decreasing daylength. This is the case in most ruminants. Others, including the fox, many small carnivores and several rodents, mate during the period when daylength is increasing. There are marked differences with change of latitude. The length of the breeding season increases for the sparrow on moving towards the equator. A similar phenomenon, although less clear, is shown by those mammals that have a sufficiently wide north-south range, for example the sheep, which is now reared throughout the world. Movement from one hemisphere to another results in a significant displacement of breeding season; the fallow deer, elk and chamois, which were introduced into New Zealand, now reproduce in April although, in the northern hemisphere, they reproduce in October. The experimental alteration of photoperiod may result in considerable changes in the breeding season. In the case of the ewe in the Paris region, for example, mating begins near the beginning of September as daylength begins to decrease. If, however, the animals are kept under conditions of artificial light and the normal annual changes in daylength are condensed into six months, breeding takes place every six months and also begins as daylength increases, rather than decreases.

There are other types of seasonal rhythm, especially in mammals. The growth of the coat in the mustelid *Mustela cicognani* is more rapid when the photoperiod is short than when it is long. In the blue hare *Lepus timidus*, the white winter coat develops as a result of shorter

106

photoperiod. Reversal of the seasonal rhythm of photoperiod for the ewe also reverses the cycle of wool growth, while temperature variations, on the other hand, have little effect.

The other type of rhythm induced by photoperiod is that responsible for the onset of diapause, of which only the ecological aspects are considered here. Diapause consists of a prolonged arrest in development which takes place at a fixed stage in the life cycle of a particular species. It shows no obvious correlation with any environmental factor, and can be broken only with difficulty. Diapause is fundamentally different from quiescence (cf. above p. 83), which begins immediately conditions of temperature, humidity, etc., become unfavourable.

The changes which take place during diapause are essentially of a biochemical nature and are under hormonal control. This explains why the reactions of arthropods during diapause are entirely different from those of normal individuals. For example, the diapausing eggs of *Cacaecia murinana* show a much greater resistance to cold, while the eggs of *Bombyx mori* can resist strong acids and doses of X-rays.

Although especially common in insects, diapause is not limited to that group. It also occurs in mites, crustaceans, rotifers and sponges. A similar process is found in plants, where seeds, bulbs and tree buds represent a cessation in development similar to that found in insects.

Diapause appears at different stages. There is an embryonic diapause in the egg of *Bombyx mori*, a larval diapause in *Hyponomeuta padella, Thaumatopoeia pytiocampa* and *Balaninus elephas*, a pupal diapause in *Mammestra brassicae* and *Rhagoletis cerasi*, and an imaginal diapause in the water beetle *Dytiscus* and the Colorado beetle. In the last species there are three phases in the life of the adult. The first, or maturation phase, begins immediately after the emergence of the adult. The insect feeds and builds up sufficient reserves for the other two phases. The diapause phase which follows should not be confused with a period of quiescence although it corresponds to either a summer pause or a period of hibernation underground during cold weather. In females, the oocytes stop growing and maturation does not take place until diapause is broken. The reproductive phase appears after diapause is completed, and corresponds to a return to active life. Death quickly follows this stage.

Photoperiod is an important factor in the regulation of diapause. It provides the stimulus which causes the animal to enter a resting state before bad weather begins. Other factors, such as temperature, are unable to synchronise life cycles with seasons because their variation is not sufficiently regular. These ideas, due to Lees (1955), are supported by Danilevskii's observations. He found that the different races within a species are always adapted to local photoperiod. Danilevskii determined the critical photoperiod which must be exceeded for continuous development to occur in the moth *Acronycta rumicis* for individuals taken from Abkhazian (43°N), Belgorod (50°N), Vitebsk (55°N) and

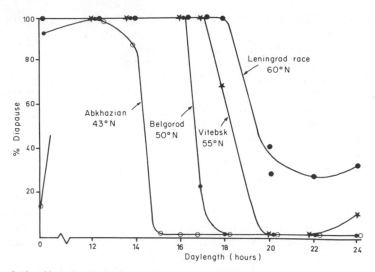

Figure 2.49   Variation in the daylength necessary to prevent diapause for several geographical races of *Acronycta rumicis* at a temperature of 23°C (Danilevskii, 1957).

Leningrad (60°N). This critical photoperiod was fourteen hours thirty minutes for individuals from Abkhazian, sixteen hours thirty minutes for those from Belgorod, eighteen hours for those from Vitebsk and nineteen hours for individuals from Leningrad. The photoperiod necessary to prevent the onset of diapause increases by approximately one hour thirty minutes for every step of five degrees latitude northwards (figure 2.49).

Two different arthropod groups can be distinguished on the basis of their response to the length of daily photoperiod. The first type shows a long-day response. In these species a long daily photoperiod allows continuous growth and development, while short photoperiods initiate diapause. Examples of this type of arthropod include *Acronycta rumicis* and many other insects. The other type show a short-day response which is the reverse of the long-day response, an example being that of *Bombyx mori*. The influence of temperature may partly mask that of photoperiod. In the mite *Metatetranychus ulmi*, nearly 100% of eggs laid enter diapause at 15°C but only 30% at 25°C, when the photoperiod is eight hours.

Daylength is not sufficient by itself to explain diapause. Lees showed that alternation of light and dark is important. The percentage of females laying diapause eggs differs in *M. ulmi* according to whether the dark period is four hours, eight hours, twelve hours or twenty-four hours in length (figure 2.50). Plants show a similar phenomenon, and here photoperiodism plays a very important part.

Some experiments have been made on variable photoperiods which

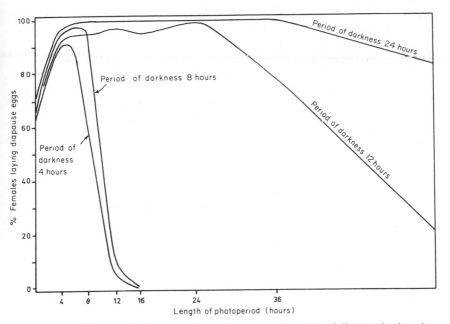

Figure 2.50   Effect of length of period of darkness on the onset of diapause in the mite *Metatetranychus ulmi*. As the period of darkness is increased from 8 to 12 hours, there is a significant increase in the number of diapause eggs laid (Lees, 1953).

resemble natural conditions more closely. Corbet (1957) found with the dragonfly *Anax imperator* that development was continuous, without diapause, when the photoperiod was increased with small increments corresponding to the conditions found in Great Britain in April. There was a diapause, however, if the photoperiod was decreased or held constant, corresponding to conditions found in July.

Photoperiod can also affect the morphological appearance of insects. The most familiar example is that of the nymphalid butterfly *Araschnia levana*, which occurs as the typical form in the spring and as the form *prorsa*, with different colouration, in summer. With a long photoperiod of eighteen hours per day, diapause is suppressed and only the form *prorsa* develops; while a photoperiod of only eight hours induces diapause and produces butterflies of the type *levana*. In the homopteran *Stenocranus minutor*, a photoperiod of eighteen hours during the nymphal period results in an imaginal diapause and produces normal adults. If the photoperiod is reduced to eight hours, dwarf adults are produced without any diapause. The photoperiod of eighteen hours represents a critical value for diapause and for size of adult.

It seems likely that morphology and diapause are both determined by endocrine function, the latter being under the control of photoperiod. As a general rule, it can be said that the phase in the life-cycle subjected to diapause is not the one directly affected by the conditions of illumination, but the previous phase. In *Araschnia levana*, for example, which undergoes diapause in the pupal stage, it is the last two larval instars which are influenced by photoperiod.

Other environmental factors may modify the effect of photoperiod. High temperatures tend to suppress diapause and low temperatures induce it, although usually within limits set by photoperiod, which is normally the controlling factor. According to Lees, high temperatures can only act during darkness, and they have a different effect from that of light. Drought can also initiate diapause, but it is necessary to distinguish between diapause and quiescence. Development is resumed in the latter as soon as water becomes available again, while this does not happen in the case of diapause.

Although temperature only plays an ancillary role in determining the initiation of diapause, it has a considerable effect on the subsequent resumption of growth and development. The length of diapause varies according to the species. Exposure to a temperature of 8°C for nine days is sufficient to break diapause in the shield bug *Eurydema ventralis*, while in the case of the saw-fly *Cephus cinctus*, an exposure of at least 100 days at 10°C is required. There is frequently an overlap between those conditions required to end diapause and those that are characteristic of the unfavourable season. The optimum temperatures for diapause development in the Palaearctic moth *Saturnia pavonia* occur between −15 and +7°C, while for species from temperate climates they are between 0 and +12°C. In the moth *Diparopsis castanea*, the larvae of which feed on cotton, the dry season is passed in a pupal diapause. This diapause lasts twenty-four weeks but, if the temperature is higher, and this is often the case under natural conditions, diapause lasts even longer.

Photoperiod induces diapause in several aquatic invertebrates as well as insects. Pourriot (1963) showed that, for some species of rotifer, the appearance of mictic females, which develop from resting eggs, is related to daylength. It is possible to produce mictic females, and thence males, in *Notommata copens* using a photoperiod of thirteen hours in length. This species reproduces during shorter photoperiods only by parthenogenesis. In Cladocera, the two factors responsible for the appearance of males and of females producing an ephippium (a case for fertilised eggs) are a temperature of 11°C and a photoperiod of twenty hours. Under these conditions, four times more females produce 'winter' eggs than under other conditions (Hutchinson, 1967). Ovarian development and moulting in an American crayfish of the genus *Cambarus* are both controlled by the length of illumination.

Photoperiod plays an important part in plants and, in many species, is the factor that controls flowering. Some plants (long-day species) require a daylength of more than eight hours for flowering to occur (spinach, *Calluna vulgaris* and spring wheat). Others (short-day plants) require a shorter daylength, and flowering takes place after daylength has been reduced to less than eight hours and the nights have become much longer (chrysanthemums). Plants from tropical climates, where day and night are of almost equal length, are short-day plants. Plants of subarctic regions flower during summer and are therefore long-day plants, and this distinguishes them from related species in alpine regions. Natural selection has resulted in the appearance of photoperiodic ecotypes. Thus several species of pine are affected by long days in mountainous areas but not at low altitudes.

b Circadian rhythms

Rhythms with a periodicity of about 24 hours are generally called circadian rhythms. They are maintained by an endogenous mechanism, a 'physiological clock', that is not fully understood. Several environmental factors, particularly temperature and light, serve to regulate this physiological clock (Pittendreigh, 1960). Circadian rhythms persist for several days when animals are subjected to constant illumination or to permanent darkness. The rhythm may persist for three months in the beetle *Bolitotherus cornutus* (Park and Keller, 1932). An inversion of the rhythm can be obtained by a reversal of the lighting regime (i.e. illumination during the night and darkness during the day) for many species such as the cockroach *Periplaneta americana*. The rhythm in activity may depend on the habitat. The carabid beetle *Steropus madidus* is active for a twelve hour period, and this activity occurs during the day on open ground but at night in woodland.

The mechanism for the physiological clock has been investigated in some animals. Harker (1956) showed that, in *Periplaneta americana*, the sub-oesophageal ganglion contains neurosecretory cells which are active throughout the twenty-four hours. These cells can maintain their rhythm for several days after the ganglion is implanted in another animal. A ganglion placed in a decapitated cockroach induced in that individual a pattern of locomotory activity that was clearly rhythmic. These neuro-secretory cells are regulated by stimuli transmitted from ocelli which are sensitive to the day and night cycle.

There are numerous examples of circadian rhythms. Bats emerge from their roosting places at dusk, and they retain this rhythm even when kept in complete darkness in the laboratory, or even when they receive a different pattern of illumination from that which corresponds to the succession of days and nights. Many species have a daily activity pattern that varies with the seasons. This has been studied in the jackdaw *Corvus monedula* over a period of a year at a latitude of 49°35′N. This bird roosts earlier and resumes its daily activity later in summer than in winter, in relation to the times of sunset and sunrise. This adjustment reduces the effect of seasonal variations in daylength on the period of activity. The activity pattern of the great tit *Parus major* shows compensation for seasonal and latitudinal variations in

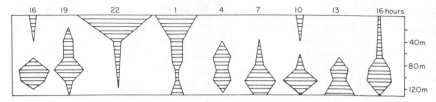

Figure 2.51   Daily vertical migrations of females of the copepod *Calanus finmarchicus* off the coast of Britain. Abundance at each depth is represented by the widths of each diagram.

daylength by going to roost in summer at latitude 68°N at almost the same time as at latitude 52°N, despite the differences in the timing of sunset.

The marine dinoflagellate *Goniaulax polyedra* is a photosynthetic and luminescent species. Three different circadian rhythms have been described for this unicellular organism. The first is in photosynthesis, the second in luminescence and the third in cell division. These three types of activity have maxima which follow one another at regular intervals of several hours. The photosynthetic peak is in the daytime, and the other two are at night. The rhythms in luminescence and cell division persist in cultures kept in darkness, demonstrating the existence of a physiological clock even at the level of unicellular organisms. Recent research has shown that synthesis and metabolism of DNA and RNA are closely correlated with the operation of these physiological clocks.

A good example of daily movement is that of many planktonic organisms which occur at the surface at night and which migrate into deeper water during the daytime as a result of their light sensitivity (figure 2.51). These vertical movements may take place through a depth of one hundred metres, or even more in some instances.

c   Lunar cycle

Lunar cycles are well known for many marine animals, but may also occur in fresh-water and terrestrial animals, according to some authors.

Lunar periodicity is well illustrated in the annelid polychaete worms. In the southern Pacific, the 'palolo' *Eunice viridis*, swarms at the surface of the sea during the last quarter of the October-November moon in such numbers that the water can be compared with vermicelli soup. The inhabitants of Polynesian islands collect them in large quantities, and these appearances are related to important religious feasts. (In many polychaetes, sexual forms have a different appearance from other individuals. The sexual form or *heteronereis*, is pelagic, swims, and swarms in large numbers when conditions are suitable). In the Cherbourg area, *Perineris cultrifera,* from thee upper shore in the Fucus and Ectocarpus zone, swarms on the flood tide during the day on the second, third and fourth days after the first quarter of the moon in

May. Individuals living at the level of LWST, however, swarm at night, during May and June, between full moon and the first quarter, with a maximum at the time of the new moon. The time of swarming in *P. cultrifera* thus appears to be related to its habitat. In the case of *Platynereis dumerilii* at Concarneau on the Atlantic coast, the heteronereis form appears at the surface during each quadrature, either in the first quarter or the last quarter of the moon, and the timing of the reproductive cycle in *P. dumerilii* can be explained by the effect of photoperiod.

The grunion *Leuresthes tenuis* spawns three or four days after the new or the full moon, corresponding to the highest tide each month between April and June on the sandy shores of California. Although this small fish is normally found in the open sea, it allows itself to be carried on to the shore by the waves during the highest tides, which occur at night. As the tide begins to recede, the grunions half bury themselves in the wet sand where the females lay their eggs, these being then fertilised by the males. The fish then return to the sea on the next wave. Because the eggs are laid during the period when tides are falling, they are not covered by the tide again for about two weeks and they complete their development in the sand without being washed away. At the next spring tide the sand is disturbed and the young fry are carried out to sea. The reasons for this remarkable synchronisation, between reproduction and development in the grunion and the tides and moon's phases, are not fully understood.

There is also evidence of a lunar cycle in the cilliate *Conchophthirius lamellidens* found on the gills of the fresh water bivalve *Lamellidens marginalis*. The frequency of conjugating individuals is much higher at the time of the new moon.

3.2 Effect of light intensity

Light of high intensity, especially in the blue and ultra-violet regions of the spectrum, has considerable ecological effects, although these are not well understood. A high proportion of ultra-violet light reduces the rate of oxygen consumption by aquatic organisms. Insects at high altitudes show a progressive darkening in colour with increase in altitude. This pigmentation is the result of several factors like cold, snow and higher humidity, but is also related to increased light intensity, especially ultra-violet light. While this increased pigmentation serves as a protection against ultra-violet light, it also absorbs some radiation and so raises the body temperature of the insect.

Activity in some insects is determined by light intensity The flight of the cockchafer *Melolontha melolontha* depends on the timing of sunset and, from April to June, becomes later each day. The fact that two populations may have different times of activity can lead to the formation of distinct biological races, although these may live in the same habitat. The small chafer *Rhizotrogus solstitialis* flies in the evening. In Serbia, however, there is a race *matutinalis* which flies in the morning. There is no morphological difference between these two forms, but they are totally separated and there is no possibility of interbreeding because their times of activity are different.

The particular conditions of temperature and illumination which determine the time of adult emergence in the mayfly *Oligoneuriella rhenana*, have been defined by Pinet (1967). An air temperature of 15°C or more is necessary for a large emergence of adults. The time of emergence changes regularly with the time of sunset as the season advances. Observations show a clear correlation between air temperature and light intensity when adults of *Oligoneuriella* begin to emerge. If temperature is high, light intensity must be low, but if the temperature is about 15°C, then illumination must be much higher for emergence to take place. This explains the lack of emergences on rainy days. These observations do not explain, however, why *Oligoneuriella* only emerges in the evening. Suitable conditions of temperature and illumination occur in the morning, but flight has never been observed. It would appear that an internal rhythm overrides environmental

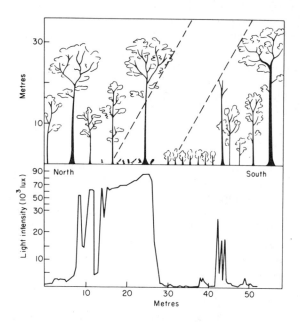

Figure 2.52   North-south section across a glade in acid oak-beech woodland in the forest of Fontainebleau showing distribution of light (in 10^3 lux) across the surface of the field layer and over bare ground at about mid-day in sunny weather during August. The zone of direct sunlight is covered by bracken and the shaded zone by scrub of hornbeam and beech (Lemée, 1967)

114

factors. It is also possible that variations in the intensity of different types of radiation, like red light, may play some part.

Light also is necessary for swarming to take place in the Scolytidae (bark beetles). Many species occupy only branches that are in direct sunlight, and these include *Pityogenes chalcographus* and *Cryphalus piceae*.

Plants can be divided into shade-loving species and those that prefer direct sunlight. The former, *sciaphiles*, are found in woodland (*Oxalis acetosella* and *Asperula odorata*) or in crevices between rocks (ferns and mosses) for example (figure 2.52). The second, *heliophiles*, occur in more open situations like heathland, waste land (rosemary, *Helianthemum*) and in fields (most crop plants). There are obviously intermediates between these two extreme conditions. In aquatic habitats, illumination decreases with depth, resulting in some selection of species. This is apparent in the sea where only red algae are able to survive below a certain level because they have abundant pigment capable of using the deeper-penetrating blue light. Light intensity also affects the degree of photosynthesis, and so acts as a limiting factor.

3.3 Effect of different wavelengths

The fact that short wavelength ultra-violet radiations have characteristic effects distinguishes them from visible radiation. They show intense bactericidal activity, especially below a wavelength of 365 nm. In addition they are essential for the production of vitamin D by the irradiation of sterols. In the sea, plankton irradiated at the surface accumulate vitamin D which is then transferred to fish through food chains.

It has been shown that ultra-violet radiation could affect population density in the pronghorn antelope, *Antilocapra americana* in the Yellowstone national park. The number of young born each year is related to the rainfall of the preceding year and to the amount of ultra-violet radiation received by the animals.

4   EFFECT OF SECONDARY CLIMATIC FACTORS

Included in this category are those components of the climate which are relatively unimportant or are not properly understood.

4.1   Wind

The first effect of wind is an indirect one. It can increase or lower the temperature according to the circumstances. The Mistral in the Rhone valley has a marked cooling effect. The Sirocco, blowing north in the Mediterranean region from North Africa, is capable of increasing the

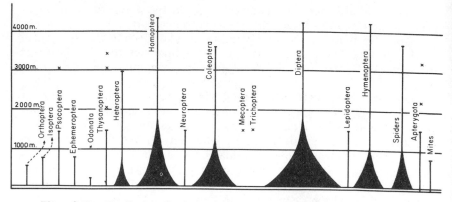

Figure 2.53   Distribution of arthropods in the aerial plankton in relation to altitude.

temperature by several degrees. It also tends to increase the rate of evaporation, and is therefore a drying wind.

In windswept localities, plant growth may be restricted and the fauna partly or completely eliminated. The most important effect of wind, especially on invertebrates, is in the dispersal of animals. It is possible to find, in the Antarctic, insects which have been transported by the wind over distances of several thousand kilometres.

Insect activity may be substantially reduced by wind. Rudolfs (*in* Uvarov, 1931) shows a marked decrease in mosquito activity, irrespective of temperature, when the wind speed exceeds 12.87 km/h. The harvesting ant, *Messor barbarus*, does not emerge from its nest while the Sirocco is blowing (Stager, *in* Uvarov, 1931). Wind is also a controlling factor in the orientation of the flight of migratory locusts.

Aerial plankton, comprising mainly small arthropods, consists of animals that are carried up to considerable heights, in large numbers, by air currents (figure 2.53).

4.2   Atmospheric pressure

It has been known for a long time that insect activity increases when atmospheric pressure is low, for example before storms. At similar temperatures, catches of insects in traps are larger during periods of low pressure (Wellington, 1954).

The experiments made by Edwards (1961) on the blowfly *Calliphora vicina* consisted of varying the atmospheric pressure within very narrow limits of about one millibar per hour. These variations resemble those occurring under natural conditions in any particular locality. Insect activity was found to increase significantly as pressure was reduced (figure 2.54). Wellington considered that the organ sensitive to pressure changes was located on the arista, the terminal part of the fly antenna.

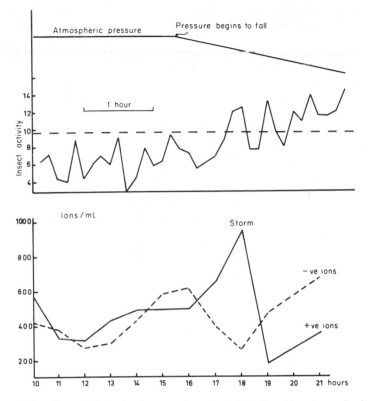

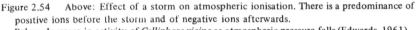

Figure 2.54   Above: Effect of a storm on atmospheric ionisation. There is a predominance of positive ions before the storm and of negative ions afterwards.
Below: Increase in activity of *Calliphora vicina* as atmospheric pressure falls (Edwards, 1961).

According to Edwards the lowering of pressure resulted in a general reaction of the whole organism, and this interpretation alone can explain the prolonged effects that were observed.

4.3   Ionisation of the air

The atmosphere contains charged particles or atmospheric ions. The density of these ions depends on local conditions, and they may be either positively or negatively charged. Oxygen ions are of particular interest to the ecologist. The oxygen atom is electrically neutral, but it can acquire an electron and become a negative ion or it may lose an electron and become a positive ion. The main causes of ionisation in the atmosphere are different types of natural radiation, especially radioactive and cosmic radiations. Secondary causes, which may be important locally, include some chemical reactions (forest fires increase

117

the numbers of positive ions in the air), electrical discharges during storms, and the stirring of water by waterfalls and waves. Ions may occur in the unstable monomolecular state, or as larger, more stable ions formed by the adhesion of several small ions to solid or liquid particles, such as water droplets or smoke and dust particles.

Marked changes occur in the ionisation of the air before a storm. Positive ions predominate before a storm, while negative ions are most common afterwards (figure 2.54).

The biological effects of atmospheric ionisation have been the subject of considerable research, but there still remains work to be done because of the difficulty in experimentation. Increased numbers of positive ions cause a rise in the activity of some insects, as Edwards has shown for *Calliphora vicina*. He also found, however, that, if exposure is prolonged, the flies adapt to the ionisation and activity returns to the normal level. Negative ions have no effect. In the aphid *Myzus persicae*, negative ions interfere with moulting while positive ions have no effect. Various climatic conditions can act as sources of ions. The Foehn, a wind blowing in the Alps from the south towards the north, contains many more positively charged ions than negative ones.

Negative ions have a beneficial effect for mammals and may even be essential for life. Experiments show that positive carbon dioxide ions reduce the frequency of ciliary movement in the trachea, causing vaso-constriction and increased respiratory movements. Negatively charged oxygen ions have the opposite effect. The effects of positive ions in mammals are similar to those of 5-hydroxytryptamine. It is thought that positive ions result in an accumulation of 5-hydroxytryptamine and that negative ions assist in its removal. Negative ions also have an effect on cytochrome-*a* resulting in more rapid oxygen movement for oxidation reactions.

4.4   The electrical field

This is a factor which has been little studied. Flight activity in the fruit-fly is temporarily reduced by sudden exposure to an electrical field of 10—62.5 volts per cm. This reduction in activity can be extended if the electrical field is reversed every five minutes. The insects react, therefore, to a change in the field rather than to the field itself. Edwards suggested that the insects were able to detect the field as a result of the accumulation of electrical charges on their bodies.

Three natural phenomena have been described here — atmospheric pressure, ionisation and the electrical field — all of which may vary simultaneously, often causing variations in the activity of insects. There is a need for further research into these phenomena, and for animal groups other than insects and mammals to be studied.

Chapter 3

ABIOTIC FACTORS: WATER AND SOIL

I THE AQUATIC ENVIRONMENT

1 WATER AS A MEDIUM

The action of water as an environmental factor is largely determined by its physical properties and movement.

1.1 Density

Changes in density, with a maximum density at 4°C, explain how ice forms at the surface of water bodies first, providing a protective layer, instead of at the bottom. Changes of density with temperature are responsible for the mixing of water in lakes at various times of the year (cf. p. 65). The high specific heat of water enables it to act as a buffer against temperature changes.

Many planktonic organisms have flotation devices, allowing them to maintain their position in open water. Examples include the swimming bells of siphonophores, the float of *Pyrosoma* and cytoplasmic gas vacuoles in radiolarians. A reduction in density may be brought by the accumulation of oil droplets in the cytoplasm (diatoms and even in large animals such as the Opah or moon fish, *Lampris guttatus*), by the reduction or disappearance of calcareous parts (the shell in heteropods and some other molluscs, the carapace of planktonic crustacea), or by increasing the water content of the tissues (over 95% in medusae). Most planktonic organisms are colourless and transparent. Expansion of the body surface may assist flotation. Examples include lateral expansion of the foot in pteropod molluscs, the bell of medusae, elongation of appendages in the copepod *Calocalanus pavo*, compression of the body in some diatoms and the phyllosoma larva of the spiny lobster and the foliaceous parapodia of the polychaete *Tomopteris*. Finally, frictional forces may be increased by a reduction in size, which increases the surface area/weight ratio thus increasing buoyancy. This explains the small size of many planktonic organisms.

1.2   Surface tension

The work of Baudoin (1955) has shown the importance of surface tension to those arthropods which live on the surface of water. Surface tension is the result of the forces of attraction between the molecules in a liquid, and acts in a plane at a tangent to the surface. It is measured in millinewtons (mN) per metre, having a value, for pure water, of 76 mN/m at 0°C and 73 mN/m at 20°C.

Arthropods are supported on water surfaces by surface tension forces acting on the foot, which is covered by hydrofuge hair. Figure 3.1 shows how the water surface is deformed by the foot. The surface tension produces a resultant force, F the carrying force, which is directed upwards and can be calculated from:

$$F = \lambda\, l \cos \alpha$$

where λ is the surface tension

   l is the perimeter of the organ in contact with the water
   α is the angle made by the meniscus with the vertical
   $(90° < \alpha < 180°)$.

In the case of a wettable body (i.e. hydrophile), α lies between 0° and 90°, the carrying force is directed downwards and the object tends to sink.

If F is greater than the weight of the animal (Archimedes force being negligible), the animal will float as the result of surface tension. Life on the water surface is only possible for fairly small animals, since weight increases as the cube of the linear dimensions, while surface tension only increases as the linear dimension. A larger species must, therefore, increase its area of contact with the surface, and this happens in *Gerris*, the largest pondskater, where the entire tarsus and tibia of the posterior legs are in contact with the water surface. In these pondskaters, the

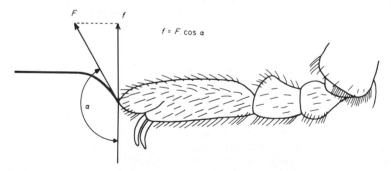

Figure 3.1   Tarsus of the fore-leg of *Velia*, a hemipteran living on the water surface. The water surface is deformed by the foot at an angle α which lies between 90° and 180°, producing a carrying force F which is directed upwards (Baudoin, 1955).

120

body is carried by the anterior and posterior legs while the middle legs function as oars. Similar adaptations are found in a range of insects including dolichopodid flies (*Hydrophorus*), bugs (*Velia*) and a caddis-fly (*Acrophylax*). Small animals, e.g. Collembola, rest on the water by the tips of the claws on their legs, and are not hindered by surface tension forces, i.e. they stand on 'tip toe'.

The scarcity of arthropods on the surface of the sea is in marked contrast to their abundance on freshwater. The only animals occurring on the surface of the sea are probably bugs of the genus *Halobates*. It is not, however, the salinity which prevents insects from colonising the sea (cf. p. 128). Baudoin showed that sea water, especially in the inshore zone, is rich in substances which lower the surface tension and are probably produced by algae. These compounds produce a foam which would trap insects, and they also lower the surface tension to 59 mN/m, which would be unable to support the weight of insects.

1.3   pH

pH also influences the distribution of aquatic organisms. Amongst plants, *Isoetes and Sparganium* grow in water of pH less than 7.5, several *Potamogeton* species and *Elodea canadensis* live in water with pH between 7.7 and 8.8, while at pH 8.4 to 9.0, *Typha angustifolia* is abundant. The acid waters of peat bogs contain *Sphagnum* mosses but molluscs are scarce because of the absence of calcium. Dipterous larvae belonging to the genus *Chaoborus* are common in lakes having acid water. Fish generally tolerate a wide range of pH from 5 to 9. Where the pH is less than 5, however, there is a high mortality, although some species adapt to a pH as low as 3.7. The productivity of fresh waters is considerably reduced below pH 5 and there is a resulting fall in fish stocks. A pH of more than 10 is fatal for fish. Maximum productivity occurs between pH 6.5 and 8.5. The following list summarises the critical values for European freshwater fish:

pH 3.0–3.5:   death of fish; some plants and invertebrates survive.

pH 3.5–4.0:   death of salmonids; roach, perch and pike may survive after acclimatization.

pH 4.0–4.5:   harmful to many fish; only pike can breed.

pH 4.5–5.0:   harmful to eggs of salmonids.

pH 5.0 9.0:   favourable zone.

pH 9.0–9.5:   harmful to salmonids and perch on prolonged exposure.

pH 9.5–10.0   prevents breeding in some species; fatal to salmonids on prolonged exposure.

pH 10.0–10.5:   roach survives for a short time.

pH 10.5–11.5:   fatal to all fish.

1.4   Water movement

The two types of fresh water habitats are *lentic* or standing waters, and *lotic* or flowing waters. The change from one type to another is often quite rapid; in torrents there are often deep pools where the water is less agitated. Water movements prevent temperature stratification and also increase the dissolved oxygen concentration of water bodies.

Animals of running waters are adapted to resist the effect of the current. Most insects have branchial (i.e. gill) respiration and do not have to surface to renew their oxygen supplies. Fish (trout, bullhead and minnow) have bodies which are almost circular in transverse section, unlike the laterally flattened bodies of still water fish (e.g. roach, perch and carp). Invertebrates are often dorso-ventrally flattened, enabling them to live under stones or attach to stones in places where the current speed is reduced. Examples include the flatworm *Crenobia alpina* of mountain torrents, and ephemeropteran larvae belonging to the genera *Rhithrogena* and *Ecdyonurus*. Hook and sucker arrangements are common. Larvae of the Blepharoceridae (Diptera) have flattened bodies with six ventral suckers. Simulid larvae have a circlet of chitinous hooks at the base of the abdomen which are attached to silken threads secreted by the salivary glands and which adhere to rocks, forming an anchor. The eggs of insects of running waters are also attached by various means. The tadpole of the frog *Rana hainensis*, from torrents in southern Asia, has a ventral sucker behind the mouth.

Animals of running water also show behavioural adaptations, one of the most marked being positive rheotaxis, where animals are orientated to face the current and move against it.

The effects of water movement are most noticeable in the intertidal zone of the sea, where waves breaking on the shore exert pressures of up to 300 kPa (300 kN/m^2). Exposed rocky shores can only support species which are able to anchor themselves securely to the substrate, like barnacles (*Balanus* and *Chthamalus*) and limpets (*Patella* and *Haliotis*).

2   SUBSTANCES DISSOLVED OR SUSPENDED IN WATER

2.1   Turbidity

The turbidity of water, due to materials in suspension, is an important environmental factor. Its main effect is to reduce light intensity and, as a result, lower primary productivity. Organisms that require clear water are absent, and dissolved oxygen is inversely related to turbidity. The effects of turbidity are shown when the water contains more than four per cent (by volume) of solid material in suspension. Fish from rivers with very turbid waters show eye reduction, explained partly by the need to protect delicate structures, and often depend on tactile organs for locating their prey, for example the genus *Neamachilus* from Himalayan torrents (Lange 1950, *in* Nikolsky, 1963). Mucus secreted by the skin of fish from very turbid water is able rapidly to coagulate suspended material. The coagulating properties of the mucus of the American lung-fish *Lepidosiren* enable it to survive in the fine sediment carried by the rivers of Chaco. This is also true of *Physoodonophis boro*, an Indian species which tunnels in liquid mud. A few drops of mucus secreted by this teleost in 500 m of muddy water result in complete sedimentation within twenty or thirty seconds, with the result that the fish is always surrounded by an envelope of clear water, even in generally turbid conditions. The pH of the mucus changes rapidly from 7.5 to 5.0 on contact with turbid water. The coagulant properties of mucus help to prevent obstruction of the gills.

Despite these adaptations, death results when the quantity of material in suspension is very large. This occurs especially during the monsoon season in Afghanistan and India, even well adapted species such as *Glyptosternum reticulatum* dying when the rapidly flowing rivers become swollen with rain water carrying large amounts of sediment (Nikolsky, 1963).

The turbidity of water may serve a protective role. Some fish spawn during the rainy season when the water becomes turbid, their floating eggs being less obvious to predators. Examples include one of the Chinese Carps, *Hypophthalmichthys molitrix* and also *Ctenopharyngodon idella* and *Aristichthys nobilis*.

2.2   Substances in solution

a   Dissolved gases

Oxygen and carbon dioxide are the two most important gases, with others, e.g. hydrogen sulphide and methane, playing an ancillary role.

Although *oxygen* is not a limiting factor in the terrestrial environment, it is often so in water. The solubility of oxygen at saturation decreases as the temperature is raised (Table 3.1). Oxygen forms about twenty-one per cent by volume of the atmosphere and about thirty-five per cent of the gas dissolved in water, its solubility in sea water being about eighty per cent of that in freshwater. Water bodies tend to be relatively poor in oxygen, however, rarely reaching saturation except in turbulent water and still water rich in plant life. In this case the liberation of oxygen during photosynthesis may even cause temporary supersaturation of the water.

The availability of dissolved oxygen in a lake depends on temperature, water movement, and on the type of organisms living there. Three categories of lake can be distinguished on the basis of these characters. *Oligotrophic* lakes are deep, with an extensive hypolimnion (cf. p. 65) containing cold water and with a relatively high oxygen content.

TABLE 3.1   The amount of oxygen which will dissolve in a litre of water at saturation and at an atmospheric pressure of 760 mm of mercury

IN SEA WATER WITH CHLORINITY 20‰ ml PER LITRE	IN FRESHWATER		TEMPERATURE (°C)
	ml PER LITRE	mg PER LITRE	
7.97	10.244	14.16	0
7.07	8.979	12.37	5
6.35	7.96	10.92	10
5.79	7.15	9.76	15
5.31	6.50	8.84	20
4.86	5.95	8.11	25
4.46	5.48	7.53	30

Productivity is low and decomposition of dead animal and plant material slow. The waters of oligotrophic lakes are clear and blue, and the characteristic fish are salmonids with a high need for oxygen. This category includes many lakes in glaciated regions such as the English Lake District, and also Lake Geneva and Lake Borget. *Eutrophic* lakes are normally not very deep, and the water at the bottom is at a higher temperature than in oligotrophic lakes. Productivity is high, decomposition rapid and the waters green. Fish present, e.g. cyprinids, are usually those with a lower oxygen requirement. Esthwaite in the Lake District is an example of a eutrophic lake. *Dystrophic* lakes may be distinguished from eutrophic lakes by the presence of humic acids which colour the water brown and make it acid. Plants are scarce, and an example is the Lake of Gérardmer.

In the sea there is a sharp fall in dissolved oxygen near the sea floor and this can be correlated with the reducing activity of micro-organisms living in the upper layers of sediments on the bottom (Fage and Brouardel *in* Peres and Devize, 1961).

The tolerance of aquatic animals to low oxygen tensions depends upon the species. Four groups can be distinguished among the fish (Wunder, 1936 *in* Nikolsky, 1963).

(i) Species requiring a high oxygen concentration. Their requirements are of the order of 7 to 11 ml/l, and they are unable to survive at a concentration of 5 ml/l. These are fish of cold, fast-flowing waters and include the trout *Salmo trutta*, the minnow *Phoxinus phoxinus* and the bullhead *Cottus gobio*.

(ii) Species which live at concentrations 5 to 7 ml/l. These include the grayling, *Thymallus thymallus*, gudgeon, *Gobio gobio*, chub, *Leuciscus cephalus* and burbot, *Lota lota*.

(iii) Species which can survive at a concentration of 4 ml/l, including the roach, *Rutilus rutilus* and ruff, *Acerina cernua*.

(iv) Those species which can tolerate poorly oxygenated water with only 0.5 ml/l including the carp and tench.

A seasonal cycle of oxygen consumption has been observed for the crustacean *Gammarus linnaeus* from Alaska which was correlated with environmental conditions. The lowest metabolic rate was found in February, when oxygen concentration in the water was almost nil. The respiratory rate increases with temperature, which also determined oxygen concentration throughout the year by its effect on the photosynthetic activity of plants.

The polychaete *Arenicola marina*, which occurs on muddy shores that are often poorly oxygenated, has a haemoglobin, erythrocruorin, which serves as an oxygen reserve, only releasing its oxygen at very low tensions. Erythrocruorin also has a protective role, as it is capable of oxidising hydrogen sulphide, a toxic gas that often forms in the mud where *Arenicola marina* occurs. The respiratory pigments present in benthic annelids are either erythrocruorin (*Arenicola, Nereis, Glycera, Tubifex*) or chlorocruorin (*Sabella*).

The level of haemoglobin in the cladoceran *Daphnia obtusa* may be increased by ten times if the animal is reared under oxygen-poor conditions for a fairly long period of time. The larvae of several species of *Chironomous* (Diptera), *Pisidium* (Bivalvia) and *Tubifex* (Oligochaeta) live in the depths of lakes where the dissolved oxygen concentration is low or zero. These animals are able to survive by reduced activity, by anaerobic respiration, or by relying on the great affinity of haemoglobin for oxygen.

Many aquatic insects come to the surface to breathe, including adult dytiscid and hydrophilid beetles, corixids, nepids and notonectids (Heteroptera), and limnaeids and planorbids (Gastropoda). Others remain permanently submerged, retaining a film of air held by a layer of dense hydrofuge hairs, e.g. dryopid beetles. Some insects take air from the aerenchyma of aquatic plants; examples include *Donacia* and *Haemonia* (chrysomelid beetles).

Carbon dioxide is thirty-five times more soluble in water than oxygen, but 600 times less abundant in air than the latter (0.035 per cent compared with twenty one per cent). It is present in water, either in solution (0.5 ml/l at $0°C$; 0.2 ml/l at $24°C$ in pure water) or in the form of carbonates or bicarbonates of alkali or alkaline-earth metals. Sea water forms an important reservoir of carbon dioxide, since it contains 40 to 50 ml/l of gas (either combined or in the free state), this being 150 times its concentration in the atmosphere.

Carbon dioxide is important as the source of carbon for photosynthesis. It also affects the pH of water and, in the combined state, acts as an alkaline reserve or buffer, especially in sea water. It also contributes towards the formation of calcareous structures (shells, skeletons, carapaces) in various invertebrates.

The most important of the other dissolved gases is *hydrogen sulphide*, a toxin which acts as a limiting factor if it accumulates in stagnant water rich in decaying organic material, for example in sewers. In the Black Sea, because of stagnation in the deeper water, oxygen is absent and hydrogen sulphide accumulates at depths greater than 150 to 200 metres, killing all forms of life. This hydrogen sulphide is produced by the reduction of calcium sulphate by the bacterium *Microspira aestuarii*.

Methane is a component of marsh gas. It is evolved through the anaerobic break-down of dead animal and plant material by the bacterium *Methanosarcina methanica* and by other bacteria belonging to the genera *Methanococcus* and *Methanobacterium*. Marsh gas from Lake Kivu in Zaire contains up to 25% methane, and from Lake Beloye (U.S.S.R.) up to 80% of methane (Dussart, 1966).

b Mineral salts

Natural waters contain variable amounts of dissolved salts. Fresh water has a salinity up to 0.5 grammes of dissolved salts per litre. Sea water has a mean dissolved salt content of about thirty-five grammes per litre. Brackish waters show a wide variation in dissolved salt content.

In *fresh waters* the most important dissolved salts are, in order, carbonates, sulphates, and the chlorides (Table 3.2). The order of importance of cations is the following; calcium 64%, magnesium 17%, sodium 16% and potassium 3%. There is considerable variation in these values however.

The chlorides are derived mainly from rain water, rather than from the sea. Sulphate comes from gypsum, from iron pyrites (which is easily oxidised) or from volcanic sources. Calcium is an important limiting factor, and a distinction may be made between '*soft*' waters poor in calcium (less than 9 mg/l Ca) and '*hard*' waters rich in calcium (more than 25 mg/l Ca).

Many molluscs and crustacea need calcium to form their shells and carapaces. The limpet *Ancylastrum fluviatile* is, however, able to live in waters containing very little calcium and flowing directly off granite. Reynoldson (1966) found that the distribution of the flatworms *Polycelis nigra, P. tenuis, Dugesia lugubris, Dendrocoelum lacteum* and *Phagocata vittata* in British lakes is indirectly related to the calcium content of the water, a high level of this element resulting in the production of a larger biomass of invertebrates and thus providing more

TABLE 3.2   Proportions of major dissolved salts in fresh water, the sea and some brackish seas

	SULPHATES	CHLORIDES	CARBONATES	SALINITY (g/l)
Fresh water	13.2	6.9	79.9	–
Open sea	10.8	88.8	0.4	35
Black Sea	9.69	80.71	1.59	19
Caspian Sea	30.5	63.36	1.24	12.86
Aral Sea	38.71	58.59	0.93	11.28

food for the flatworms (see also p. 9, the example of lakes in Wisconsin).

Calcium decreases the permeability of cell membranes, while magnesium has the opposite effect. This reduction in permeability may possibly explain how marine species entering fresh waters can survive if these waters contain sufficient calcium to prevent the loss of electrolytes, especially sodium. One example is the polychaete *Marifugia cavatica*, which is found in caves in Yugoslavia, where the calcareous tubes made by this worm form a solid mass up to a metre in depth. The isopod *Sphaeroma hookeri* is able to enter rivers in the Mediterranean region if the calcium level is sufficiently high, and forms colonies which are genetically isolated from the marine population. The intertidal polyclad *Procerodes ulvae* can become acclimatised to very dilute brackish waters if they contain at least 5 mg/l of calcium. The prawn *Palaemonetes varians* lives at a range of salinites from over 20‰ to nearly fresh water provided that there is sufficient calcium present. This element limits the production of urine by the green gland in the prawn, and thus prevents an excessive loss of sodium ions which the animal would be unable to survive.

The chemical composition of *sea water* is more stable than that of fresh water. Tables 3.3 and 3.4 show the main constituents of sea water containing 34.4816 grammes of dissolved salts per kilogramme. Nearly all the elements have been recorded in sea water, including thirteen non-metals and at least forty metals. Some rare elements can be concentrated by living organisms, e.g. strontium by some radiolarians, bromine and iodine by corals, and vanadium by ascidians.

The mean salinity of sea water is about 35‰, but it varies between 33 and 37‰ in the open sea. It reaches 41‰ in the Red Sea, and falls to 19‰ in the Black Sea and 12‰ in the Baltic. Dilution occurs in

TABLE 3.3 Weights of the major ions present in sea water containing 34.4816 g of dissolved material per litre (weights shown in grammes) (after Lyman and Fleming, 1940 *in* Pérès and Devèze, 1963).

Cl^-	(chloride)	18.9799
SO_4^{--}	(sulphate)	2.6486
HCO_3^-	(bicarbonate)	0.1397
Br^-	(bromide)	0.0646
F^-	(fluoride)	0.0013
H_3BO_3	(boric acid)	0.0260
Na^+	(sodium)	10.5561
Mg^{++}	(magnesium)	1.2720
Ca^{++}	(calcium)	0.4001
K^+	(potassium)	0.3800
Sr^{++}	(strontium)	0.0133

TABLE 3.4 Percentages of common salts dissolved in sea water

Sodium chloride	77.8
Magnesium chloride	9.7%
Magnesium sulphate	5.7%
Calcium sulphate	3.7%
Potassium chloride	1.7%
Calcium carbonate	0.3%
Others	1.1%

seas which are almost enclosed and receive large quantities of fresh water, and also at river mouths.

The term *brackish waters* includes all waters other than fresh and sea water. Brackish waters may reach very high salinities. Examples include 230 g/l in the Dead Sea and 170 g/l in the Great Salt Lake.

One characteristic of brackish waters is their variability. The term *homeohaline* waters is applied to those with stable salinity while *poikilohaline* waters have variable salinity. Aguesse (1957) distinguished four categories of brackish waters on the basis of mean annual salinity, from observations made in the Camargue on the preferences of different invertebrates:

(i) *oligohaline* waters, having 0.5 to 5 g/l of dissolved salts;

(ii) *mesohaline* waters, having 5 to 16 g/l of dissolved salts;

(iii) *polyhaline* waters, having 16 to 40 g/l of dissolved salts;

(iv) *saline* waters, having more than 40 g/l of dissolved salts.

He then distinguished four types of waters based on changes in dissolved salt content during the annual cycle:

(i) *oligopoikilohaline* waters, where the maximum and minimum salinities occur in the same category as the mean salinity;

(ii) *mesopoikilohaline* waters, where the maximum salinity falls in the category immediately above that of the minimum, irrespective of the mean salinity;

(iii) *polypoikilohaline* waters, where the maximum salinity is situated two categories above that of the minimum;

(iv) *subpoikilohaline* waters, where salinity varies only slightly as the result of human interference.

By combining these two classifications, it is possible to define 16 categories of brackish waters on a dynamic basis. Using this classification there are, for example, oligohaline – mesopoikilohaline waters, etc.

A more widely used classification of brackish waters is the Venice system for marine waters described in 1958 (Symposium, 1959).

c Effect of salinity on living organisms

Salinity is an important factor in determining the distribution of living organisms. Many groups of animals are entirely marine (sipunculids; echiuroids; phoronids; brachiopods; chaetognaths; mollusc groups including Amphineura, Monoplacophora, Scaphopoda, and Cephalopoda; stomochordates; pogonophorans and tunicates), or almost entirely marine (sponges; coelenterates; polychaetes and bryozoans).

Brackish faunas include several animals capable of living at high salinities. The crustacean *Artemia salina* occurs in saline pools containing up to 350 g/l of dissolved salts, together with ephydrid fly larvae, the flat worm *Macrostoma hystrix* (which is also found in fresh water) and the flagellate *Dunaliella viridis*. In the Camargue, the dytiscid beetle *Potamonectes cerisyi* occurs only in saline pools, where it is able to multiply rapidly.

In estuaries and brackish seas, such as the Baltic, the marine stenohaline component of the fauna is rapidly reduced below salinities of 30‰. The first species to disappear are ascidians, and about a half of the marine species disappear when the salinity has fallen to 20‰. The remaining fauna can tolerate salinites as low as 4‰. The barnacle *Balanus improvisus* occurs in almost fresh waters in company with the worm *Mercierella enigmatica* (a euryhaline species characteristic of hyperhaline waters), rotifers and the hydroid *Cordylophora lacustris*.

Brackish waters contain fewer species than fresh water or the seas, since only euryhaline species are able to survive there, but much higher numbers of individuals. In the lake at Vaccarès in the Camargue, where salinity varies between 2 and 7‰, species of freshwater fish which can tolerate low salinities, such as the carp, tench, pike and zander, occur with marine species like the mullet (*Mugil cephalus*), which are able to live at low salinities (Lévêque, 1957).

The effect of salinity variation in brackish waters on the composition and abundance of animals was studied by Aguesse (1957) in the Camargue. Figure 3.2 shows changes in the abundance of cladocerans and copepods over two successive years, during which time there were considerable variations in the salinity of the pools.

The influence of salinity on species diversity has been shown by the work of Zenkewitch (1963) on the Mediterranean, Black Sea and Sea of Azov. Over 7000 species were recorded from the Mediterranean (salinity 35‰), but only 1200 of these were found in the Black Sea (19‰) and about 100 in the Sea of Azov (12‰).

Stenohaline species include some radiolaria and reef-building corals, and euryhaline species include the jelly-fish *Aurelia aurita*, mussel *Mytilus edulis*, crab *Carcinus maenas* and the urochordate *Oikopleura dioica*. Tolerance of changes in salinity may be modified by temperature. The hydroid *Cordylophora caspia* is more tolerant of low salinities if the temperature is slightly raised. The crab *Carcinus mediterraneus* and several other decapods can occupy water at lower salinities if the temperature is increased.

Size variation within species may be related to salinity. The crab *Carcinus maenas* has dwarf forms in the Baltic, and also large forms in estuaries and lagoons. The bivalve *Cardium glaucum* has giant forms in Mediterranean pools. Mussels grow to a larger size in lagoons, while sea urchins are smaller than in the sea.

The size of euryhaline species is often dependent on the salinity of their habitat, the length of *Artemia salina* varying from 10 mm at a salinity of 122‰ to between 24 and 32 mm for a salinity of 20‰. The shape of the body and appendages and also the distribution of pigmentation similarly changes with salinity. Bertin (1925) found that,

129

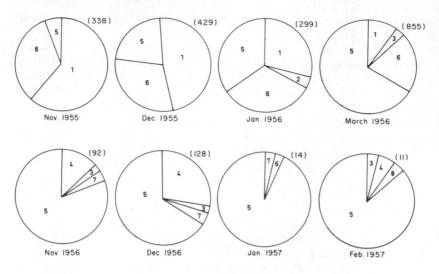

Figure 3.2   Cladoceran and copepod fauna of the Relongues marshes in the Camargue. Above: Saline polypoikilohaline cycle; Below: Oligohaline mesopoikilohaline cycle. 1. *Daphnia magna*; 2. *Macrothrix hirsuticornis*; 3. *Alona rectangula*; 4. *Chydorus sphaericus*; 5. *Megacyclops viridis*; 6. *Arctodiaptomus wierzejskii*; 7. *Calanipeda aquaedulcis*; 8. *Onychocamptus mohammed*. Figures in parentheses represent the number of individuals in a litre of water.

in the stickleback *Gasterosteus aculeatus*, the development of bony plates on the sides of the body is related to several factors, including temperature and salinity. In the mosquito larva *Culex pipiens*, reduction in size of the anal papillae is almost directly related to increased salinity. These papillae are the only part of the integument which is permeable to water and dissolved salts.

Nitrogen and phosphorus play an important part in the synthesis of living materials, and may act as limiting factors in photosynthesis. Where the sea water is stratified, the amounts of these elements present in the assimilable forms of nitrates and phosphates tend to be reduced in the light zone. Nitrates and phosphates accumulate in the depths, where they are formed through the bacterial decomposition of plant and animal remains. There is a seasonal cycle in the distribution of these nutrients, with a minimum in spring and summer and a maximum in winter when the photosynthetic rate is reduced. This cycle is more marked near the surface than in deep waters.

Upwelling water movements inject nutrient-rich waters into the light zone with a resultant increase in the rate of productivity and the abundance of zooplankton (Figure 1.1, page 10).

II   ABIOTIC FACTORS IN THE SOIL

The soil is a complex habitat occupied by an abundant and varied fauna. Only the more important of the abiotic factors operating in the soil are described here.

1   SOIL WATER

Pedologists classify the water present in the soil into four categories:

(i) *hygroscopic or imbibed water*, forming a thin film covering the soil particles; held firmly by soil colloids and not available to animals or plants;

(ii) *non-available capillary water*, occupies spaces less than 0.2 μm in diameter; similarly held too firmly to be available to soil organisms;

(iii) *available capillary water*, occupies spaces measuring from 0.2 to 8.0 μm: normally available to plants during dry periods and allowing the activity of bacteria and small protozoa, including some flagellates, e.g. *Bodo*, and the Testacea (Bonnet, 1964).

(iv) *gravitational water*, temporarily occupies the larger air spaces in the soil. Drains away under the influence of gravity unless drainage is impeded. Rapidly-draining water occurs in the larger air spaces of sandy soils, while slowly-draining water remains for a longer time in the smaller, non-capillary air spaces of soils.

To estimate the available water in the soil, the permanent wilting point is determined. This is the water content (as a % of total dry weight) of the soil at the point when the leaves of plants first show signs of permanent wilting. Wilting point can be determined either by a biological method (a plant is grown in a pot, the soil is allowed to dry and the soil moisture content at which the plant wilts is calculated), or by a physical method. The simplest of the latter methods is to place a sample of about ten grammes of dry soil in a dessicator for five days at a relative humidity of 98.8 per cent, using a two per cent aqueous sulphuric acid solution. The water content of the soil, *hygroscopic water*, is determined by drying the sample in an oven at 105°C. The wilting point for the soil is then 1.5 times the hygroscopic water content.

The permanent wilting point is high for peaty soils (up to fifty per cent), about fifteen per cent for silty clay soils, and 1.5 per cent for coarse sand. As a general rule, soils with finer particles have a higher permanent wilting point, clays holding more water than sands, and colloidal humus particles even more than clays.

The *soil moisture potential (pF)* is the logarithm to the base ten of the numerical value of the negative pressure of the soil moisture,

131

expressed as centimetres of water. The value increases as the soil dries out. The pF equivalent of 1000 cm of water is thus $\log_{10} 1000$, i.e. 3.0.

There are four important points on the pF scale:

(i) *field capacity, pF* 1.8, corresponding to the point at which the fast-draining water has been removed and slow-draining water begins to drain away;

(ii) *retention capacity*, or moisture equivalent, corresponding to a pressure of one third of an atmosphere and a pF of 2.5;

(iii) *temporary wilting point, pF* 3.9 to 4.2, is the point at which plants absorb water with difficulty from the soil;

(iv) *permanent wilting point*, corresponds to a pF of 4.2.

The concept of available soil water is important to botanists. Hygrophytes are plants which need large amounts of available water and, therefore, soil at a low pF value. Xerophytes are able to survive when the soil pF approaches wilting point.

The pF concept can also be applied to the water requirements of soil animals. Bonnet (1964) found that some species of Testacea, including *Centropyxis halophila* and *Geopyxella sylvicola*, are characteristic of mineral soils where water, although not abundant, is readily available, i.e. the pF is low. Other species, *Assulina muscorum, Heleopora sylvatica* and *Nebela collaris*, occur in moist soils, rich in organic matter, which behave, however, as physiologically dry habitats (high pF).

The experiments of Vannier (1967) showed that two phases can be distinguished in the behaviour of soil collembola and oribatid mites. One phase corresponds to normal activity, while the other, escape behaviour, begins to show when the soil pF reaches a particular value which varies with the species. For the majority of Arthropleona (Collembola) and all oribatid mites, the escape phase begins at wilting point. Escape behaviour begins in *Isotomiella minor* (Collembola) at a pF between 4.2 and 5.0, and for *Tectocepheus sarekensis* (Oribatidae) when the pF is between 5 and 6. On the other hand, a collembolan *Dicyrtoma atra* begins to show escape behaviour at pF less than 2.5, that is before the slower draining water has been totally removed from the soil. The presence of micro-arthropods in the soil cannot, therefore, be directly related to the water content of the soil. In temperate regions, wilting point is rarely reached in the soil, and changes in water content do not affect the behaviour of soil animals. In mediterranean climates, however, the soil pF reaches high values in the summer, and collembola tend to disappear from the soil. It is clear that microarthropods remain active at higher pF values than most plants can survive. This is also true for some bacteria which, in dry tropical or semi-arid regions, continue to remain active at a pF of 5.0 or even 5.5.

Termites are very susceptible to dessication and must live in an atmosphere of at least fifty per cent relative humidity. They move into the soil to depths of up to twelve metres in search of water to maintain the humidity of their termitaria.

Earthworms, which are rather more aquatic than terrestrial, require a supply of water. If the soil dries out, they either die, or move deeper into the soil, or aestivate in mucous-lined cells, in which state they are able to survive a loss of over half their body water. They become active again when the soil becomes moist.

Some nematode species are found in very moist soils, while others are characteristic of soils subjected to periodic drying out. The role of moisture in the distribution of nematodes is shown in the following example. In *Brachystegia* forests in Zaire, there are seventy-three nematodes per gramme of soil where the soil moisture content is thirty per cent, but only fifty per gramme in soils of *Stylosanthus* cultures where the moisture content is only twenty-two per cent.

2 TEXTURE AND STRUCTURE OF THE SOIL

The texture of the soil depends on its mineral composition. Soil particles can be classified according to the following table:

Coarse sand	from 2.0 mm to 0.2 mm in diameter
Fine sand	from 0.2 mm to 0.02 mm
Silt	from 0.02 mm to 0.002 mm
Clay	less than 0.002 mm.

Particles larger than 2.0 mm in diameter form gravel.

Soil structure is also important. If soil colloids are dispersed and their particles isolated irrespective of size, then the structure is particulate. If the colloids are flocculated into aggregates which are more or less stable, then the soil has a granular structure. Flocculation takes place when Ca^{++} and Mg^{++} ions are abundant. The soil structure is important, especially with regard to the aeration of the soil. Granular soils are permeable and well aerated, while particulate soils are more or less impermeable, because of their colloidal nature, and so relatively poorly aerated. The porosity of the soil depends on both texture and structure.

Earthworms are more numerous in silty or sandy clay soils than in sands, gravels or clays. The most favourable soils for the development of soil-dwelling beetles have a high proportion of fine particles (clays and silts) which hold the necessary water. Larger particles are unsuitable as they allow too rapid drying out of the soil.

Particle size is important in the distribution of animals in underground waters and on sandy beaches. The polychaete *Arenicola marina* lives in sandy muds containing about twenty-four per cent of water and with a mean particle size of about 247 μm. The mystacocarid *Derocheilocaris remanei* lives in mainly siliceous sands, where the

133

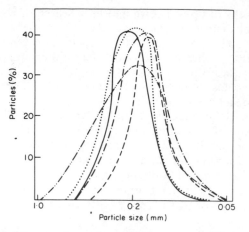

Figure 3.3 Particle size range in sands containing mystacocarids in France and Italy (Delamare Deboutteville, 1960).

interstices are free from clay. These sands are fine, uniform with a mean particle size of 0.2 mm, and occur on the coasts of Algeria, France and Italy (Figure 3.3).

The shape of sand grains is also important. Siliceous sand grains, having sharp edges, enclose larger interstitial spaces between the grains than calcareous sands, where the edges are rounded. The larger spaces are more suitable for colonisation by interstitial animals. The diameter of these animals is generally less than or equal to the spaces which they

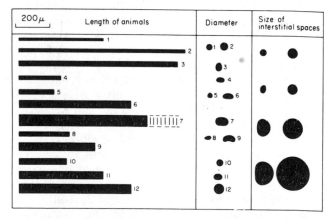

Figure 3.4 Comparison between the size of some animals from siliceous sands of the Bassin d'Arcachon and the interstitial spaces between the sand grains where they occur. 1 and 2: nematodes; 3: oligochaetes; 4: turbellarians; 5 and 6: archiannelids; 7: annelids; 8 and 9: gastrotrichs; 10: tardigrades; 11: harpacticoids; 12: mystacocarids (Renaud Debyser, 1963).

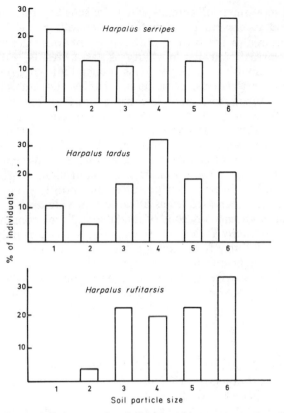

Figure 3.5   Preferences of three species of *Harpalus* with respect to soil particle size. 1 and 2: coarse sand; 3: medium sand; 4 and 5: fine sand; 6: clay.

occupy (Figure 3.4). Interstitial animals .also have an elongated form and are circular in section, an adaptation to locomotion in the interstitial spaces. These animals are found among the annelids, flatworms, ciliates and gastrotrichs.

The effect of particle size on soil animals is not fully understood. Lindroth (1949) studied the preferences of some species of the carabid beetle genus *Harpalus*. His results (Figure 3.5) show that *H. serripes* is relatively indifferent, while *H. tardus* and, particularly, *H. rufitarsus* prefer soils with a fine particle size (fine sand, clay and silt) and avoid coarse sand.

3   AERATION OF THE SOIL

The porosity of the soil regulates the circulation of water, air and of many animals. A compacted and not very porous soil may hinder

135

the vertical movement of animals which are sensitive to temperature or moisture, and so prevent their continued survival. The Colorado beetle hibernates nearer to the surface in heavy, moist soils than in sandy silts or brown earths, and its mortality is, as a result, rather higher. Burrowing animals are clearly less dependent on porosity and these include earthworms, some insect larvae and ants. The lack of oxygen in compact soils may become a limiting factor, and carbon dioxide contents of 1.1% at a depth of fifteen centimetres and 9.4% at seventy centimetres have been recorded. Surface and litter animals are less able to tolerate high carbon dioxide levels than soil animals. Termites are tolerant of carbon dioxide, and wireworm larvae (*Agriotes*) are attracted by the carbon dioxide from plant roots. Earthworms are more numerous in heavy, poorly aerated and poorly drained soils and many can tolerate increased levels of carbon dioxide.

The fauna of the marshy soils of the Siberian taïga consists mainly of collembola and enchytraeids. These animals are twice as numerous, and their biomass is thirty-five times greater, when the water contained in the soils is not stagnant but well aerated. There are many species of soil protozoa which survive under semi-aerobic conditions. Some species can obtain their oxygen through the reduction of compounds containing oxygen (Pochon, 1958 *in* Bachelier, 1963).

4 SALINITY

There are several different types of halomorphic or salt soils. There are *solontchaks* (saline soils), with a *pH* less than eight and sodium forming less than fifty per cent of the cations in solution. There are *solonetz* (alkaline soils), containing an excess of sodium as sodium carbonate and with *pH* reaching 9.0. Halomorphic soils have a very characteristic flora and fauna.

The plants are *halophytes* and among the most characteristic are some Chenopodiaceae (*Salicornia, Salsola, Suaeda*). The salt tolerance of the various species varies widely, explaining the zonation of plants in salt marshes in terms of the salinity gradient. In the Camargue, an area which has been especially well studied, the most tolerant species is *Arthrocnemum glaucum* (glasswort), which during the summer period, can tolerate salt concentrations of over twenty per cent, and which forms tussocks separated by areas of bare mud covered by salt efflorescences (this is an *Arthrocnemetum glauci* association). Zones where some of the salt has been leached out by rain and which are raised a little support *Salicornia fruticosa, S. radicans, S. herbacea* and other chenopods like *Suaeda maritima* and *Halimione portulacoides*, which together form the *Salicornietum fruticosae* association. These plants can only tolerate salt levels of 1.5% during the wet season and ten per cent during the summer. Areas containing little or no salt support either a psammophytic flora on dunes, including *Artemisia glutinosa* and *Teucrium maritimum*, or plants of waste ground, forming the *Thero-brachypodion* and having the sea lavender *Limonium vulgare* and *Brachypodium phoenicoides* as characteristic species. The numbers of terrestrial invertebrates in the Camargue increase as the soil salinity decreases (Bigot, 1965). In the three plant associations forming the halosere, the following species of invertebrates were found:

(i) in the *Arthrocnemetum glauci*, there were 120 invertebrate species including nine characteristic species;

(ii) in the *Salicornietum fruticosae*, there were 211 invertebrate species including eleven characteristic species;

(iii) in the *Thero-brachypodion*, there were 295 species including sixteen characteristic species.

For this part of the halosere, where the effect of the salt is well marked, there were 414 invertebrate species present. In the Camargue as a whole, where there are many habitats with little or no salt, there are over 1700 species of animals. The characteristic species of the *Arthrocnemetum glauci* included several carabid beetles (*Cicindela circumdata, Dyschirius cylindricus, Pogonus pallidipennis* and *Tachys scutellaris*), all of which are strongly halophilous species.

Lumbricid worms are scarce where the soil salinity exceeds 0.07 M for very long (Barley, 1961). They react mainly to the cations in the salts present. *Eisenia foetida*, a species of manure heaps, is more susceptible to sodium chloride than either ammonium, lithium or potassium chloride, in that order. In the case of *Allolobophora caliginosa*, a species of garden soils, the order of susceptibility is potassium, ammonium, sodium and then lithium chloride. Earthworms react to solutions of M/100 concentration, and even M/500. They do not survive long in solutions of M/10 concentration.

5   pH

The effect of *pH* in soil is an indirect one, and it is the result of the action of several soil factors. *pH* depends mainly on the humus and clay colloids present, which are electronegative and surrounded by the cations H^+, Ca^{++}, Mg^{++}, Na^+, and K^+. The *pH* is also affected by the type of plant cover and by climatic conditions, especially temperature and rainfall. The *pH* of calcareous soils is about 8; it may reach 9.5 in saline soils and may be as low as 4.0 in peat bogs and podzols.

Protozoa are found over a wide range of *pH*, from 3.9 to 9.7, depending on the species. *Paramecium* occurs at *pH* values between 5 and 9. The Testacea (Protozoa) include acidophilic species e.g. *Centropyxis vandeli*, requiring soils with a *pH* less than 6, neutrophilic species, e.g. *Difflugia lucida*, found in soils with *pH* between 6 and 7, and basophilic species which require soils with a *pH* of over 7, e.g. *Centropyxis halophila*. Neutrophilic species are the most numerous of these protozoans. *C. vandeli* requires very dry soils on acid rocks, *D.*

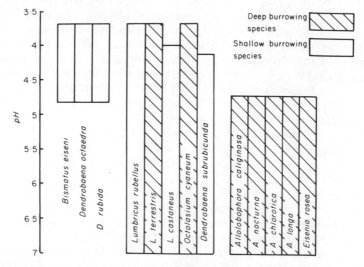

Figure 3.6   Classification of earthworm species according to *pH* of litter or soil (Satchell, 1955).

lucida is associated with mosses, and *C. halophila* is restricted to littoral saline soils.

Earthworms do not occur in soils with a *pH* less than 4.4, while between 4.5 and 8.0 their preferences are masked by interactions with other factors, especially calcium, present in the soil. *Eophila icterica*, taken from soil with *pH* 7, can live under laboratory conditions in a range of soils from *pH* 4.2 to 8.0, but is three times more active at *pH* 8 than at *pH* 4.2. Satchell (1955) classified earthworms on the basis of their *pH* preference into acid tolerant, ubiquitous and acid intolerant species (Figure 3.6).

The distribution of snails shows a correlation with soil *pH* because of their need for calcium. Highest species diversity is found for soils with a *pH* of 7 or slightly above.

6   CALCIUM

Calcium is essential for many soil animals. Some lumbricids secrete calcium from their calciferous glands in the form of calcium carbonate granules, which then pass through the gut. *Allolobophora caliginosa, Eisenia rosea, E. terrestris* and *Lumbricus castaneus* are scarce when calcium is absent, while *Eophila icterica* is common in sandy soils and absent from calcareous soils at Cotentin. Several species of millipede also require calcareous soil. There are calcicole and calcifuge species of testacean protozoa. *Centopyxis plagiostoma, Geopyxella sylvicola* and *Bullinularia gracilis* are indicators of the presence of Ca^{++} ions (Bonnet, 1964).

Plants may be classified as calcicoles or calcifuges according to their preferences, and the marked contrast between the vegetation of calcareous and sandy soils has been known for a long time. In the Paris area, crops on calcareous soils harbour a number of weeds including cornflowers, poppies, various umbellifers (e.g. *Caucalis lappula*) and the hemp-nettle *Galeopsis ladanum*, and yet these plants are absent or scarce on sandy soils. In the forest of Fontainebleau, calcareous sands are colonised by an association which includes *Silene otites*, while siliceous sands are colonised by the grass *Corynephorus canescens*. In mountain areas alpine meadows are invaded by *Carex firma* and *C. sempervirens* on calcareous soils, and by *C. halleri* on siliceous soils.

Chapter 4

ENVIRONMENTAL FACTORS: FOOD

Food is an important component of the environment. Depending on its quality and abundance, it can modify fecundity, longevity, speed of development and mortality in animals. Changes in diet may cause structural, physiological and ecological adaptations.

I THE QUALITY OF FOOD

The quality of food is mainly the concern of the physiologist, but its effect on the abundance, longevity, growth rate and fecundity of living organisms is of interest to the ecologist.

1 EFFECT ON FECUNDITY

When females of the polyphagous beetle *Silpha atrata* are restricted to a vegetable diet, they only lay about thirty-nine eggs, while those receiving a diet of animal origin lay up to 235 eggs (Heymons *in* Bodenheimer, 1955). The composition of the food of the Colorado beetle affects its fecundity. When females were fed on old leaves from potato plants, about half of them had stopped laying within three days, and they had all ceased after eleven days. This effect is reversible, and fecundity returns to normal if young leaves are provided for food (fig. 4.1). The larva of the honey bee may develop into either a fertile queen or a sterile worker according to whether it receives royal jelly or not.

The diet of the red grouse *Lagopus scoticus* may contain between fifty and a hundred per cent of the heather, *Calluna vulgaris*, depending on the season. Reproductive success in this species depends on the availability of an early growth of young heather shoots before egg laying begins. It has also been shown from experiments that birds select heather rich in calcium, phosphorus and nitrogen as food, ignoring plants deficient in these elements, which are essential for the female during egg laying. The amount of heather available in a particular area is not, therefore, necessarily a reliable measure of available grouse food. A

140

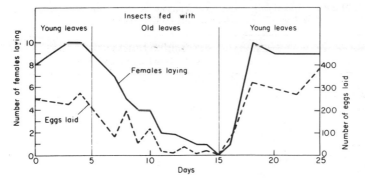

Figure 4.1   The effect of altering the quality of food on fecundity in the Colorado beetle.

high density of grouse in an area may be due to a high calcium, phosphorus and nitrogen content of the heather. In this example it is not the quantity of food that acts as a limiting factor, but the quality of that food and, particularly, the quantity of certain elements present (Moss, 1967).

Ewes fed on kale have a shorter oestrus period and reduced fertility compared with control animals fed on hay. Deficiency in phosphorus and copper lowers fertility in this species. The more rapid growth and earlier sexual maturity of arctic mammals can be interpreted as an adaptation to the shorter period of plant growth. Their food is also richer in vitamins (cf. p. 324).

2   EFFECT ON LONGEVITY, GROWTH RATE AND MORTALITY

The quality of food has a similar effect on longevity in animals. Andrewartha (1935) showed that longevity and fecundity of the insect *Thrips imaginis* were increased at 23°C if pollen was added to the diet. The duration of larval development in the moth *Ephestia elutella* depends on the type of food provided (table 4.1).

Although the cockchafer is polyphagous, it shows a clear preference for those plants which are beneficial for growth. The dandelion, ox-eye daisy (*Chrysanthemum leucanthemum*) and yarrow (*Achillea millefolium*) allow more rapid growth than clover (*Trifolium pratense*) or birdsfoot-trefoil (*Lotus corniculatus*) or than certain grasses, such as the cocksfoot (*Dactylis glomerata*), which are unable to support the larvae.

Xylophagous cerambycid beetle larvae living in the inner bark and sapwood, which is rich in starch and soluble sugars, are able to complete their development in a year. Their development may, however, last several years in the heartwood, which is lacking in starch and soluble sugars. *Lyctus* larvae can only develop in wood containing starch, which is the only carbohydrate that can be assimilated by this

141

TABLE 4.1   Duration of larval development in the moth *Ephestia elutella* reared at 22°C and 70% relative humidity (Waloff, 1948)

FOOD	% SURVIVING	COMBINED DURATION OF LARVAL AND PUPAL STAGES (DAYS)
Wheat	87	50
Wheat germ	100	42
Tobacco	10	120
Figs	10	134
Cocoa	9	87
Beans	6	–
Flour	25	137

insect. If starch is absent (for example in wood treated with hot water), most of the larvae die.

In temperate regions, it has been shown that earthworms avoid litter formed from coniferous needles, and are also unable to survive in litter containing alder or false acacia leaves (Bachelier, 1963).

3   VARIATIONS IN DIET

The diet of any species is rarely constant throughout the year and in every locality. Seasonal variations will depend on the food available and the activity of the animals. The Corsican wild sheep feeds on shoots of the strawberry tree during autumn and winter, and on herbaceous plants during the summer (Pfeffer, 1967). The American fox, *Vulpes fulva*, feeds mainly on fruit and insects during the summer and autumn, with small mammals comprising less than a quarter of its diet. However, in winter and spring it feeds mainly on rodents.

Diet also depends on age and stage in development. The fry of the Caspian Sea roach, *Rutilus rutilus caspius*, feed on small planktonic organisms such as algae and rotifers; as they grow larger, planktonic crustacea are added to their diet, and then benthic insect larvae. Finally, the adults feed largely on molluscs. Changes in the structure of the gut and in the form of the mouth and teeth are correlated with these changes in diet (Nikolsky, 1963). Hymenoptera with parasitic larvae often feed on nectar from flowers in the adult state. Herring fry (7 to 12 mm long) feed on ciliate protozoa (*Tintinnopsis*) flagellates (*Peridinium*) and copepods, while the adults eat pteropods, euphausids (*Nyctiphanes*), amphipods and copepods (Hardy, 1924).

Diet may also depend on the physical characteristics of the habitat. It has been demonstrated that the particle size of the substrate partly determines the diet of the cumacean *Cumella vulgaris*. This crustacean occurs in large numbers in the intertidal zone in substrates where the mean diameter of the particles is less than 160 μm. In fine sand (less than 150 μm) and in mud it is a deposit feeder, while in coarse sand (more than 150 μm) it is an epistrate scraper, sucking the surface of the sand grains one after the other.

Variations in diet according to sex have been recorded for the bream from the Sea of Azov. The female feeds mainly on the crustacean *Hypaniola kowalevskii* and the male on the polychaete *Nereis succinea*. In the abyssal teleost family Ceratidae (angler fish), the females of many species live as active predators, while the males are parasitic on the females. In the chalcid wasp genus *Coccophagus* (Aphelinini) male larvae of some species live as hyperparasites in scale insects, which are already hosts to female larvae, at the expense of the latter (the primary parasites).

Some animals, such as the squirrel of temperate regions, store food. In North America the rodent *Ochotona* of mountainous areas cuts the vegetation and allows it to dry in the sun during the summer. This dry material is then stored under rocks and used as a food reserve in winter (Howell *in* Dice, 1955).

4 FEEDING TYPES

The following feeding types can be distinguished:

(i) *polyphagous species*, which feed on a wide range of food species. Many carnivorous mammals and predatory insects are polyphagous. Some polyphagous insects eat a number of different plants. In North America, the caterpillar of *Pyrausta nubilabis* attacks more than 200 plant species. The tachinid fly *Compsilura concinnata* uses several hundred different insect species. Polyphagy is not, however, incompatible with food preferences. The wild sheep of Corsica will eat over one hundred different plants, but searches especially for the strawberry-tree and the broom (*Cytisus triflorus*) and ignores conifers (Pfeffer, 1967).

(ii) *oligophagous species*, which live at the expense of a smaller number of often related species. The Colorado beetle feeds on potato and several other wild plants belonging mainly to the family Solanaceae.

(iii) *monophagous species*, which depend solely on a single host or food plant. Monophagy is common among insect parasitoids, the hymenopteran *Aphelinus mali* choosing only the woolly aphid. The weevil *Anthonomus pyri* is restricted to the pear and the silk worm to the mulberry tree. Examples of monophagy are less common amongst the vertebrates. One of the most striking is that of the Everglades kite (*Rostrhamus sociabilis plumbeus*), restricted to a small area of Florida, which feeds exclusively on the snail *Pomatia caliginosa*.

Many species are able, through change in diet, to adapt to another food species. The most common examples are found amongst polyphagous insects which fed originally on wild plants and have now become adapted to cultivated plants. The vine had few natural enemies in the wild state. However, today under cultivation it is attacked by insects which originally fed on willow herbs, *Galium* etc. This type of adaptation may be so complete that monophagous biological races are formed which are narrowly restricted to their new host. For example, the flea beetle *Haltica lythri*, which feeds on Lythraceae and Onagraceae, has developed a race restricted to the vine, the sub-species *ampelophaga*.

II THE QUANTITY OF FOOD

Food, which is the only source of energy available to animals, is obviously a limiting factor if the supply is insufficient. The amount of food required is, weight for weight, much greater for small species than for large ones (table 4.2). Relatively more food is required by homeotherms than poikilotherms, as the former need to maintain a constant body temperature.

By comparison, some benthic marine fish eat only the equivalent of ten times their body weight in a year.

The quantity of food available may affect the size of the individual. The size of adults of the parasitic wasp *Ichneumon nigritarsis*, is closely related to that of the chrysalis parasitised by the wasp larva. There is a significant reduction in size in flatworms that are starved. The body weight of the carnivorous fresh-water fish *Lepomis macrochirus* can vary from 112 grammes in a pond where there is sufficient food, to 0.5 grammes in a pond where food is in short supply as the result of overpopulation.

Starvation in insects leads to a reduction in the number of ovarioles and, therefore, in the number of eggs laid, although the size of the egg is not affected. If the blowfly *Calliphora erythrocephala* is undernourished, it develops only 93 ovarioles instead of the 243 of normal flies. Under similar conditions the mosquito *Anopheles maculipennis* shows a reduction in number of ovarioles from 384 to 142. The number of eggs laid by the bed-bug *Cimex lectularius* depends on the quantity of blood ingested. The fecundity of the Colorado beetle depends on the amount of potato foliage eaten.

TABLE 4.2   The quantity of food ingested daily as proportion of body weight for some birds (Ferry, 1964)

SPECIES	BODY WEIGHT IN GRAMMES	DAILY QUANTITY OF FOOD AS % OF BODY WEIGHT
Blue tit	11	30
Robin	16	15
Song thrush	90	10
Kestrel	200	7.7
Buzzard	900	4.5

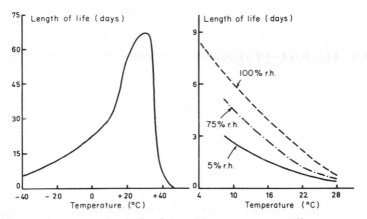

Figure 4.2   Left: mean length of life of starved house sparrows at different temperatures. Right: mean length of life for the caterpillar of *Panolis flammea* in relation to temperature and relative humidity. Values shown are for starved animals (Zwolfer, 1931).

Animals that obtain sufficient food develop more rapidly than those that are undernourished. The third larval instar of the water beetle *Dytiscus* takes fourteen days if the larva is fed with twenty-five tadpoles per day, but is extended to fifty-one days if the food is reduced to only one tadpole per day.

If homeotherms are starved, the length of life increases as ambient temperature approaches body temperature (fig. 4.2). The length of life decreases, however, with increasing temperature in starved poikilotherms.

The basic problem that arises from ecological studies of bird migration is how birds which migrate from northern Europe to the Mediterranean region, or even to tropical Africa, are able to coexist with the indigenous bird fauna without competing with them (Blondel 1969). In order for migratory behaviour to develop in a species, its food resources must become exhausted in autumn and there must be available food in the region to which it migrates. Migrations tend to flood with birds at the most favourable time of the year, those habitats which are not fully exploited by their resident species. In the case of passerines the resident species are always relatively scarce. The result is that migratory birds tend to use seasonal surpluses of food. These surpluses develop because the non-migratory indigenous species are maintained at low densities by harsh conditions during part of the annual cycle, for example in the winter or the dry season, and this explains the surplus food. 'Migrations are simply a means of allowing birds to wait until their own ecosystem becomes available again for them to breed' (Blondel, 1969).

Chapter 5

BIOTIC FACTORS

The term biotic factors includes all those interactions which take place between the different organisms found in a habitat. These interactions were called *coactions* by Clements and Shelford (1939). They are of two types:

(i) those occurring between individuals of the same species: intraspecific interactions,

(ii) those occurring between individuals of different species: interspecific interactions.

I INTRASPECIFIC INTERACTIONS

Intraspecific interactions are varied, and include the following types:

1 THE GROUP EFFECT

This expression (Grassé and Chauvin, 1944; Grassé, 1965) describes changes that take place when animals of the same species are brought together into groups of two or more individuals. The group effect occurs in many orders of mammals and insects. One important result is an increase, sometimes considerable. in the growth rate.

The group effect is shown by many species which are only able to live and reproduce normally when they are present in sufficiently large numbers. The cormorant, *Phalacrocorax bougainvillei*, the guano bird of Peru, can only breed in colonies of at least 10 000 individuals at a density of three nests per square metre. This 'principle of the population minimum' explains why it is impossible to prevent the extinction of species that become very rare, and is probably the reason why the North American whooping crane, despite efforts made over a number of years, is still represented by only thirty or forty individuals. It has been estimated that a herd of African elephants needs at least twenty-five individuals to survive, and that a reindeer herd requires a minimum of 300 to 400 head.

The principle of the population minimum also explains why the birth rate is reduced when numbers are low, as females have less chance of mating than when numbers are high. Searching for food, and defence against enemies are made easier by living in a group. Wolves are able to kill much larger prey when hunting in packs than by themselves. The bison, musk ox and many other ruminants can defend themselves more effectively against predators when grouped together in herds.

1.1 Locust phases

One of the most familiar and striking group effects is the phase phenomenon discovered by Uvarov (1931) in the migratory locust *Locusta migratoria*. This species and *Schistocerca gregaria* have been extensively studied. Solitary nymphs are green in colour and not very active. If locusts are crowded together, however, nymphs are produced which show increased activity and gradually become mottled with black and orange. In addition, morphological differences appear. The pronotum in gregarious adults has a median keel which is straight or concave when seen in profile, while solitary adults have a convex keel. The differences between gregarious and solitary forms were so marked in *Locusta migratoria* that, for a long time, the solitary form was considered to be a distinct species, *Locusta danica*. Gregarious individuals have a much larger appetite than solitary forms, are more active and tend to group together into swarms. The nymphs also grow more quickly and are heavier. Solitary forms develop more ovarioles than gregarious individuals, and their life cycle includes a diapause which is not found in gregarious forms. As locusts become more gregarious, this high reproductive potential of the solitaries (measured by the number of ovarioles) diminishes. This explains the disappearance of locusts after migratory swarms have saturated a region. Changes in phase are due to sensory phenomena, for example tactile stimuli or the sight of other individuals. Hormones are also involved, especially those secreted by the corpora allata. The group effects develop at densities higher than two individuals per 2600 ml.

The existence of phases has been demonstrated in some moths, beetles, psocids, aphids, crickets and in the cockroach *Blatella germanica*. In all these examples, modifications have been observed in fecundity, speed of development, and often in characteristic morphological and physiological changes in the animals.

The discovery of the group effect shows that the result of increased density is just as important in ecology as predation or competition, yet the latter were studied almost exclusively for a long time. The close proximity of many individuals in the same habitat is not necessarily harmful and may even be advantageous, especially from the point of view of growth (Chauvin, 1957).

2 THE OVERCROWDING EFFECT

This term, due to Grassé (1929), describes the effects produced when a habitat becomes overpopulated. There are, obviously, intermediate stages between the group effect and the overcrowding effect, but in general the consequences of overcrowding are harmful, while those of grouping are beneficial.

The work of Park (1941) provides an illustration of the effect of overcrowding in the beetle *Tribolium confusum*. There is an optimum density at which the number of eggs per female reaches a maximum, and this corresponds to the group effect. If the density is increased above this optimum value, the culture medium (flour) becomes significantly contaminated by faeces and other moderately toxic secretions of the beetles. This phenomenon, known as conditioning, results in disturbances in the life cycle of the beetle such as a reduction of fecundity in the female, and an increase in the larval developmental period. These effects are reversible, and cease as soon as the *Tribolium* are placed in fresh flour. A similar conditioning effect has been described by Crombie (1942) for *Rhizopertha dominica*, a grain beetle.

These phenomena caused by overcrowding have been called self-regulatory processes by Chapmann (1928). Crombie (1943) showed, for *Tribolium confusum*, that cannibalism by adults on their eggs increases with population density in the following way:

Number of adults per gramme of flour	1.25	2.5	5	10	20	40
% eggs eaten	7.7	17.0	20.0	39.7	70.2	98.4

Pearl (*in* Prenant, 1934) showed that, for *Drosophila*, as soon as there were two females in a culture bottle (nutrient medium surface area 300 mm^2) interaction resulted in a decrease in the mean number of eggs per female. A rise in population density increased the chances of disturbance, and therefore reduced the time available for egg laying and feeding, with a consequent lowering of fecundity.

Another example is that of the grain weevil *Sitophilus oryzae*. In this species, the recorded number of eggs laid per female each day reaches a

Table 5.1 The relationship between population density and fecundity in *Sitophilus oryzae* at 25°C and 90% relative humidity

Number of weevils	4	8	32	128	128	128
Number of grains	800	400	400	200	50	25
Number of eggs laid daily per female	6.60	3.50	3.02	1.60	0.90	0.50
Number of eggs per grain	0.08	0.17	0.60	2.56	6.14	7.55
Number of adults appearing per day and per female	–	3.50	2.20	0.70	0.20	0.08
% of grains without eggs	92.0	84.2	61.0	5.0	2.0	0.0

maximum when the ratio of number of individuals (males and females combined) to the number of grains equals 1/200. The number of eggs decreases when fewer grains are at the disposal of the females. This effect is due, firstly, to 'saturation' of wheat grains, females hesitating to lay in grains already containing eggs or larvae; and, secondly, to an overcrowding effect, as insects hinder one another and so interfere with egg laying (table 5.1).

3   INTRASPECIFIC COMPETITION

Competition may be intra- or interspecific. In the first instance, competing individuals remain together so that they continue to interbreed, thus ensuring the transmission of those genetic factors characteristic of the population. In this way intraspecific competition is fundamentally different from interspecific competition.

Intraspecific competition is evident in territorial behaviour where an animal defends a nesting site or a specific area of its habitat. This is the case in many birds (cf. p. 171) and fish (for example the stickleback). In the mating season, the male marks out a territory and, with the exception of the female or females, prevents other members of the species from entering it.

Many hypotheses have been put forward to explain the function of the territory. The existence of a territory may increase the chances of survival by dividing the resources of the environment and so avoiding intense competition amongst individuals. It also provides for protection against predators, the resident animals having a good knowledge of possible escape routes and so being able to avoid their enemies more easily. It has been observed that the existence of territories can influence the evolution of a species by increasing the opportunity for interbreeding, young pairs formed often tending to differ genetically from animals occupying neighbouring territories. In this way, mutations that may lead to the differentiation of new species are more easily retained.

The maintenance of a social hierarchy with dominant and subordinate individuals is another example of intraspecific competition. In the common chafer, the white three year old grubs hinder the development of the one or two year old grubs. This explains why flights of adults take place only every three years, while in other insects (for example elaterid beetles of the genus *Agriotes*) where the larval stage also lasts three years, there are emergences of adults every year because of the lack of competition in the larval stages.

Competition for food between individuals of the same species increases with a rise in population density. Those consequences that relate to population dynamics will be discussed in chapter 8 (p. 201). Intraspecific competition resulting from high plant densities causes significant adaptive changes in individuals, a phenomenon only rarely observed in animals. Agriculturalists and foresters have, for a long time, recorded these changes. Many theoretical and practical studies are at present being made on these changes because of their potential importance, particularly for increasing crop yields. One particular change that has been observed is a decrease in the mean weight of individual plants, and in the number of seeds found on those plants although the weight of each seed has tended to remain constant (fig. 5.1). Some data for the effect of competition are given below for well known examples. In lucerne, for example, the diameter of the root base is related to the production of dry matter, this being normal for a perennial species where the root is the storage organ. If d is the diameter of the root base (in mm) and n the number of plants per linear metre, the relationship $d = 6.20 - 0.02\,n$ holds for the first autumn after sowing, and $d = 10.00 - 0.06\,n$ for the second autumn, showing that the relationship between d and n varies with age. In the Paris region, the relationship $E = 112 - 1.76\,n$ was found for the hybrid maize

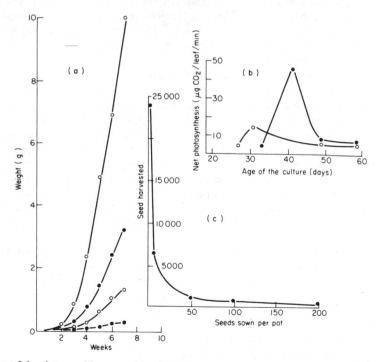

Figure 5.1   Intraspecific competition in plants. A: Weight increase for individuals from four white mustard populations with densities of 125, 12.8, 3.2 and 1.5 plants per dm² respectively; B: Variation in the net rate of photosynthesis for the third leaf appearing in two populations of white mustard, one with 3.5 plants per dm² (black dots) and the other with 12.8 plants per dm² (white circles); C: Seed production as a function of density in *Capsella bursa pastoris* (after Palmblad. 1968).

INRA 300, relating E, the number of cobs per 100 plants, to n, the density of plants per square metre.

Intraspecific competition occurs particularly for water and light. In the case of competition for light, plants interfere with one another by cutting out light when they are too dense (fig. 5.1). As density increases, numbers are greatly reduced by high mortality, which is a result of competition. This effect is well known for some crops (carrots and beetroot), and in woods where young trees are more numerous than mature trees (cf. table p. 188).

Competition for light has an effect on the form of trees, isolated trees being very different from trees growing in a forest group. A tree such as the oak, when isolated, has a globular standard form, the lower lateral branches receiving as much light for development as those higher up in the crown. In a forest, however, the leaves of the lower branches are shaded by those above them on the same tree or on neighbouring trees. A self-pruning process operates, causing the death and shedding of branches whose leaves do not receive sufficient light, and where the assimilation/respiration balance is negative. Lower branches tend to disappear as the tree grows in height, and it adopts a closed canopy form with a straight, cylindrical trunk lacking branches for much of its height.

In some instances, intraspecific competition may result in a population being divided into several distinct groups living in different habitats. The finch *Passerculus sandwichensis* has one race that lives in coastal salt marshes, while the other lives on dry hillsides. Competition may cause the geographical movement of a part of the population. For example, several specialised seed-eating birds of the taïga, e.g. the slender-billed nutcracker and the waxwing, periodically erupt into eastern Europe when food is in short supply throughout their normal range.

It can be said, in conclusion, that competition for food between individuals of the same species becomes more intense as population density increases.

II INTERSPECIFIC INTERACTIONS

When two species interact, they may have either no effect, or a harmful, or beneficial effect upon one another. The range of possible interactions are:

(i) *neutralism*: the two species are independent, neither species affecting the other.

(ii) *competition*: each species inhibits the other. Competition is the result of the search for food, shelter, nesting sites and other resources. Both species are called competitors.

(iii) *mutualism*: each species is unable to survive, grow or reproduce without the other being present. The two species live in symbiosis.

TABLE 5.2   Summary of interactions between different species (after Odum, 1959)

INTERACTION	SPECIES OCCURRING TOGETHER		SPECIES SEPARATED	
	SPECIES A	SPECIES B	SPECIES A	SPECIES B
Neutralism	0	0	0	0
Competition	−	−	0	0
Mutualism	+	+	−	−
Cooperation	+	+	0	0
Commensalism (A commensal with B)	+	0	−	0
Amensalism (A amensal with B)	−	0	0	0
Parasitism (A parasite, B host)	+	−	−	0
Predation (A predator, B prey)	+	−	−	0

0: species not affected in their development
+: development of species made possible or benefited
−: development of species hindered or made impossible

(iv) *cooperation*: both species form an association which, though not obligatory (i.e. each one being able to live alone) brings benefits to both. The term proto-cooperation is sometimes used in preference to the term cooperation since it is argued that the latter implies a voluntary choice. Nesting colonies of many species of birds, certain terns and herons for example, illustrate a form of cooperation which provider for more efficient defence against predators

(v) *commensalism*: an association of one species (the commensal), which benefits, with another species (the host), which gains no advantage. Commensal organisms show reciprocal tolerance of one another. Phoresis, the transport of a small animal by a larger one, is a form of commensalism. An example of phoresis is the transport of several species of mite by scarab beetles like *Geotrupes*.

(vi) *amensalism*: in this type of interaction, one species (the amensal) is inhibited in its growth or reproduction, while the other (the inhibitor) is not affected.

(vii) *parasitism*: a parasite, generally the smaller, may inhibit growth or reproduction of its host, and depends directly upon the latter for its food. The parasite may or may not cause the ultimate death of its host.

(viii) *predation*: one species (the predator) kills another species (the prey) in order to obtain food.

The different kinds of interaction are summarised in table 5.2. Too much reliance should not be placed on this classification, which is a simplification of the many kinds of interaction that may take place. In addition, interactions between species may change with time. For example, the great spotted woodpecker competes for food with the tits, but in spring woodpecker and tits are commensals, abandoned nests of the woodpecker being occupied by tits. The classification is used simply to show the main examples of interspecific interactions.

1 COMPETITION

Interspecific competition can be defined as the active search, by members of two or more species, for the same environmental resource. Some authors have a wider conception of competition, which, according to them, corresponds with all adverse interactions between two species. They then distinguish between direct and indirect competition. In the former, a species affects another simply by its presence, for example by secreting toxic substances into the environment, and they may or may not depend on the same food supply. This direct competition, in the strict sense, is called amensalism by Odum (1971), and his view is adopted here by excluding it from true competition. Competition takes place when two species seek the same environmental resource: food, nesting site, etc. Kendeigh (1961) used the term 'indirect competition' to describe this process. The idea of competition was restricted by Labeyrie (1960) to the use of the same limited energy resource by different individuals, as opposed to the situation when two individuals seek the same shelter and nesting sites.

It is necessary, lastly, to distinguish between active competition (the interference of Elton and Miller, 1954; contest competition of Collier, 1973) when one of the species, by its behaviour, prevents the other from gaining access to food or necessary space (this is common in higher vertebrates, particularly with the complex behaviour of birds, but is also found in insects like dragonflies); and passive competition (the exploitation of Elton and Miller; scramble competition of Collier, 1973) which is more frequent and where behaviour does not prevent other competitors from obtaining the required resources (this is illustrated by the experiments of Nicholson, 1957 on the blow fly *Lucilia cuprina*).

This abundance of definitions shows that competition is a complex process, the details of which are still not properly understood. Its importance was already recognised by Darwin, and for some authors it forms, together with the concept of the ecological niche arising from it, one of the fundamental problems in ecology.

1.1 The idea of the ecological niche

Observation and experiment have shown that competition is greater between closely related species. In extreme instances it can be shown

TABLE 5.3 Food taken by two species of British cormorant (after Lack, 1945)

PREY TAKEN	PERCENTAGE COMPOSITION OF THE FOOD OF	
	P. aristotelis	*P. carbo*
Sand eels	33	0
Clupeoids	49	1
Flat fish	1	26
Shrimps/prawns	2	33
Labrid fish	7	5
Gobies	4	17
Other fish	4	18

that two species with virtually identical needs cannot survive together, as one of them is eliminated after a certain length of time. This statement has been called Gause's hypothesis or the competitive exclusion principle.

It occasionally happens that two related species, having apparently similar food requirements, live together without entering into competition. Some apparent exceptions to Gause's hypothesis can be found among the birds. The cormorant *Phalacrocorax carbo* and the shag *P. aristotelis* nest together on the same cliffs in Britain and fish in the same waters. Their prey is not, however, the same. *P. carbo* is a deep diver and feeds mainly on benthic animals (flat-fish, shrimps), while *P. aristotelis* fishes in surface waters feeding on clupeoid fish and sand eels (Lack, 1945). Despite appearances, therefore, these two species do not compete for food (Table 5.3).

Among insects, three species of ichneumonid wasps of the genus *Megarhyssa*, found together in North America, lay eggs, by means of an ovipositor, in larvae of *Tremex columba*, which lives in dead wood. The mean length of the ovipositor differs sufficiently, however, between one species and the other. (40 mm for *M. greenei*, 80 mm for *M. macrurus* and 115 mm for *M. atrata*) for competition to be avoided, each species of *Megarhyssa* choosing only *Tremex* larvae situated at a fixed depth (Heatwole and Davis, 1965).

These examples lead to the idea of the ecological niche which was first introduced by Elton in 1927. Using Odum's illustration, the habitat of a species corresponds to its 'address', while its ecological niche represents its 'profession' within the community to which it belongs. In other words, information about the ecological niche provides answers to the questions: what does it feed on? where does it rest? what feeds on it? and how does it reproduce?

1.2 Theoretical studies on the ecological niche

If the extent of the ecological niche of two species in a particular habitat is represented by two variables N_1 and N_2, there are four possible situations:

(i) that the two ecological niches N_1 and N_2 are totally separated;

(ii) that the two niches overlap partly and have a common portion $N_1 \cap N_2$;

(iii) that the two niches are adjacent to one another;

(iv) that the niche N_2 is completely contained within niche N_1.

When two closely related species are *allopatric*, i.e. when their ranges are distinct from one another, their ecological niches may be distinct, condition (i) above, or else may partly overlap, condition (ii). In this case geographical separation is sufficient to prevent competition. When the species are *sympatric*, i.e. they live together within a fairly large area, their ecological niches may be partly superimposed, condition (ii), or one may be totally included within the other, condition (iv). A third type of possible distribution is that of *contiguous allopatry*, where the two areas come into contact and the ecological niches adjoin one another, condition (iii).

It is possible to distinguish (Hutchinson, 1967) between a fundamental niche and a realised niche. The former is represented by the total niche which would be occupied by the species in the absence of any restraint by competition with other species. It corresponds to the maximum expansion that a species can attain. The realised niche corresponds to that portion of the fundamental niche which is actually occupied in the habitat. These theoretical ideas are expanded and justified by the many observations described in the following sections.

1.3 The effect of competition on geographical distribution

The effect of competition on geographical distribution is clearly illustrated in the case of species accidentally introduced into regions where they have not previously been found. When the introduced species are more robust and more prolific than the indigenous species, they tend to replace the latter. This is true for the grass *Spartina townsendi* which has gradually replaced the related species *S. stricta* on coastal mud-flats of western Europe. Two species of hymenopterous sawflies, *Cephus tabidus* and *C. pygmaeus* whose larvae develop in wheat stems, were accidentally introduced into the U.S.A. at the end of the nineteenth century. *C. tabidus* has been gradually restricted to the northern part of its range by *C. pygmaeus*, which lays its eggs earlier and whose larvae eat the eggs of other species. The Colorado beetle *Leptinotarsa decemlineata* which originally fed on *Solanum rostratum*, has become perfectly adapted to two new hosts, the potato *Solanum tuberosum* and a wild plant *Solanum carolinense*. The latter is also the host plant for *Leptinotarsa juncta*, which has gradually been replaced by the Colorado beetle over much of its range and is now restricted to Florida. Many Australian marsupials have retreated or disappeared as a result of competition with rabbits or sheep. Endemic species of earthworm in tropical regions have been replaced by palearctic species.

In Britain, the indigenous squirrel *Sciurus vulgaris* has been slowly replaced in the midlands and south by a species introduced from America, *Sciurus carolinensis*.

1.4 The effect of competition on distribution between several habitats in the same locality

Species that are ecologically closely related and live in the same region are often allopatric, despite appearances, as they occupy different habitats. Numerous examples illustrating this result of competition can be found in the terrestrial environment. When three species of woodpecker of the genus *Dendrocopos* occur together, each exploits a different part of the tree. The great spotted woodpecker searches for food on the trunk, the middle spotted woodpecker on the larger branches and the lesser spotted woodpecker on the small branches. The brown or sewer rat *Rattus norvegicus* has driven the black rat *Rattus rattus* out, and the latter now lives in lofts in houses while the former is found in cellars and sewers. In the savannah of Lamto in the Ivory Coast, the avifauna is so rich that all available niches appear to be occupied. Some closely related species are separated by their habitat. Thus on the savannah are found *Halcyon chelicuti* and *Tockus nasutus*, while *H. senegalensis* and *T. semifasciatus* occur on the edge of the forest, and *H. malimbicus* and *T. hartlaubi* in the forest itself. Where two species with almost identical diets occur together, they either occupy only a part of the habitat or both species may be fairly scarce. One example is that of *Cypsilurus parvus*, which is found in the crowns of palms, while *Cheatura ussheri* searches hollows in the trunk of the same tree. Another example is that of *Oriolus brachyrhynchus* and *O. nigripennis,* which live together in the forest canopy, both being fairly scarce. Alternatively, one species may be far more abundant than the other, as in the case of *Pogoniulus leucolaima* and *P. subsulphureus*. In the last two examples, competition for food is avoided by low population densities.

Some examples can also be found of animals living in neighbouring aquatic habitats. Connell (1961) showed that the distribution of barnacles (Cirripedia, Crustacca) of the genera *Chthamalus* and *Balanus* on the Scottish coast is partly determined by competition. *Chthamalus* adults are found attached to rocks above the level of mean high water neap tides, while those of *Balanus* are found below this level. Planktonic larvae of the former settle further down the shore as far as mid tide level, but are smothered by the more rapidly growing *Balanus* larvae before they can complete their metamorphosis. In Lake Lucerne, Switzerland, the copepod *Mesocyclops leuckarti*, which has a largely carnivorous diet, occurs in water below the level occupied by *Cyclops strenuus*, another copepod with similar diet. In Lake Maggiore in Italy, however, the two species are separated in time, *M. leuckarti* having maximum abundance in summer and *C. strenuus* in winter (Hutchinson, 1967).

Blondel (1969) also observed that the closely related species of warbler, *Sylvia cantillans* and *S. melanocephala* avoid competition by a time separation. The egg laying period of these two birds differs by as much as two weeks. In Sénégal and in other tropical regions, these birds reproduce throughout the year and this reduces competition.

When competition operates, the most specialised species, with the most restricted ecological niche, persists and replaces the others. This is confirmed by work on the relationships between several American burrowing rodents of the family Geomyidae (Miller, 1964). The various species of Geomyidae are allopatric and mutually exclusive, and so do not form mixed populations. Their distribution is mainly determined by the depth and texture of the soil, as well as by interspecific competition. *Geomys bursarius* has the most exacting requirements, being confined to deep sandy soils and excluded from soils that are too hard, i.e. compact clays and gravels. The least exacting species, *Thomomys talpoides*, is, on the contrary, found in a variety of soils. *Cratogeomys castanops* and *Thomomys bottae* have intermediate requirements. The relationships between these species have been described using a system of ecological niches based on two variables, the depth and texture of the soil. When two species do occur together, the one with the broader niche is suppressed by that with the narrower niche. Thus, *G. bursarius* is found along rivers where the soil is sandy, between the territories of *T. talpoides* and *T. bottae*, which it has driven out. In the same way *T. bottae* occupies deep soils on lower ground, driving *T. talpoides* to less suitable soil on higher ground.

Beauchamp and Ullyott (1932) showed a similar process operating in flatworms. When the two species studied were allopatric, *Planaria montenegrina* occupied only the upper part of rivers, where water temperatures varied between 6.5°C and 16°C, while *P. gonocephala*, which tolerates temperatures varying between 6.5°C and 23°C, lived along the whole length of the rivers. Where the two species were sympatric, however, *P. montenegrina*, having a narrower niche, replaced *P. gonocephala* in the upper reaches of rivers, and the latter species was found only in waters where the temperature ranged from 13°C to 23°C.

These examples show that the presence of more competitive species can reduce the fundamental ecological niche of species. A similar process occurs among plants. Many apparent calcifuge plants will tolerate a wide range of calcium levels in pure culture. However, under natural conditions, their habitat is restricted by competition with other species. This has been demonstrated experimentally. In the Swiss Alps at Grisons, two distinct plant associations are found. One, the *Nardetum*, is characterised by the grass *Nardus stricta* and develops on siliceous soils having a *pH* less than six. The other, the *Seslerio-semperviretum*, with *Sesleria caerulea* as the main species, occurs on

157

calcareous and dolomitic soils with *pH* greater than seven. However, experiments made with monocultures show that at least four species from the *Nardetum* can grow fairly well on calcareous soil, and eight species of the *Seslerio-semperviretum* on siliceous soil. Although no marked changes in vitality were observed in monocultures of *Nardus stricta* and *Sesleria caerulea* on various soils, in mixed culture *N. stricta* gradually eliminated *S. caerulea* on siliceous soils, and *S. caerulea* eliminated *N. stricta* on calcareous soils. The reason is not yet known for this result, but it probably has a physiological basis.

The opposite effect, i.e. the extension of an ecological niche in the absence of competitors, is found in island faunas. Islands often afford simpler habitats than the mainland and the number of available niches is more restricted. Under these conditions, many species have a broader niche than on the mainland. Examples are birds like the Wake Island rail (*Rallus wakensis*), *Porzanula palmeri* from Laysan Island (both probably now extinct) and *Gallirallus* from New Zealand. These birds are omnivorous, but usually attack eggs of other species, a diet unknown in related birds from the mainland. The frequent disappearance of island species following the introduction of competing foreign species can be explained by the stability of the habitat not allowing the modification of niches, although this is the normal result of interspecific competition on the mainland.

1.5 Effect of competition on the morphology and productivity of plants

The result of interspecific competition on plants is similar to that of intraspecific competition, i.e. adaptive changes and reduction in fertility and numbers. It has often been shown that a dominant species will gradually eliminate a subordinate species or, at least, greatly reduce

TABLE 5.4   The yield of three different groups of wheat and peas. The groups have almost identical yields, suggesting the existence of some kind of self-regulating mechanism

VARIETY OF PEA	YIELD OF WHEAT in kg of dry matter per m^2	YIELD OF PEAS	TOTAL WHEAT + PEAS
A	0.478	0.459	0.937
B	0.473	0.480	0.953
C	0.315	0.585	0.900

its vitality. In mixed cultures of wheat (variety Étoile de Choisy) with each of three varieties of forage pea, it was found that the more vigorous and coarse variety *C* almost smothered the earlier wheat. As a result of competition for light, the wheat became too shaded to maintain a sufficient rate of photosynthesis (table 5.4).

1.6 Effect of competition on the evolution of species and communities

Competition for light may act as a factor in the evolution of plant associations. The germination of a forest tree species underneath another species having more exacting light requirements is a regular feature. In the forest at Fontainebleau the birch (*Betula verrucosa*) is the first to colonise bare soil in burnt areas. Scots pine appears under the cover of the young birch after some years. After twenty years, the birch becomes dominated by the pine, cannot obtain sufficient light, and so dies. The oaks *Quercus sessiliflora* and *Q. pedunculata* gradually establish themselves under the pines. Finally, the beech, which is a shade producing tree, gradually eliminates all the other trees. The elimination of birch by pine is assisted by the birch bracket fungus (*Piptoporus betulinus*), a fungus which severely attacks trees stunted by competition. In North American deciduous forests the first woody plants are 'pioneer' species requiring a great deal of light, like *Populus tremuloides* and *Betula payrifera*, and these are gradually eliminated by 'climax' species like *Acer saccharum*. It can be seen from table 5.5 that numbers of individuals of *Acer saccharum* decrease as a result of intraspecific competition, but that the relative importance of this species increases as the proportion of bare ground increases.

TABLE 5.5   Changes in the number of trees per hectare and the proportion of bare ground (figures in parentheses) in a North American deciduous forest

SPECIES		YEARS					
		1933	1938	1944	1949	1955	1960
'Pioneer'	*Populus tremuloides*	20	11	3	3	2	3
	Betula payrifera	35	23	17	12	8	9
'Climax'	*Acer saccharum*	322	315	292	280	266	215
		(39)	(44)	(49)	(53)	(56)	(58)

Competition plays a part in the evolution of species, especially when they are sympatric over part of their geographical range. In this case, competition is responsible for the development of the different kinds of morphological, ecological and behavioural adaptations that allow the two species to exploit the habitat in different ways. This is the reason that two species, which are partially sympatric, show more differences from one another, over that part of their range where they occur together, than in areas where they are not in contact. This 'character displacement' has been known since Darwin, and has been demonstrated in insects, fish, amphibians, birds and mammals, as well as for some flowering plants. Character displacement forms the basis for adaptive radiation, one of the causes of the evolution of species. In the

Galapagos Islands, passerine birds of the family Geospizidae, better known as Darwin's finches, have colonised many habitats and become adapted to a wide range of diets. Two related species, *Geospiza fuliginosa* and *G. fortis* feed on seeds, their beak lengths being related to the type of seeds taken. Lack (1947) studied the beak length of these birds in four islands: Daphne, where *G. fortis* occurs by itself, Crossman, where *G. fuliginosa* occurs by itself, and Charles and Chatham, where both species are found together. He showed that, on the islands of Daphne and Crossman, the average beak length was ten millimetres for both species. On the islands of Charles and Chatham, however, the two species have evolved away from one another. *G. fortis* has a longer beak and *G. fuliginosa* a shorter beak. Those individuals of *G. fortis* with a beak length of less than ten millimetres, and those of *G. fuliginosa* with a beak length of more than nine millimetres, had been eliminated by natural selection as a result of competition for food (fig. 5.2).

Finally, competition plays an important part in population dynamics. Gause's work has provided a foundation for this aspect, which is discussed further in chapter 8. Two illustrations are given, in conclusion, showing how diversification of ecological niches enables the habitat and its food resources to be exploited to the full.

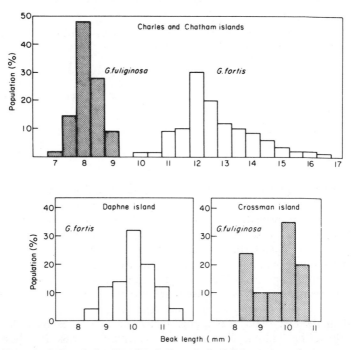

Figure 5.2   Beak length in *Geospiza fortis* and *G. fuliginosa* from four islands of the Galapagos archipelago (Lack, 1947).

In warm water fish ponds in China, where fish production reaches 5000 kg per hectare, traditional pisciculture uses five or six species chosen on the basis of their diets, so that every ecological niche is occupied. The species include a surface-feeding herbivore *Ctenopharyngodon idella*; two plankton feeders, *Hypophtalmichtis molitrix* and *Aristichthys nobilis*; *Cirrhina molitorella*, a fish feeding on benthic invertebrates, including larvae and worms, and organic detritus; a mollusc feeder, *Mylopharyngodon piceus*; and an omnivore, *Parabramis pekinensis*. In addition, another species of the same family (Cyprinidae), the common carp *Cyprinus carpio* is sometimes added. The latter is omnivorous, feeding mainly on detritus and the faeces of other fish.

The work of Lévêque (1957) in the Camargue has shown that the important limiting factor for waders, gulls and terns is the abundance of food. No species frequents the same feeding grounds in the same proportions, thus allowing for ecological isolation (fig. 5.3). There are, in addition, differences in methods of fishing in brackish water, as the depth of the localities where the birds search for food is determined by the length of their legs. The following order has been established:

	Kentish plover (*Charadrius alexandrinus*)
	Redshank (*Tringa totanus*)
Fishing in	Oyster catcher (*Haematopus*)
increasing	Avocet (*Recurvirostra avosetta*)
depth of water	Shelduck (*Tadorna tadorna*)
	↓ Flamingo (*Phoenicopterus ruber*)

The large number of ecological niches formed in these ways makes it possible for many species to nest in the Camargue.

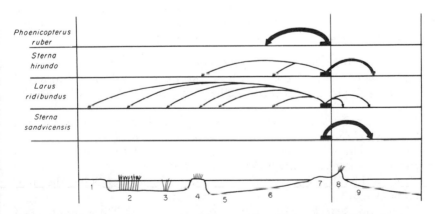

Figure 5.3   Diagram of the habitats where some birds of the Camargue feed. The thickness of the arrows is proportional to the amount of food taken from each habitat. 1. dry or cultivated land; 2. rice fields; 3. marsh; 4. saline grassland; 5. slightly brackish pond (from 5 to 15%); 6. brackish ponds (more than 15%); 7. nesting islands; 8. dunes and beaches; 9. sea (simplified from Lévêque, 1957).

2   PREDATION AND PARASITISM

A predator is a free-living animal which obtains its food by killing another animal, death being almost instantaneous. Both young and adult predators may feed in this way. By contrast, a parasite feeds on living tissue and does not necessarily cause the death of its host. Parasites spend at least part of their life cycle entirely within (endoparasites) or attached to or feeding at the surface (ectoparasites) of the host. They show a close physiological dependence on the host, and an immunological response is initiated in the latter. Parasitoids, mainly hymenopterous and dipterous insects, resemble parasites in that the young larvae are located within a host, which is sometimes paralysed, feeding on non-vital living tissue. The older larvae, however, finally cause the death of the host and in this respect are more similar to predators. The adult parasitoid is free-living.

2.1   The idea of the parasitic complex

The parasitic complex is the group formed by a species together with its parasites. The thorough investigation of the parasitic complex of a species is indispensable to studies of numerical changes for that species. Two well known examples are described here.

a   The parasitic complex of Choristoneura murinana

This moth, which belongs to the family Tortricidae, has a caterpillar which is a pest of the silver fir *Abies alba* in Europe. The diagram (fig. 5.4) shows the importance of the main parasites of this species during the larval and pupal stages. The parasites are mostly ichneumonids, chalcids and braconids belonging to the Hymenoptera. The pupal stage is more heavily parasitised than the larval stages. Those parasites restricted to *C. murinana* include one ichneumonid, *Cephaloglypta murinianae* and a braconid, *Apanteles* sp.

b   The parasitic complex of rape beetles

The fauna of the rape crop has been thoroughly studied by Jourdheuil (1960). This plant is attacked by phytophagous beetles of the following groups:

(i) flea beetles and chrysomelids of the sub-family Halticinae, like *Psylliodes chrysocephala* and *Phyllotreta nemorum*;

(ii) weevils of the genus *Ceutorhynchus*, like *C. napi, C. pleurostigma, C. assimilis, C. quadridens* and *C. picitarsis*;

(iii) nitidulids of the genus *Meligethes, M. aeneus* being the most frequent species.

Predators attack these phytophagous beetles throughout their inactive stages (eggs and pupae buried in the soil). These predators are mainly carabid beetles and *Cantharis livida*.

Parasites include nematodes, which may attack several stages (larvae,

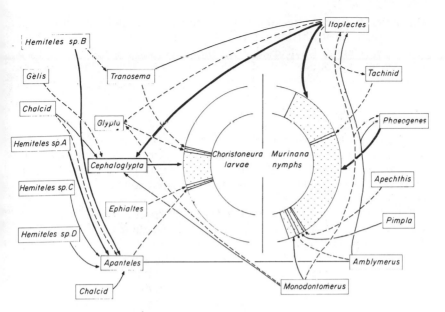

igure 5.4   Parasitic complex for *Choristoneura murinana*. Percentage parasitism for larvae and
pupae is shown by the dotted portion of the two semi-circles. Arrows of dashed lines
correspond to a rate of parasitism less than 1%; thin continuous lines to 1–5%; medium
continuous lines to 5–10%; thick continuous lines to more than 10%. The names of some
hyperparasites are given in the column on the left side of the diagram. The parasites are a
tachinid, some braconids (*Apanteles* sp.), some chalcids (*Monodontomerus aereus* and
Amblymerus subfumatus), all the others being ichneumonids.

pupae and imagos), and, especially, Hymenoptera which most
frequently attack the larvae. These hymenopterous parasites belong to
the superfamilies Ichneumonoidea and Chalcidoidea. They are nearly all
monoparasitic, i.e. only one parasite developing in each host. They can
be classified in the following way:

(i) Internal parasites of beetle larvae.
   Univoltine cycle – only one generation per year. These include
   Ophioninae (Ichneumonidae) of the sub-tribe *Thersilochina*, with
   seven species.
   Polyvoltine cycle – many generations per year. Three species of
   Thersilochina and nine species of Braconidae.

(ii) External parasites of beetle larvae.
   These include Braconidae (three species), Pteromalidae (two species),
   one Eurytomidae and one Euplophidae.

   The superfamily Ichneumonoidea is represented by a small number
of genera comprising endoparasitic species restricted to beetles. These
are essentially braconids of the genera *Diospilus* and *Microctonus*, and
ichneumonids of the sub-tribe *Thersilochina*. These Ichneumonoidea,

which attack only beetles of crucifers and which are polyphagous, form a small community which is restricted to these plants.

The superfamily Chalcidoidea, on the other hand, is represented by a heterogeneous group of ectoparasitic species attacking a small number of beetles.

2.2   The idea of coincidence

For a parasite to develop, it is necessary that the receptive stage of the host and the infective stage of the parasite appear at the same time. The parasite cannot develop if the host is not present, is inaccessible or in a non-receptive stage. This phenomenon has been called coincidence by Thalenhorst (1950).

A particular study has been made for the purposes of pest control, of coincidence between phytophagous insects and their plant hosts. One example given is that of the apple blossom weevil. This insect lays its eggs in young flower buds, and the larvae develop in the fruit. Observations on the development of flower buds of the apple have made it possible to describe the various stages in development by letters A to H. The time of appearance of these stages depends on temperature, and may vary slightly from year to year. It has been shown that the period of oviposition of the weevil is closely related to certain stages in bud development. There is, in other words, a phenological coincidence between oviposition of the weevil and the development of its host. (Phenology is that set of observations which relate the action of temperature to the date of appearance of biological phenomena. These observations are particularly precise in plants. The flowering date of certain plants is recorded regularly each year for many localities, and it is possible, from this data, to draw up phenological maps.) Maximum oviposition for the apple blossom weevil is at bud stage C, which is characterised by a reduction division in the pollen grain mother cells and, morphologically, by the appearance of the green inner bud scales. The few eggs that are laid in stage D (characterised by the appearance of the pink flower buds) produce larvae which hatch too late and are rejected by the flower when it opens. Coincidence between oviposition of the weevil and development of the buds is sufficiently close that identification of the phenological stages of the buds serves to guide agricultural advisors when indicating the timing of insecticidal treatments.

Other examples have been given by Jourdheuil (1960) for the insect fauna of the rape crop. The larva of the weevil *Ceutorhynchus napi*, which lives on this plant, is attacked by the anthomyid fly *Phaonia trimaculata*. Phenological variation, however, limits the efficiency of the fly predator, whose larvae appear at the moment when many weevil larvae are burying themselves in the soil to pupate. The many phytophagous beetles which depend on rape for food lay their eggs at a predetermined stage in the development of the plant. The choice of earlier or later varieties makes it possible to cancel the coincidence between insect and plant host, and so hinder the development of the insect.

Phenological coincidence is to be expected between two species of animal when the existence of one depends on the other. The fly *Epistrophe balteata*, a parasite of aphids, enters a diapause in winter in central Europe and in summer in Israel; these are the periods when the aphid host disappears in those regions. Coincidence is very difficult in cases where a parasite has a cycle that includes several intermediate hosts. The large number of eggs laid by these species compensates for the enormous loss which results from the frequent lack of coincidence with the various hosts.

3 COMMENSALISM AND MUTUALISM

It is often difficult to distinguish between these two categories. Animals that invade and are tolerated in the nests of other species are commensals. The insects of mammal burrows, bird's nests and the nests of social insects comprise many species. A hundred and ten species of beetle have been found in marmot burrows in the Alps.

In general commensals are not dependent on a particular species. It is possible to distinguish *pholeoxenes*, which are found accidentally in burrows and nests; *pholeophiles*, which occur more frequently in nests and burrows than outside them; and *pholeobies*, whose entire life is spent in nests and burrows. Many commensals are attracted by the higher temperatures of nests and burrows, and this is true for the staphylinid beetle *Microglotta nidicola*, which is unable to copulate unless the temperature is perceptibly higher than that of the external environment. The increased relative humidity in the burrows of steppe rodents attracts visitors to these habitats. Organic matter, such as the faecal material found at the entrance of rabbit burrows, similarly attracts certain insects.

Commensalism is common amongst marine animals. A crab from the islands of Hawaii, *Lybio tessellata*, holds a small sea anemone of the genus *Bunodeopsis* in each of its claws, using it to capture its prey. The polychaete *Nereis fucata* lives in a shell occupied by a hermit crab, *Eupagurus prideauxi*, and collects the debris which forms its food from between the mouthparts of the crab. An anemone, *Adamsia palliata*, is associated with *Eupagurus prideauxi*, and defends it with the aid of its stinging tentacles and by extending its basal region to form a flexible covering for its partner.

Symbiosis or mutualism is well known in Lichens, formed by the association between an alga and fungus. In the animal kingdom, one of the best examples of symbiosis is that of termites, which harbour in the gut either flagellates or bacteria, depending on the species. This

165

symbiosis enables the termites to break down wood and they, in turn, provide the micro-organisms in their gut with a suitable environment.

Symbiosis plays an important part in ecology. For example, some plants live in association with nitrogen-fixing bacteria, and these are responsible for the nitrogen enrichment of the soil. An alder tree two metres high fixes from a quarter to half a kilogramme of nitrogen per year with the aid of symbiotic bacteria living in its root nodules. Trees possessing nodules for nitrogen fixation are often those species which pioneer poor soils. This is true for the alder, which is an early coloniser of alluvial moraines in Alaska. Symbiosis with fungi, forming mycorrhiza on their roots, explains the presence of forests on podsolic soils, where nitrates form slowly and are rapidly removed by leaching.

4 AMENSALISM

This phenomenon, involving the inhibition of growth of a species (the amensal) by substances secreted by another species, is particularly well known in the plant kingdom. Amensalism corresponds to direct competition in the sense of Kendeigh (1961), and to the antibiosis or antagonism of various authors. Secretion of a toxic substance by the roots of *Hieracium pilosella* enables this composite to eliminate annual plants and to develop pure stands over large areas. Many fungi and bacteria synthesise antibiotic substances which inhibit the growth of other bacteria. Examples of amensalism in the aquatic environment are numerous. Marine dinoflagellates of the genus *Gonyaulax*, responsible for the phenomenon of 'red water' throw out substances which diffuse through the water and may cause the death of the entire fauna over large areas. Planktonic Cyanophyceae like *Aphanizomenon flosaquae* frequently multiply and accumulate in large masses at the surface, where they form 'water flowers', poisoning the aquatic fauna and also cattle which come to drink. Other algae can cause this intoxication effect. The substances produced are of varied chemical composition and include peptides, quinones, etc. These substances act at low concentrations, and have been called ectocrine substances. While their role is important, it is still not clearly understood.

Chapter 6

THE CHARACTERISTICS OF POPULATIONS

A population is a group of individuals of the same species which occupies a particular area at a given time. The members of a population generally interbreed with one another, except in the case of a few partheno-genetic and apomictic species. Each population has a group of genes forming a common gene pool and these are continually being redistributed within the population. Since suitable habitats for a species are often separated from one another, each species will consist of a collection of populations which are more or less isolated from one another. The word population is sometimes used to describe a group of individuals belonging to several related species within a fixed area, for example, the rodent population of a field, or the carabid population of a marsh. The term 'peuplement' is used in the French literature to describe groups of individuals belonging to several species.

The aim of population ecology is to understand and to explain the fluctuations in populations which take place under natural conditions, and also, when these are exploited by man. Population dynamics is of considerable importance from both the theoretical point of view and for its applied aspects. These include the prediction of outbreaks of pest species, game management, the dynamics of exploited fish populations and biological control. In order to interpret population fluctuations, it is necessary to study the action of environmental factors, both biotic and abiotic. Since changes may also be brought about by genetic factors, it is also necessary to understand population genetics.

I THE SPATIAL DISTRIBUTION OF INDIVIDUALS

Some knowledge of the properties and characteristics of a population is necessary to an understanding of their dynamics.

In the case of a *regular distribution*, the variance (s^2) is zero, since the number of individuals in each sampling unit is constant and equal to the mean. In the case of a *random distribution*, the mean ($\bar{x}$) and the variance (s^2) are equal. For a *contagious distribution*, the variance

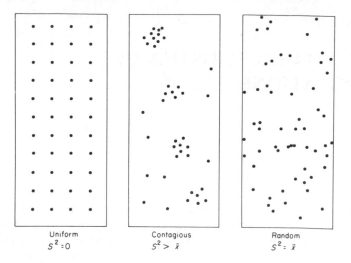

Uniform	Contagious	Random
$S^2 = 0$	$S^2 > \bar{x}$	$S^2 = \bar{x}$

Figure 6.1 The different possible types of distribution for individuals in a population. $\bar{x}$ mean; s^2: variance.

(s^2) is greater than the mean, and increases as the animals become more aggregated (fig. 6.1).

The random distribution corresponds to what mathematicians call the Poisson distribution, for which the expected frequencies are calculated from the series given by:

$$Ne^{-m} \left(1, m, \frac{m^2}{2!}, \frac{m^3}{3!}, \cdots \frac{m^x}{x!}\right)$$

where N is the number of sampling units in a sample, and m the mean number of organisms per sampling unit. Comparison of experimental data with expected frequencies shows whether a distribution is random or not. It is also possible to compare values of the variance and mean. If the ratio $s^2/\bar{x}$ tends towards zero, there is a uniform distribution; if the ratio tends towards one, the distribution is probably random, and if it is greater than one, it is a contagious distribution (for the calculation of limits for the ratio $s^2/\bar{x}$, see Milne (1964)).

A uniform distribution is rare in nature, and is usually the result of intense competition between individuals. The stickleback, a solitary carnivorous fish which defends a territory, appears to be uniformly distributed. Another good example of uniform distribution is that of the bivalve *Tellina tenuis*, which is found in littoral sands in the English Channel. The reasons for the uniform distribution of this species are not known (Holme, 1950).

A random distribution is found only in uniform habitats and for those species where there is no tendency to aggregate. The distribution of *Tribolium* in culture media is often random. The same is true for aphids in a field during the initial stages of invasion, while population density is low. The eggs of insects are often laid at random and, as a result, the young which hatch from the eggs are also distributed at random. As they become older, however, they tend to become aggregated. This is true for wireworms, the chafer *Amphimallon majalis* and the caterpillar of *Pieris rapae*. Changes in the density of these populations affect their distribution so that it is no longer random. For example, as the aphids multiply, their distribution over the field becomes contagious, and this is also true for wireworm larvae and caterpillars of *Pieris*.

The contagious or aggregated distribution is by far the most widespread. It is often due to small but significant variations in the environment, or to changes in the behaviour of the animals. The aggregations themselves may be distributed at random or clumped. A well known example is that of trees in a wood. In a single-species stand, the trees at first appear to be aggregated but, as time passes, this changes to a regular distribution accompanied by a decrease in density as a result of intraspecific competition. A regular spatial distribution is characteristic of Scots pine and beech forests. In mixed woods the suppressed species are generally distributed in clumps (contagious distribution), the dominant species having a regular distribution.

II DENSITY

The density of a population is the number of individuals present per unit area of volume. The estimation of density is important, as the effect of a species on an ecosystem depends to a large extent on its density. There are a number of methods for estimating animal numbers but they may be grouped under four main headings. (A detailed account of these methods is beyond the scope of this book. Further information can be found in Southwood (1966) and in the I.B.P. Handbooks.)

1 DIRECT COUNTING

There are several ways of making direct counts of a population. In an 'open' habitat, i.e. one with a cover of short vegetation and no trees to hide the animals, mammals can be counted directly. This method can also be used for birds, where nests or nesting pairs can be counted, and it is particularly suitable for sea birds living on steep slopes and cliffs without vegetation. Fisher and Vevers (1944) have used the data from actual counts of gannets (*Sula bassana*), which nest in widely dispersed colonies on the North Atlantic shores of Europe and America, to show

how numbers have changed:

1834: 334 000 individuals;

1894: 106 000 individuals; showing a considerable decline in numbers consistent with overcropping by man, which was followed by the protection of this species;

1939: 165 000 ± 9,500 individuals, showing recovery of the colonies as a result of protection.

Global censuses are infrequent as they are too lengthy and expensive. Partial censuses have been made over fixed areas, and these give some idea of the dynamics of a local population. A regular census has been taken of caribou in northern Canada, of Saïga antelope in the Volga basin, and of ungulates in some African national parks. Among birds, colonies of the heron, *Ardea cinerea*, have been recorded in England, and changes in numbers of birds on the Sept Iles reserve in Brittany have also been monitored.

Where a habitat is very open, larger areas can be covered from planes flying at low altitude and slow speeds. In the U.S.A. aerial photography has been used to count numbers of pronghorn antelope, *Antilocapra americana*, over an area of 100 000 km² of the desert plateau, on the borders of the states of California, Nevada, Oregon and Idaho. In 1949 there were 29 940 individuals in this area (Springer, 1950). The census of overwintering duck populations in the Camargue has been made possible by aerial observation. The number of shelduck *(Tadorna tadorna)* in Europe has also been estimated at between 75 000 and 100 000 with the aid of aerial photographs, taken while the birds gathered for moulting on the German North Sea coast between Bremerhaven and Cuxhaven (Goethe, 1961). The density of guano birds off the Peruvian coast reaches twelve per square metre (Dorst, 1963).

Direct counts can be made of insects. Dreux (1962) estimated that the mean density of grasshoppers in the Alps was in the region of two individuals per square metre.

The sampling route method has been widely used for mammals. It entails going over a carefully marked-out route under similar conditions, and recording all animals seen in a strip on either side of the route. The width of the strip depends on the size of animal. This method was used by Boulière and Verschuren (1960) to estimate the mean density per square kilometre for a number of large mammals in Albert National Park in Zaire. Results are shown below in Table 6.1, animals being counted in zones 300 metres wide for the first five species and 150 metres wide for the remaining five.

Among smaller herbivorous mammals, the density of hares was estimated at eighty-three, eighty-three and one hundred and eleven individuals per square kilometre, respectively, along the three different routes. In more closed habitats the width of the counting zone was reduced to twenty-five metres. Verschuren, using this width, counted

TABLE 6.1   Counts of large mammals in the Albert National Park

| SPECIES | mean number of individuals per km² | | |
	ROUTE I (steppe)	ROUTE II (steppe and bush savanna)	ROUTE III (park savanna with trees)
Elephant	0.22	0.91	3.36
Hippopotamus	1.50	0.45	0.07
Buffalo	25.00	10.02	5.31
Topi	14.50	1.00	0.17
Waterbuck	0.54	–	5.28
Buffons kob	40.60	5.60	2.29
Reedbuck	0.16	–	–
Bushbuck	–	0.24	–
Wart-hog	4.30	0.43	1.39
Giant forest hog	–	0.11	–
Total mean number per km²	86.7	18.71	17.87

black-fronted duikers (*Cephalophus nigrifrons*) in dense *Hagenia* scrub of the Virunga mountains in Zaire.

Blondel (1965) estimated the density of nesting birds in a 28 ha plot of Mediterranean maquis, containing holm oak, rosemary and broom, not far from the Chaîne des Alpilles. A network of marked routes made it possible to walk through the area, the paths being closer together in the wooded areas than in the clearings. The nesting period lasted from 15th March until 1st July. Population density was estimated by counting birds singing, the assumption being that a singing male was

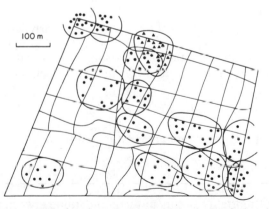

Figure 6.2   Territories of two species of passerine birds: the Dartford warbler, *Sylvia undata* (dots), and the nightingale, *Luscina megarhynchos* (triangles), in an area of 28 ha of Mediterranean maquis. Pathways are indicated by plain lines. Each triangle or dot represents a visual or audible contact with a bird (Blondel, 1965).

171

marking the territory containing its nest. The area was frequently walked over early in the morning on fine days, and every visual or audible contact with a bird was recorded on a large scale map. This map (fig. 6.2) shows results obtained for the two least abundant species. Territories of the more abundant species touched but never overlapped. Four nesting species were found: the nightingale with 12 pairs, the blackcap with 15 pairs, the subalpine warbler with 39 pairs, and the Dartford warbler with one pair. The mean density of nesting birds was 21 pairs per 10 hectares. By comparison, there were 68 pairs per 10 hectares in a deciduous forest at Cîteaux in Burgundy, 100 pairs in a Czechoslovak spruce forest, 20 pairs in a coniferous forest in Finland, and only 0.6–3 pairs in the semi-desert areas of the northern Sahara.

2   METHOD OF CAPTURE AND RECAPTURE

The principle of this method is theoretically simple. If a individuals are captured from a population of N individuals, marked by an indelible method (spots of colour, rings, some form of radioactive marker, etc.) and then released; and if, at the end of a certain time, b individuals are captured and, of these, c are already marked (i.e. recaptured), then

$$a/N = c/b \quad \text{or} \quad N = ab/c$$

It is assumed in making the calculation that the population is stable, without immigration or emigration, without births or deaths, and that the capture of an animal for the first time does not modify its behaviour and does not make its subsequent recapture easier or more difficult. This method has been used, with modifications, for many animals including dragonflies, Lepidoptera, bats, *Cepaea nemoralis*, finches and fish.

The method is only suitable if $a, b, c,$ and N are fairly large. Otherwise the errors become greater. It is possible to obtain the upper and lower limits of the estimate with the aid of a simple statistical calculation. If N_1 and N_2 are the limiting values for ninety-five per cent probability, the relationship

$$a/N_1 = c/b + 2\sigma \quad \text{and} \quad a/N_2 = c/b - 2\sigma$$

applies, where σ is the standard error estimated from

$$\sigma = \sqrt{c(b-c)/b^3}$$

whence it is possible to calculate N_1 and N_2.

If significant mortality occurs in the population between the times of marking and recapture, more complicated mathematical methods must be used. These, besides giving a population estimate, may sometimes indicate the mean length of life of the animals (Dreux, 1962).

172

3   SAMPLING

When the above methods are not practicable, it is necessary to resort to sampling when estimating the size of a population. The main difficulty lies in selecting a sample that is as representative as possible of the whole population. Depending on the type of habitat and the species studied, it is necessary for the ecologist to choose a method which leads to the smallest possible number of errors. Some of the techniques used to sample the three main types of habitat, terrestrial, freshwater and marine, are reviewed below.

3.1   Sampling terrestrial animals

a Arthropods, especially insects

Entomologists for a long time have used *sweep nets* to sample from herbaceous plants. A more recent tool is the *sélecteur* or sampler described by Lamotte *et al* in Lamotte and Bourlière (1969). This consists of a hinged metal cylinder with sharp edges which enables the operator to sample the faunas of specific strata in the vegetation. A similar method is the *vacuum sampler* described by Southwood (1966). These contrast with the sweep net, which gives an incomplete sample of animals.

The sweep net has been criticised for not sampling the whole fauna, but it can give comparable results if used in a standard way, and can be used to follow changes in a population for certain species. Balogh (1958) recommended the use of a square net to avoid unequal sampling at different levels. Compared with the sélecteur, only the sweep net allows the capture of relatively large animals. It can also give a more accurate picture at lower densities (Roth, 1963).

According to Whittaker (1952), a series of fifty sweeps, each 120 cm in length, using a net 33 cm in diameter, were required to collect that number of animals which corresponded to those actually occurring in a square metre. Thus, the net sampled only ten per cent of the fauna. The sweep net is not effective over very short vegetation, and not efficient in tall vegetation because of the difficulty of penetration, the lower layers not being sampled at all.

A quantitative method for sampling short grassland consists of collecting all the animals present in quadrats five metres square for larger animals, and from quadrats of one metre square (or even less) for smaller animals. The vegetation is first removed and carefully searched, stones turned and the soil surface disturbed. This method is laborious (four people required for a square metre, twelve for twenty-five square metres and twenty-four for one hundred square metres), but it has been used in Africa (Gillon and Gillon, 1965) and in the Camargue (Bigot, 1965) amongst others.

Results from samples can be expressed either in numbers of

173

individuals or by weight (biomass) for the different species or, as often happens, when it is impossible to identify all individuals, for the different groups.

Insects on trees can be collected by beating branches held over a sheet or tray; however, only those that fall and cannot fly are sampled. In addition, the quantity of foliage sampled is unknown. An alternative method encloses the branches in a large cloth bag in which the arthropods can be anaesthetised and counted. For equal weights of foliage, this method collects fifteen times more insects than are obtained by beating, and the relative proportions of some groups are very different. A similar method, using a hinged metal box, has been described by Dempster (1961). Beating does, however, give quantitative results for less mobile forms like caterpillars.

Some accurate techniques for sampling insects living on foliage of trees have been developed in Canada (Morris, 1955 and 1960) and Switzerland (Auer, 1961). For example, the larch moth *Zeiraphera griseana*, an important pest of larch in alpine forests, has been studied at Engadine. Auer (and also Morris in Canada) came to the conclusion that there were no significant differences in numbers of caterpillars between the edges of the forest exposed on the west, the east, the north or the south. There were, however, significant differences in the vertical distribution of caterpillars on the same tree and from one tree to another, even when they were of a similar size. These conclusions led to the design of a large-scale sampling programme. The larch forests were divided into a number of plots in which the trees were numbered. Trees to be sampled were chosen at random to avoid any choice on the part of the collector. The crown of each tree was divided into strata of equal height, and teams of climbers removed branches from each stratum. Every sample (between one and three kg) from a tree was accompanied by a detailed report, giving physical and biological data for the habitat. The results of laboratory examination of the samples were recorded on special forms and on punched cards. It was then possible to study variations in numbers occurring in the course of successive generations, and to examine the causes. Figure 6.3 shows results obtained from the studies at Engadine.

For the study of soil microarthropods (mainly collembola and mites), sampling units are taken with a soil corer, usually with surface area 5x 5 cm and between 2.5 and 5.0 cm deep. The fauna are extracted either by dynamic methods, which depend on the behaviour of the animals, or by mechanical methods, when the animals play only a passive role (Vannier and Cancela da Fonseca, 1966; Murphy, 1962). The Berlese funnel is the commonest dynamic method used. It has been modified, since its introduction in 1905, by many workers, such as Tullgren, Macfadyen and Vannier. In the apparatus designed by the latter, the soil cores of known volume are placed over funnels of diameter twelve centimetres, and kept at a temperature of 30°C, the

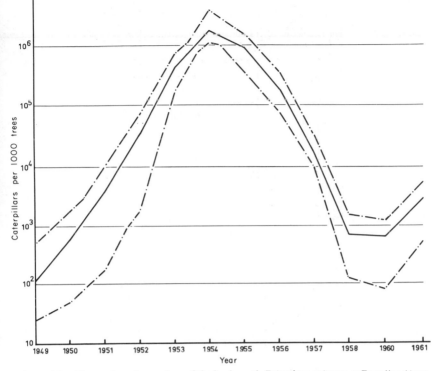

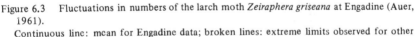

Figure 6.3   Fluctuations in numbers of the larch moth *Zeiraphera griseana* at Engadine (Auer, 1961).
Continuous line: mean for Engadine data; broken lines: extreme limits observed for other localities.

lower part of the funnel being at air temperature. A thermal gradient, which favours movement of animals out to the core, is set up as a result of ventilation of the lower part of the funnel. A diagram of a programmed automatic extractor (Vannier, 1964) is shown in figure 6.4. It can be seen that, at any particular time, the water content of the soil core can be estimated. In addition, the apparatus will record the order in which various groups of the soil microfauna emerge in relation to soil moisture content. It can be shown that groups always emerge in the same order, and these reproducible results are evidence of the reliability of the apparatus, which, when used with reasonable care, is the only rapid means by which large collections of collembola and mites can be made and the development of populations studied. Flotation methods (Murphy, 1962) are inconvenient and laborious, and extract dead animals as well as living ones. These methods are, however, more suitable for the extraction of larger soil animals like earthworms and molluscs.

175

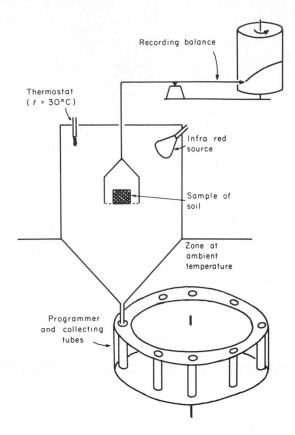

Figure 6.4   A programmed automatic extractor for soil microfauna (Vannier, 1964).

b   Trapping

Mammal populations are often studied by trapping. If all the individuals in a rodent population have the same probability P of being captured, and if P is the total number of animals in the population, then, the number of individuals caught at the time of the first trapping,

$$C_1 = p \cdot P \text{ individuals}$$

and, at the time of the second trapping,

$$C_2 = p \cdot P'.$$

where $P' = P - C_1$. This gives

$$P = \frac{C_1^2}{C_1 - C_2}.$$

The method consists either of capturing all the animals until numbers are exhausted, or calculating P graphically. The points representing numbers of animals captured lie on a straight line which cuts the x axis at a point representing the value of P for the population. This method is equally applicable to fish population estimates, and sometimes for insects. In the case of mammals, results depend on the type of trap used, the attractiveness of the bait, climatic conditions and, lastly, behaviour of the animals with respect to the traps.

Many different kinds of traps are used for the capture of arthropods, especially insects. Simple containers buried at soil level, i.e. pit-fall traps, catch non-flying insects. The addition of a fixative such as 50% alcohol kills the animals and ensures their preservation. If, however, fixative is not added, it is possible to keep the animals alive and then to release them after marking. This method was used by Grum (1959) who, using eighty pots, 60 mm in diameter, distributed over 400 square metres, was able to study the movements of the carabid beetles *Carabus arvensis* and *Pterostichus niger* at Pologne. Doane (1961) designed an improved trap capable of retaining elaterid beetles, which are active jumpers (fig. 6.5). Baits can be used in traps, but it is then difficult to use the results in quantitative way.

Sticky traps, such as the glass plates or squares of metal foil coated with adhesive used by Ibbotson (1958) and Roth (1963), may be used for the collection of small flying insects or those that are transported by wind. Suction traps can be used in some instances to obtain quantitative data. Light traps using either visible or ultra-violet light, which is more attractive to most insects, may be used to trap flying insects. All species are, however, not equally attracted, and in some instances there are differences between males and females. The height

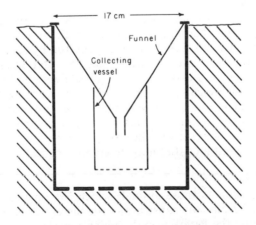

Figure 6.5   Doane's pitfall trap, with an inner removable container for the collection of elaterid beetles.

of the light trap above the ground, the weather, the wavelength and intensity of light all affect the results.

3.2 Sampling fresh-water animals

Quantitative sampling of invertebrates is difficult. The two tools most frequently used are the pond net and the plankton net. A net which has been used for sampling restricted, thinly populated, aquatic habitats, consists of a closing net, fixed in the open position at the sampling site, left for a fixed time and then closed by a remote control system. For the shallow waters of the Camargue, a brass tube eight centimetres in diameter and forty centimetres high, with a sharpened lower rim which was pushed into the substrate, has been employed. This enclosed a vertical column of water of known volume which could be removed, the fauna being extracted by filtering through a fine net. The fauna is disturbed by the movement of the apparatus but, if this is always used in the same way, the error will be standardised and results comparable. The benthic fauna and other large animals can be removed with a bucket or hand grab.

Samples of fish can be taken with nets, including seine nets, by draining part of a river or by poisoning water with formalin or rotenone, which results in the dead fish floating at the surface just long enough for collection. The electric fishing machine is one of the best methods for collecting fish in freshwater. For a full account of sampling in freshwter, reference should be made to the I.B.P. Handbook No.17.

3.3 Sampling marine animals

a The plankton
Marine plankton has an uneven distribution, and contains animals of a wide size range (less than 1μm to several centimetres), and this causes sampling errors. It is possible to collect plankton by removing a known volume of water with the aid of a Nansen water sampler, which is lowered at the end of a cable to the required depth and then closed by means of a special weight or 'messenger'. Water can also be filtered *in situ* through a plankton net towed for a fixed distance. A recording device measures the volume of water filtered. These methods do not entirely avoid sampling errors, but they can be standardised so that results from different localities may be compared.

b The benthos
The benthos comprises those animals which live on the sea bed. They can be collected from soft substrates with the aid of various grabs, the commonest being the Petersen grab, which samples an area of 0.1 m^2. On hard substrates, only the littoral zone has been investigated. The

development of skin-diving techniques has made it possible to extend the zone investigated down to a depth of sixty metres. Further information on the sampling of marine animals is given in the I.B.P. Handbook No.16.

4   INDIRECT METHODS OF ESTIMATING NUMBERS

Desert rodents with individual burrows can be estimated by counting holes. The method of 'reopened holes' is also used. All holes found are blocked, and the number reopened or redug are recorded (Petter, 1961). This technique has been used to estimate numbers of the vole *Microtus arvalis* in Belgium.

Molehills and beaver lodges can be counted, but the correlation with actual numbers is often poor. In the case of rabbits, droppings found over a fixed area can also be counted and used to calculate an index of frequency. Data from dead animals have been used; for example, numerical fluctuations of arctic mammals have been studied from numbers of pelts. The analysis of actual captures by predators can give information about their prey. Small mammals, for example, can be studied from the analysis of pellets regurgitated by birds of prey.

Attempts have been made to estimate insect abundance from plant damage. It is possible, for example, to count exit holes made by adults whose larvae have developed in the trunks of trees, or to estimate, from the tunnels under elm bark, the numbers of individuals present in various stages from the egg to the adult in the beetle *Scolytus scolytus*. In Canada, Morris (1955) used collections of excreta falling from trees to estimate the densities of caterpillars of *Choristoneura fumiferana*. Head capsules left after successive moults, recovered at the same time, gave information about the stage of development as well as the species present. Since the foliage intercepts some cast skins and droppings, and wind disperses others, there is some inaccuracy in this method. The collection of moulted skins of dragonflies (Corbet, 1957), and cicadas (Dybas and Lloyd, 1962) can similarly be used to give estimates of insect density.

III   THE INTRINSIC RATE OF NATURAL INCREASE

The growth rate of a population depends on two opposing factors: births and deaths. If emigration and immigration are ignored and the population is assumed to be a closed one with no interchange with neighbouring populations, then the *growth rate of the population* (dN/dt) is:

$$dN/dt = B - D$$

where N is the number of individuals in a population at time t, and B and D are numbers of births and deaths. Since B and D are proportional to the size of the population, then $B = bN$ and $D = dN$ where b and d are the *birth rate (natality)* and *death rate (mortality)* per individual in the population. If $r = b - d$ then:

$$dN/dt = (b - d)N = rN \tag{1}$$

On integration equation (1) gives:

$$N_t = N_o \cdot e^{rt} \tag{2}$$

where N_o is the population size at time $t = 0$ and r is the intrinsic rate of natural increase. The calculation of r for experimental populations is laborious and can only be made for stable populations, having a constant proportion of individuals in each age class. Some values of r are shown in table 6.2. The value of r varies widely between species and, even within a species, it is not a constant parameter but varies with environmental factors such as temperature.

Equation (2) represents exponential growth and corresponds to the 'biotic potential' for a species. This is its capacity for increase in the absence of any constraints. With the exception of a few special cases (laboratory culture, colonisation of new habitats), it is clear that this population growth is not achieved and that the situation is very different in nature. Exponential growth is restricted by the carrying capacity of the environment, K, which is the maximum number of individuals that a particular habitat can support. The equation for the logistic curve put forward by Verhulst in 1845 describes this situation:

$$dN/dt = rN \left(\frac{K - N}{K} \right),$$

TABLE 6.2   Value of the intrinsic rate of increase, r for some animal populations

SPECIES AND ENVIRONMENTAL CONDITIONS	INTRINSIC RATE OF INCREASE (r) PER FEMALE PER YEAR	FINITE RATE OF INCREASE (e^r) PER YEAR (Number of times population would multiply per female per year)
INSECTS:		
Sitophilus oryzae at 29°C	39.6	$1.58 \cdot 10^{17}$
Sitophilus oryzae at 23°C	22.4	$5.34 \cdot 10^9$
Sitophilus oryzae at 33.5°C	6.2	493
Tribolium castaneum at 28.5°C and 65% relative humidity	36.8	$9.60 \cdot 10^{15}$
MAMMALS:		
Microtus agrestis	4.5	90
Rattus norvegicus	5.4	221
Man	0.0055	1.0055

which gives

$$N = \frac{K}{1 + e^{a - rt}}.$$

The integration constant, a, defines the position of the curve in relation to the origin, ie. the value of N at time, $t = 0$. The correction factor $(K - N) / K$ is an expression of the environmental resistance of the habitat to the growth of a population. Although natural populations rarely achieve an exponential rate of growth, the calculation of r is of interest and especially when comparing closely related species. The ecological optimum for a species represents those conditions where r reaches a maximum value. In the case of two sympatric species of flatworm occurring in the Lyonnaise region, the value of r varies with

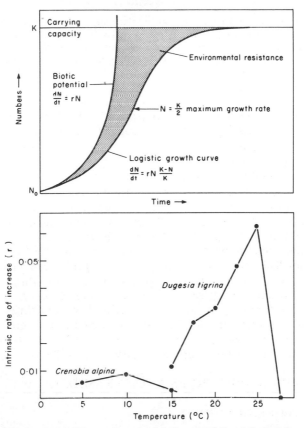

Figure 6.6   Above: exponential growth curve (biotic potential) and the logistic growth curve. Below: variation in the intrinsic rate of natural increase, r, as a function of temperature, t, for two species of flatworm.

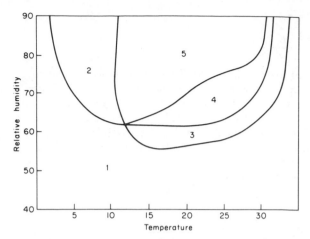

Figure 6.7   Effect of temperature and relative humidity on two mites, *Acarus siro* and *Cheyletus eruditus*. In zone 1, neither species can survive; in zone 2, *Acarus siro* alone survives; in zone 3, *Cheyletus eruditus* alone survives; in zone 4, *Cheyletus eruditus* has a higher rate of increase than *Acarus siro* and controls its prey efficiently; in zone 5, *Acarus siro* has a higher rate of increase than its predator which can no longer control its prey (after Solomon, 1962).

temperature. Where there is interspecific competition between flat-worms, *Crenobia alpina* succeeds in cold waters where it has a higher value for r and *Dugesia tigrina* eliminates its competitors in waters where the temperature exceeds 20°C as its rate of increase is higher under these conditions (fig. 6.6). Similarly, natural selection acts by either forming geographic races or producing species having a value of r which varies with physical environmental factors (cf.p. 221).

Solomon (1962) analysed the effect of temperature and humidity on the growth rate of two species of mite. These were *Acarus siro*, which lives in flour and its predator, *Cheyletus eruditus*. He found that the two species had very different optimum temperatures and humidities. *A. siro* survived best at higher temperatures and under these conditions increased its numbers by seven times in a week. *C. eruditus* preferred lower temperatures and its numbers increased by four times in a week. As a result of this variation in rate of increase with environmental factors, *C. eruditus* was only able to reduce numbers of *Acrus siro* in a narrow zone of temperature and humidity (zone 4, fig. 6.7).

IV   SURVIVORSHIP CURVES AND PYRAMIDS OF AGE

1   SURVIVORSHIP CURVES

The principles underlying the construction of survivorship curves were established by specialists in human demography, and were introduced

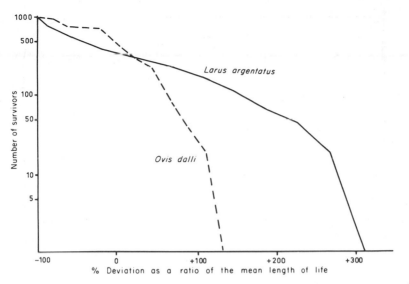

Figure 6.8   Survivorship curves for the Dall mountain sheep *Ovis dalli* and the herring gull *Larus argentatus* (Deevey, 1947).

into the study of animal populations by Pearl (1925). Work on the Dall mountain sheep, *Ovis dalli*, in the mountainous area of Mt. McKinley National Park, Alaska is used here as an example (fig. 6.8). Since the animal is protected, hunting is forbidden and mortality is solely due to natural causes. Murie (1944) made a study of 608 skulls of dead animals, ageing them from the form of the horns. Two kinds of graphical representation are possible for this data: (i) the age at death in years can be plotted on the abscissa, and the number of survivors, using a logarithmic scale, on the ordinate; or (ii) time can be plotted on the abscissa in the form of the difference between each age class and the mean age at death, expressed as a percentage of the mean. For example the mean age at death of *Ovis dalli* is 7.09 years, and the difference from the mean for the age class two to three years is $2.00 - 7.09 = -5.09$, or, as a percentage,

$$\frac{-5.09}{7.09} \times 100 = -71.8\%.$$

The second method of representation, shown in figure 6.8, is more useful, enabling curves for two species with very different life spans to be compared.

There are three types of survival curve (fig. 6.9):

(i) In the case of *Drosophila*, man and many mammals, most individuals have the same life span and die during a very short period of time. These curves are convex.

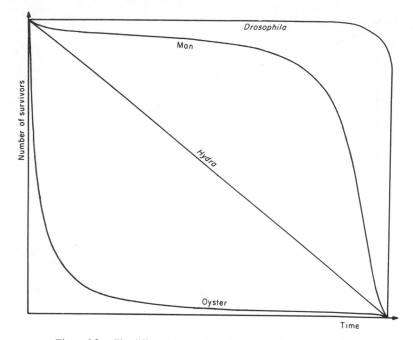

Figure 6.9   The different types of survivorship curves (Deevey, 1950).

(ii) In the case of the fresh water *Hydra*, the coefficient of mortality remains constant throughout life, and the curve is a straight line.

(iii) In the case of oysters, fish and many birds and invertebrates, the curve is concave, showing a high mortality in the early stages.

The curves also show that, in some species like *Ovis dalli*, the mean life span is little different from the maximum longevity. In others, like the gull *Larus argentatus* and the American robin *Turdus migratorius*, the maximum life span is much greater than the mean longevity.

Survival curves for insects are obtained by plotting the length of different stages on the abscissa, and not by taking equal time intervals (fig. 6.10). The density of a population will often influence the form of the survivorship curve. The black-tail deer *Odocoileus hemionus columbianus*, living in the chaparral (a plant formation similar to maquis) region of California, has been studied by Taber and Dasmann (1957). A high density population (twenty-five individuals per km²) was found   in a managed area, where an open scrub and herbaceous cover was maintained by repeated burning, as this provided a greater quantity of new growth for the animals. A lower density population (ten per km²) was recorded in an unmanaged area of old bushes unburned for ten years. Recently burned areas could support up to thirty individuals per km², but the population was unstable and a

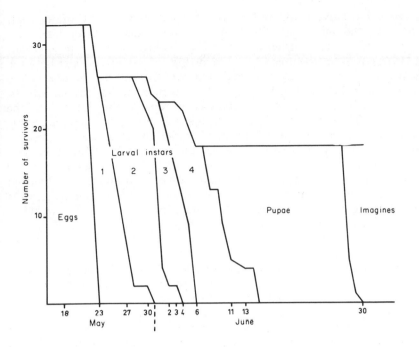

Figure 6.10   Development of an egg batch (of 32 eggs) from one female Colorado beetle (Grison, 1963).

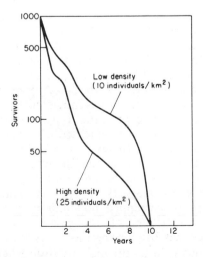

Figure 6.11   Survivorship curves for two populations of the deer *Odocoileus hemionus* in California (Taber and Dasmann, 1957).

185

survivorship curve could not be constructed. Figure 6.11 shows that, although the maximum life span was the same in both instances, juvenile mortality is higher in the case of the higher density population.

Differences in the form of survivorship curves often exist between sexes—in man, in many mammals (e.g. *Odocoileus hemionus*) and in insects (e.g. *Tribolium*). Shortage of available food has similar effects.

The study of survivorship curves is of considerable interest to the ecologist. They indicate at what age a species is most vulnerable. Some form of intervention during the most vulnerable stage has the greatest effect on the future development of the population. These principles are applied, for example, in the management of game reserves or in the control of insect pests. Since mortality appears to be more influenced than natality by environmental factors, the study of mortality factors is necessary to explain fluctuations in populations.

2 PYRAMIDS OF AGES

Since natality and mortality vary considerably for different age groups, the potential for increase in a population will depend on the relative proportions of the different age classes present.

Three periods can be distinguished in the life of an animal: the pre-reproductive, the reproductive and the post-reproductive periods. In many species the pre-reproductive period is by far the longest. This is particularly true for mayflies where the adult life is very short, and for the American cicada *Cicada septemdecim* which has a larval life of seventeen years.

There are three kinds of age pyramid. The first has a broad base, indicating a high proportion of young individuals, and is characteristic of rapidly increasing populations. The second is an intermediate type with a fairly large percentage of young, and the third has a narrow base with more older individuals than young. This is characteristic of a declining population (fig 6.12). For many species that are hunted or fished, the age pyramid is arbitrarily modified by man (fig. 6.13).

Research on laboratory animals has indicated that populations show homeostasis, that is, if some individuals are regularly removed, their replacements tend to reach the same age distribution as that of the individuals removed (Watts *in* Chauvin, 1957). Lotka (1925) suggested the same thing on theoretical grounds. The work of Petrides (1950) on the muskrat supports these conclusions. In a population that had been heavily trapped for many years, there were 48% adults and 52% young animals; whereas, in an untrapped population, there were 85% adults and 15% young. Reduction of the population by trapping had apparently increased natality among the survivors, and the population returned spontaneously to its original structure when trapping ceased.

Fish provide useful material for investigation as their scales often give

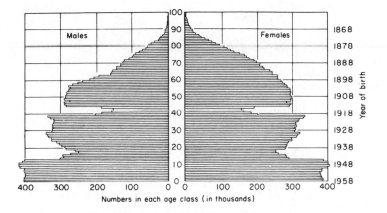

Figure 6.12   Pyramid of ages for the population of France on the first of January, 1959.

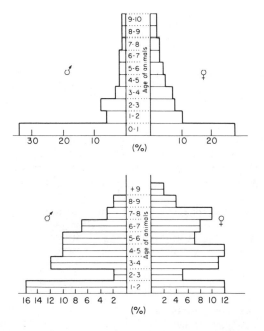

Figure 6.13   Above: pyramid of ages for a population of the deer *Odocoileus hemionus* in the chaparral of California. The pyramid shows each sex separately (Taber and Dasmann, 1958). Below: pyramid of ages for a population of wild sheep from the Davella reserve in Corsica. The scarcity of two- and three-year old animals corresponds to a forest fire which caused the death of many animals (Pfeffer, 1967).

187

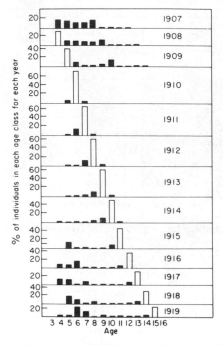

Figure 6.14   Age classes for herring caught in the North Sea between 1907 and 1919. The age class formed by individuals born in 1904 was dominant for several years. This class is represented by a colourless histogram (Hjort, 1926).

a precise measure of age. In the case of North Sea herring studied by Hjort (1926), fish younger than five or six years were not caught in nets and only older individuals were properly sampled. Figure 6.14 shows that fish hatched in 1904 dominated the population until 1918 when, even at the age of fourteen years, they were still the most abundant age

TABLE 6.3   The age classes of a beech forest at Soignes, Belgium

YEAR RECORDED	AGE OF TREES	NUMBER OF TREES PER HECTARE	MEAN ANNUAL MORTALITY	MORTALITY RATE
1866	1	10 000	–	–
1897	31	2 274	250	2.5
1903	37	1 949	54	2.4
1919	53	861	68	3.5
1928	62	758	21	2.4
1932	66	698	15	2.0
1936	70	620	19	2.7
1940	74	546	18	2.9
1944	78	509	9	1.6
1948	82	498	3	0.6
1962	96	414	5	1.0
1968	102	367	8	1.9

TABLE 6.4 Diameter classes in a beech forest in central Europe, the diameter being that of the tree at a height of 1.30 metres, which is approximately proportional to age

CLASSES GREATER THAN 10 cm IN DIAMETER	NUMBER OF INDIVIDUALS PER HECTARE			
	BEECH	PINE	SPRUCE	
11–15 cm	20	12	–	
16–20 cm	16	8	4	small
21–25 cm	–	4	–	timber
26–30 cm	4	4	–	80
31–35 cm	4	4	–	
36–40 cm	4	4	8	
41–45 cm	16	4	–	medium
46–50 cm	4	4	–	timber
51–60 cm	24	–	4	72
61–70 cm	16	16	4	large
71–80 cm	–	4	4	timber
81–90 cm	4	4	–	64
91–100 cm	4	8	–	
greater than 100 cm	–	–	–	
Total	116	76	24	216

group. This phenomenon of a dominant age class has been observed in fish with a very high reproductive potential.

For plants, there are usually fewer old than young individuals. This is the result of competition (cf. page 150). A beech forest (the forest of Soignes in Belgium), for example, shows an almost constant mortality rate of about two per cent over a period of one hundred years (table 6.3)

The age class distribution may not always be even, as is the case in the virgin climax beech forests of central Europe, where there are three or four dominant age classes (table 6.4). This suggests that the present day population includes individuals descended from several waves of extensive regeneration spread over a period of time. Figure 6.15 shows a pyramid of ages for a population of Scots pine in the forest of Fontainebleau.

V SEX RATIO

Another important factor in the development of populations is the proportion of the sexes, or sex ratio. This is rarely unity. One sex is frequently more abundant than the other. In vertebrates there is often a slightly larger number of males at birth. In a muskrat population in Wisconsin, the sex ratio at birth was unity, but after three weeks the ratio of males to females was 140 to 100. Males tend to be more numerous than females in the Anatidae (ducks and geese). Small

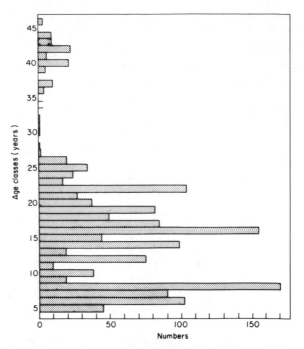

Figure 6.15   Pyramid of ages for a mixed population of Scots pine in the Solle valley, forest of Fontainebleau. Area covered by population: 3.76 ha.; numbers: 1440 trees of 5 or more years (Lemée, 1967).

populations of the American rabbit *Sylvilagus* show a strong predominance of males. On the other hand, large populations of the squirrels *Sciurus carolinensis* and *S. niger* show a predominance of females.

VI   GENETIC POLYMORPHISM IN POPULATIONS AND ITS ROLE IN ADAPTATION

Genetic variability is an essential feature of natural populations. It is almost true to say that no two individuals are exactly identical from the genetic point of view. This polymorphism can be readily demonstrated for selected species. The banded snail *Cepaea nemoralis* has a basic shell colour of either yellow, pink or brown and a variable number (0 to 5) of black bands. These characters are determined by allelomorphic genes whose action is not influenced by environmental conditions. This polymorphism, however, has an adaptive function since there is a correlation between shell colour, number of bands and the habitat of the snail (fig. 6.16). In beech woods, snails lacking bands 1 and 2 and with a pink shell predominate. In hedgerows and on rough grassland,

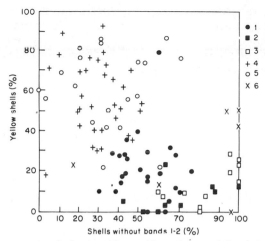

Figure 6.16   An example of adaptive polymorphism: the correlation between habitat, shell colour and banding in *Cepaea nemoralis*. 1: mixed deciduous woods; 2: oak woods; 3: beech woods; 4: hedgerows; 5: rough herbage; 6: short turf (after Cain and Sheppard, 1954).

yellow, fully banded snails predominate. Visual selection by bird predators, song thrush and blackbird, is responsible for this difference, snails with a yellow, banded shell being less obvious against the uneven background of rough grassland. The work of Dobzansky (1947) on *Drosophila pseudoobscura* across the southern United States showed

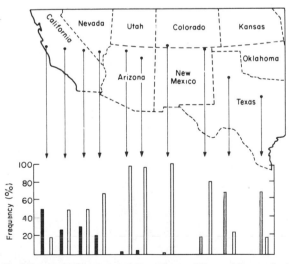

Figure 6.17   Variation in the frequency of three genes of *Drosophila pseudoobscura* along an east-west transect across the southern U.S.A. Black columns: *standard* gene; white columns: *arrowhead* gene; hatched columns: *pike peak* gene. The *standard* gene is absent from the east, and the *pike peak* gene from the west (after Dobzansky, 1947).

geographical variation in the frequencies of the genes, *standard, arrowhead* and *pike peak* (fig. 6.17). Genetic polymorphism can equally well be demonstrated for many plants. It may occur in either morphological or physiological characters and it provides a mechanism for adaptation to new environments.

Chapter 7

POPULATION FLUCTUATIONS UNDER NATURAL CONDITIONS

'In looking at Nature, it is most necessary to keep the foregoing considerations always in mind—never to forget that every single organic being may be said to be striving to the utmost to increase in numbers; that each lives by a struggle at some period of its life; that heavy destruction inevitably falls either on the young or old, during each generation or at recurrent intervals. Lighten any check, mitigate the destruction ever so little, and the number of the species will almost instantaneously increase to any amount'
Darwin, *The Origin of Species.*

1 THE CAPACITY FOR INCREASE

Some examples of the capacity for increase of living organisms have been given already (table 6.2, p.000). This potential has been described by Darwin in *The Origin of Species.* The bacterium *E. coli* completes a division in twenty minutes and, at this rate, the organism could cover the surface of the earth with a continuous layer in only thirty-six hours. A single *Paramecium* could produce, in a few days, a mass of protoplasm equal to 10 000 times the volume of the earth. Among the Metazoa, the tapeworm *Echinococcus* produces nearly one billion eggs. A pair of sexuales of the aphid *Phylloxera* could theoretically produce, in one year, between 10^{11} and 10^{18} offspring. Amongst the vertebrates, a fish like the cod lays many millions of eggs. The offspring from a pair of birds laying only five or six eggs per year could amount to ten million individuals in fifteen years. Darwin made a similar calculation for the elephant: 'at the end of seven hundred and forty or seven hundred and fifty years, there would be nineteen million living elephants, all descended from the same pair'.

This increase in numbers, if unchecked, follows a geometric progression, and it is this capacity for increase which Chapman (1931) called the *biotic potential*. The potential rate of growth must now be compared with the actual rate of growth for a population. For example, the following data have been given for flamingoes in the Camargue, where the population numbers about 8000 individuals. In 1957, 2500 young were hatched, representing a population increase of 31%. There were no births in 1958, and 600 young hatched in 1959, an increase of 7.5%. If it is assumed that one pair lays one egg per year, this would give a potential rate of increase of fifty per cent, a much higher figure than the observed rate of increase.

Data for *Pieris brassicae* illustrate some of the factors which limit the growth of populations. The following mortality factors act on this butterfly:

> diseased larvae: 59%
> larvae parasitised by *Apanteles glomeratus*: 34%
> larvae eaten by birds: 4%
> diseased pupae: 3%
> pupae parasitised by *Pteromalus puparum*: 0.14%
> numbers of imagines emerging as % of original number of eggs laid: 0.32%.

A careful study of the causes of mortality must be made in order to predict changes in a population. Life tables have been calculated for a number of insect pests of forest trees. Morris (1957) has given the data in table 7.1 for *Choristoneura fumiferana*, a pest of the Canadian spruce.

In this table x is the age of the animal, represented by the stage in its development, l_x corresponds to the number of individuals surviving at the beginning of each age class, x (here, for example, there were 200 eggs to begin with), and d_x corresponds to the number dying during the age interval x. The table shows that only two imagines developed from a batch of 200 eggs, a mortality of 99%. The most vulnerable stage is that of the later larva, as mortality here reaches 90%.

TABLE 7.1 Summary of life table for *Choristoneura fumiferana* (spruce budworm)

x	l_x	MORTALITY FACTORS	d_x	d_x as a % of l_x
Eggs	200	parasites	10	5
		others	20	10
		total	30	15
Young larvae	170	dispersal	136	80
Later larvae	34	parasites	13.6	40
		disease	6.8	20
		others	10.2	30
		total	30.6	90
Pupae	3.4	parasites	0.35	10
		others	0.55	15
		total	0.90	25
Imagines	2.5	miscellaneous	0.5	20

Sex ratio 1:1, mortality 99%, survivors 1%.

For a species with a sex ratio of one to one, the following mortality rates, prior to attaining reproductive ability, would be necessary to maintain the population at a steady state:
a mortality of 90% with a fecundity of 20 progeny per female;
a mortality of 99% with a fecundity of 200 progeny per female;
a mortality of 99.9% with a fecundity of 2000 progeny per female.

II FLUCTUATIONS IN NATURAL POPULATIONS

1 IRREGULAR FLUCTUATIONS OVER LONG INTERVALS

These changes are unpredictable. The American shad, *Alosa sapidissima*, is a good example. In 1900 the catch of this fish reached almost 2000 tonnes per year. As a result of intensive fishing, water pollution and the construction of barrages, the catch decreased by 98%; then, just as this species appeared to be almost extinct, numbers increased, and 2500 tonnes were caught in 1944. At the present time, numbers of the shad are again beginning to decrease (Walford, 1950 *in* Vibert and Lagler, 1961).

Variation in the catches from mackerel fisheries off the coast of North America provides another example of irregular fluctuations. The most marked changes were between 1884 and 1886, when catches fell from over 60 000 tonnes to 10 000 tonnes per year.

An increase in numbers of the sea-star *Acanthaster planci* has taken place since 1966 in the Pacific Ocean, especially on the Great Barrier Reef off north-east Australia, as well as in the Mariana Islands and Tahiti. This species which, until then, had not been very abundant and which consequently had hardly been noticed, soon reached a density of one per square metre. This sea-star has caused considerable damage, as it feeds on the polyps which make up the living part of the coral reefs. It has probably cleaned out forty kilometres of reefs in less than three years on the island of Guam. The skeletons of some corals have been eroded by wave action to such an extent that some people have predicted, perhaps a little pessimistically, the early disappearance of many atolls. The fishing returns of those peoples in the Pacific who depend largely on the sea for a living have been considerably reduced as a result of fish abandoning the reefs. None of the hypotheses put forward to explain the sudden increase of the sea-star seem to provide a satisfactory explanation. One suggested cause is the elimination of one of its predators, the gastropod *Charonia tritonis*, whose shell is collected for its mother-of-pearl. Other suggestions include increases in the D.D.T. content of sea water, and the influence of radioactive fallout.

Unexpected fluctuations are also found in terrestrial animals. In 1946 in the forest of Fontainebleau, two carabid beetles *Nomius pygmaeus* and *Agonum quadripunctatum* were quite common. Only two examples of the first species had been found previously, one in 1864 and the other in 1935; similarly, only a few examples had been found of the second species. The two species have not been recorded in this forest since 1946.

2 REGULAR FLUCTUATIONS

2.1 Fluctuations with a cycle of several years

The most familiar example is that of some mammals and birds in arctic regions. Cyclic oscillations in fur-bearing animals are well known from the records of the Hudson Bay Company for numbers of pelts brought in annually by trappers. These oscillations have a cycle of 9.6 years for the snowshoe hare and the lynx. The peak abundance of the hare generally precedes that of the lynx by one or two years. Since the lynx is largely dependent on the hare for food, the oscillations appear to be related with those of its prey (figure 7.1). Cyclic oscillations with a mean period of four years occur in tundra species like the snowy owl, *Nyctea scandiaca*, the arctic fox, *Alopex lagopus*, and also in the lemming (*Dicrostonyx*) which forms their main prey. In the taïga, the great grey shrike, *Lanius excubitor*, the red fox, *Vulpes fulva*, and the field voles (*Microtus*) which serve as their prey, also have cycles of four years.

Some authors consider that the regularity of the 9.6 year cycle for the snowshoe hare and the lynx is related to the sunspot cycle. This idea would seem to be supported by the fact that a marine animal, the Canadian Atlantic salmon (*Salmo salar*) shows the same nine or ten year cycles of abundance. Odum (1971) has, however, discussed the considerable objections to this theory. Dorst (1963) gives an example of fluctuations for which the cause is understood. There is an influx of warm water towards the south off the Peruvian coast. This warm water

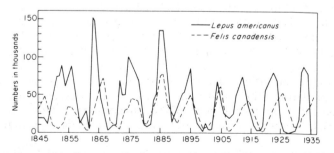

Figure 7.1   Cyclic fluctuations in numbers of the snowshoe hare *Lepus americanus* and lynx *Felis canadensis*, the data being numbers of pelts received by the Hudson Bay Company (Mac Lulich, 1937).

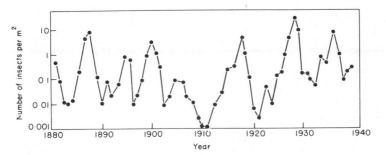

Figure 7.2   Annual fluctuations in numbers of the moth *Bupalus piniarius*, whose caterpillar lives on pine in the Letzlingen region of Germany (after Schwerdtfeger, 1941).

covers the cold water regularly, about every seven years, in 1917–18, 1925, 1932, 1939 etc. When this happens, the water temperature increases immediately by five degrees or more, salinity changes, plankton die, and the water becomes filled with decomposing material. The fish eventually die, and sea birds disperse over considerable distances as they have nothing to eat. This mortality is important, resulting in anaemia in underfed birds, in epizootic diseases and in increased parasitism. Cormorants are much more susceptible than gannets and pelicans as the former fish nearer the surface, while the other birds fish in deeper water which is unaffected by the superficial warm water layer.

The abundance of the crossbill *Loxia curvirostra* in Finland shows cyclical changes with a maximum about every three years. This corresponds to periods of abundance of spruce cones, the seeds of which provide their food. In the Alps, the larch moth has a ten year cycle, during which the population varies in a ratio of the order of one to 10 000 (figure 6.3). Another example from Germany is the oscillation in numbers of the moth *Bupalus piniarus*, whose larvae feed on pine. Peaks of abundance are separated by irregular intervals of approximately ten years (figure 7.2).

2.2   Seasonal fluctuations

Andrewartha (1935) has shown that numbers of *Thrips imaginis* in Australia reach a maximum each year towards the month of December, that is, during the Australian summer (figure 7.3). Emlen (1940) has described the development of a Canadian quail population. Every year maximum numbers for this species are reached towards the months of August and September, when the young birds, who form a large part of the population, appear. The remaining individuals are two or three year old adults.

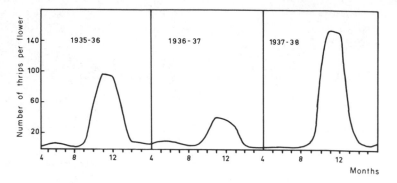

Figure 7.3 Seasonal variations in the abundance (in Australia) of *Thrips imaginis* (Andrewartha and Birch, 1954).

Marked seasonal variations are shown by lake plankton. Fluctuations in phytoplankton in a lake in Ohio were correlated with changes in zooplankton (figure 7.4). In 1948, for example, a large cladoceran population kept the phytoplankton at a very low level during the summer, while in 1949 the phytoplankton showed a marked summer increase as a result of a reduction in the numbers of Cladocera. Bates (1945) found that numbers of mosquitos in Colombia show seasonal changes of varying size depending on the species. Some are almost stable, while others undergo considerable variation (figure 7.5).

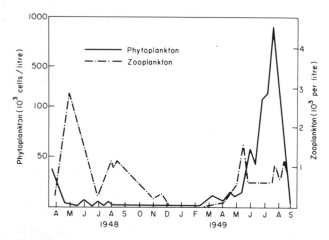

Figure 7.4 Seasonal variations in the abundance of zooplankton and phytoplankton in the Atwood lake reservoir, Ohio (after Wright, 1954).

198

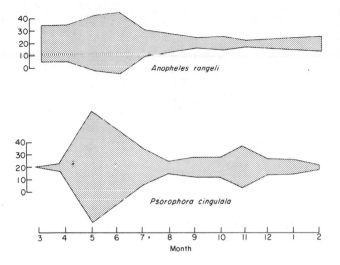

Figure 7.5   Seasonal fluctuations in the abundance of two species of mosquito in Colombia. The number of insects shown corresponds to nightly catches in a trap (after Bates, 1945).

3   IRREGULAR FLUCTUATIONS, OFTEN AT FAIRLY SHORT INTERVALS WITHIN CERTAIN LIMITS

In England, numbers of the heron *Ardea cinerea* tend to decrease after severe winters. The populations recover, however, within a year or two, so that average numbers are maintained. In Holland, the tits *Parus major* and *Parus ater* show irregular changes in abundance, but again, these are within certain limits.

4   THE RAPID INCREASE OF A POPULATION INTRODUCED INTO A NEW HABITAT

When a new habitat is colonised by a species, numbers increase rapidly. In Australia, a few rabbits that were introduced have given rise to many millions of descendants. A study has been made of changes in the numbers of a type of pheasant introduced on the island of Protection in Washington State. A census was taken each year in the spring and autumn from 1937 until 1942. Two males and five females were originally released in 1937, and these gave rise to a population which increased slowly at first, but then very rapidly. A fall in numbers every autumn was due to natural mortality. A steady state would probably have been reached if the pheasants had not been completely wiped out in 1942.

199

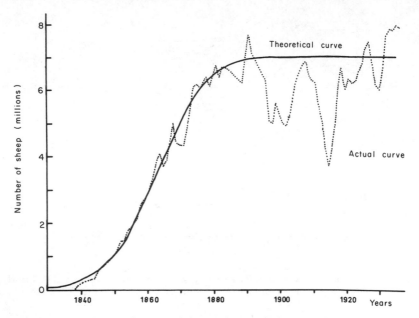

Figure 7.6   Growth curve for a population of sheep in South Australia. The corresponding logistic growth curve is also shown (Davidson, 1938).

Davidson (1938) studied the records of sheep introduced into South Australia. Numbers follow a growth curve which approximates to the logistic curve, but with fairly large fluctuations once a maximum level has been reached (figure 7.6).

200

Chapter 8

THE REASONS FOR POPULATION FLUCTUATIONS

Population changes are the result of the complex interactions which take place between a wide range of environmental factors. These fluctuations can be studied in three different ways. Firstly, by the use of simple laboratory models which attempt to reconstruct natural variations. Secondly, by formulating hypotheses based on the effect of one or more factors which are likely to be important and subsequently testing the validity of these models. Thirdly, by analysing natural population changes as accurately as possible and determining which factors are mainly responsible.

I EXPERIMENTAL STUDIES USING SIMPLE MODELS

Gause (1931, 1935), Lotka (1934) and Volterra (1931) have been responsible for a number of theoretical studies aimed, firstly, at determining the limits for growth of a population in terms of the amount of available food; and, secondly, at evaluating the effect of interspecific competition on each of two populations in the presence of one another. Their object was to formulate a mathematical expression for the intensity of the 'struggle for existence'.

1 POPULATION GROWTH

A population of the beetle *Tribolium confusum* is used here as an example. The experimental growth curve (figure 8.1) shows close agreement with Verhulst's logistic curve (cf. p. 180). The population reached a maximum of 650 individuals when supplied with sixteen grammes of flour, and 1750 individuals with sixty-four grammes of flour. There is, therefore, evidence of a maximum biotic capacity for this environment, food acting as a limiting factor. The population increase is less than the increase in available food. The rate of increase declines slowly as the population grows, i.e. the rate is inversely proportional to the density.

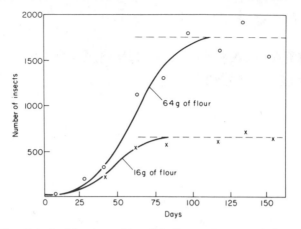

Figure 8.1   Growth in the laboratory of two *Tribolium confusum* populations, one with 64 g of flour and the other with 16 g (Gause, 1931).

2   GROWTH OF TWO POPULATIONS COMPETING FOR THE SAME NICHE

The first experiment designed to show competition was due to Gause (1934). He used two species of ciliate protozoa, *Paramecium caudatum* and *P. aurelia*, cultured in test-tubes at fixed pH and fed regularly with the bacterium *Pseudomonas aeruginosa*. This bacterium does not multiply in the selected medium. When separate, each species of *Paramecium* exhibits a typical sigmoid growth curve, the population being maintained at a level that is almost directly related to the amount of food provided (figure 8.2). When together, the two species compete with one another and, after two weeks, *P. caudatum* is eliminated. Under these experimental conditions only that species with the higher rate of growth (here *P. aurelia*) can survive. Gause also showed that, in a closed system, *P. caudatum* was sensitive to the metabolic substances produced, while *P. aurelia* showed only a slight response to them. If the experiment is repeated with a medium that is renewed regularly, the result is reversed, *P. aurelia* gradually disappearing. These experiments serve to illustrate the complexity of the processes involved in interspecific interactions.

If *Paramecium caudatum* and *P. bursaria* are used in these experiments, they coexist and an equilibrium is reached. In fact *P. bursaria* gathers at the bottom and on the sides of the culture tube, while *P. caudatum* remains in the open part of the culture. These two ecological niches are sufficiently distinct to avoid competition, although the food source is the same for both species.

Similar results to those of Gause have been obtained with higher animals. For example, Crombie (1942) allowed two species of flour beetle, *Tribolium confusum* and *Oryzaephilus surinamensis*, to compete

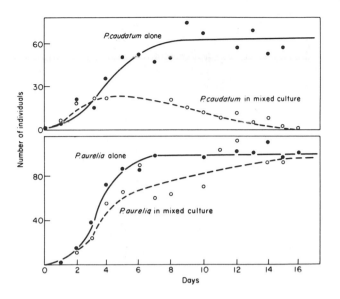

Figure 8.2   Competition between two related species, *Paramecium caudatum* and *P. aurelia* occupying the same ecological niche (Gause, 1934).

in a renewable medium. The former species eliminated the other after some months in a uniform medium. If, however, thin glass tubes were added to the culture medium to diversify the ecological niches, the two species were able to coexist. Frank (1957) in his experiments, used two planktonic species of *Daphnia, D. magna* and *D. pulicaria*, fed on unicellular green algae of the genus *Chlamydomonas*. In mixed culture only *D. pulicaria* survived, *D. magna* having disappeared after forty-five days. The main reason for this disappearance was an increase in the proportion of males of *D. magna* (up to eighty per cent), which resulted in a much lower rate of increase for this species.

The eventual elimination of one of any two competing species is not always the rule. Many experiments on *Drosophila* have shown that coexistence is possible under particular conditions of temperature and humidity. Among the plants, competition between the grasses, *Anthoxanthum odoratum* and *Phleum pratense*, leads to a stable equilibrium, with the more abundant species giving up some space to the less abundant one. These results may be explained in terms of the selection coefficient (cf. p. 220).

3   GROWTH OF TWO POPULATIONS CONSISTING OF A PREDATOR AND ITS PREY

In Gause's experiments on predator/prey interactions, the prey was the ciliate *Paramecium caudatum* and the predator another ciliate *Didinium*

203

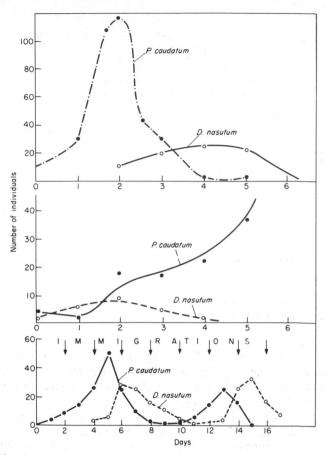

Figure 8.3   Growth curves in mixed culture for *Paramecium caudatum* and its predator *Didinium nasutum*.
Above: in a uniform environment without immigration.
Centre: in a heterogeneous environment without immigration.
Below: in a uniform environment with periodic immigrations (Gause, 1934).

nasutum. Three types of experiment were carried out. In the first, the culture tube contained only the predator and its prey. In this limited space the *Didinium* destroyed all the *Paramecium* very quickly, and finally died of hunger. In the second experiments, the culture tube contained, in addition, a layer of oatmeal on the bottom. The *Didinium* multiplied slowly at first, and the *Paramecium* population was reduced. The latter then found refuge in the oatmeal layer, where they were able to multiply protected from their predators, which finally disappeared. The third experiments were more complicated. Innoculations of *Didinium* were made at regular intervals into a culture containing

Paramecium plus a layer of oatmeal. Regular oscillations in numbers of both species took place. These occurred only when the destructive capacity of the predator was restricted or when regular immigrations of the latter took place (figure 8.3). Huffaker (1958) studied the predator-prey interaction between a predatory mite, *Typhlodromus*, and its prey, another mite belonging to the genus *Eotetranychus*, which lives on oranges. He showed that cyclic oscillations occurred when the two were cultured together under certain conditions. This kind of oscillation has been called reciprocal, since it depends on reciprocal interactions between the predator and its prey.

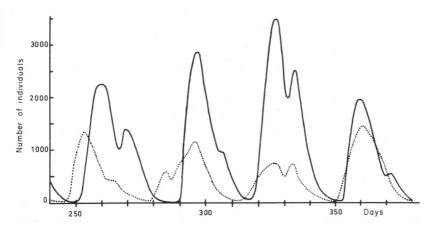

Figure 8.4   Fluctuations in an adult population of *Lucilia cuprina* where the density is regulated by the amount of food given to the larvae, the adults receiving an excess of food (Nicholson, 1954).
Continuous line: larvae receive 50 g of food.
Dotted line: larvae receive 25 g of food.

The experiments of Nicholson (1954) showed cyclic fluctuations within a single species, the blowfly *Lucilia cuprina*. In a first set of experiments, the adults received ample food, but the larvae were restricted to 25 g or 50 g of meat according to the experiment. Regular fluctuations were seen in the population. When larvae became too numerous, they entered into competition and hindered one another, the result being an increase in mortality. When the density was reduced, the larvae once again developed normally and numbers rose (figure 8.4). In the second set of experiments, the larvae received ample food, while the adults received only a limited supply (0.5 g of meat per day). This lack of food resulted in a lowering of fecundity in the adults, followed subsequently by a decrease in numbers, and regular fluctuations were set up (figure 8.5).

205

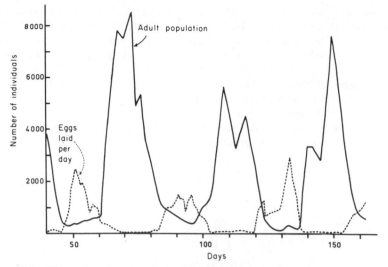

Figure 8.5 Fluctuations in a population of *Lucilia cuprina* where the adults are given a limited quantity of food, the larvae receiving an excess of food. The dotted line shows the numbers of eggs laid after a delay of two days, this being the time taken for the development of eggs after feeding (Nicholoson, 1954).

II THEORIES

The theories put forward to explain changes in abundance of a species can be arranged into two groups. Some ecologists consider that the only factors capable of regulating populations are those which act in a density-dependent way, i.e. the proportion of the population affected by the factor is proportional to population density. Many mathematical models have been put forward to support this viewpoint. Examples include the work of Lotka, Volterra, Nicholson and Solomon. Other ecologists argue that density-independent factors (climate, irregular spatial distribution, etc.) are the most important with density-dependent factors (biotic factors) having only a minor role. These include Bodenheimer, Uvarov and Andrewartha and Birch.

MATHEMATICAL THEORIES

There are many mathematical theories, each relating to a particular set of conditions. The more important theories are reviewed below.

1. The growth of an isolated population may be represented by Verhulst's logistic curve (cf. p. 180) which shows that the rate of increase diminishes as numbers increase, that is, as density-dependent factors begin to operate. However Verhulst's equation depends on the assumption that numbers are maintained at a steady level once the carrying capacity is reached. This is rarely true and fluctuations in

numbers about a mean value are more common. Other more elaborate models, depending upon age structure of the population, may be used to explain these fluctuations.

2. The situation where two species occupy the same niche and thus compete with one another, has been described in a model put forward by Volterra. This model predicts that in every instance, one of the two species will be eliminated and this agrees with the experiments of Gause on *Paramecium*. It also explains the scarcity, or even the absence, in nature of any two species with the same niche, coexisting together.

3. A rather more detailed study was made by Gause and Witt on the development of two species with overlapping niches. Their conclusions indicated that, depending on the conditions, either, one of the two species would be eliminated or, the two species would continue to coexist in a stable manner.

4. Lotka and Volterra have suggested a model to describe the fluctuations in numbers observed in predator/prey interactions. Assuming that environmental conditions are constant and that only the hunger of the predator and the reproductive potentials of predator and prey are involved, then the rate of increase of prey in the absence of predator will be:

$$\mathrm{d}H/\mathrm{d}t = a_1 H$$

and the rate of increase of predator in the absence of prey will be:

$$\mathrm{d}P/\mathrm{d}t = -a_2 P$$

where H and P are the densities of prey and predator, a_1 is the intrinsic rate of increase for the prey and $-a_2$ represents the death rate of the predators in the absence of prey. When the predator and prey are put together in a limited space, the predator population increases at a rate which depends on the prey density and, conversely, the increase of the prey is slowed as the density of the predator increases. Thus

$$\mathrm{d}H/\mathrm{d}t = (a_1 - b_1 P)H \quad \text{and} \quad \mathrm{d}P/\mathrm{d}t = (-a_2 + b_2 H)P$$

where b_1 is the coefficient of escape and b_2 is the coefficient of predation. These two equations when combined, become on integration:

$$a_2 \log H - b_2 H + a_1 \log P - b_1 P = \text{constant}$$

Fluctuations in numbers of predator and prey show a periodicity (figure 8.6) and the following laws were put forward by Volterra:

(i) *Law of the periodic cycle*: the changes in numbers of predator and prey are periodic. The duration of each cycle depends only on the coefficients of increase for the two species, and on the initial conditions.

(ii) *Law of the conservation of averages*: the mean numbers of individuals of predator and prey are independent of the initial conditions, and remain steady provided that the coefficients of increase, the coefficient of escape and the coefficient of predation remain constant.

(iii) *Law of the disturbance of averages*: if numbers of both predator and prey are reduced in proportion to their original numbers, the mean number of prey will increase while that of the predator will decrease.

5. Thompson (1923) put forward a mathematical hypothesis relating to the effect of entomophagous parasites. He makes the following assumptions when setting up his model:

(i) that the rate of parasitism is independent of host density;

(ii) that generations of host and parasite coincide;

(iii) that mortality due to the action of external factors is the same for both host and parasite.

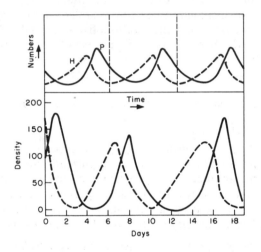

Figure 8.6 Above: fluctuations in numbers of prey, H, and predator, P, according to predictions made by Volterra's model.
Below: actual fluctuation observed in a culture of *Paramecium aurelia* (continuous line) feeding on the yeast *Saccharomyces exiguus* (dotted line). *Paramecium* counts per 15 cm³ and yeast counts per 0.1 cm³ (after Gause, 1935).

Beginning with these premises, Thompson deduced equations which gave the percentage of individuals parasitised after n generations. These show that the rate of parasitism increases slowly at first, while the host population increases exponentially. The rate of parasitism is then suddenly increased, and the host population is reduced almost to zero. In one of his examples Thompson gives the following figures for an initial population of 1000 individuals of the host and ten parasites.

Generations	1	2	3	4	5	6	7
% parasitism	0.15	0.23	0.33	0.51	0.77	1.16	1.76

Generations	8	9	10	11	12	13	14
% parasitism	2.70	4.20	6.50	9.60	18.80	34.20	78.00

6. The theory of Nicholson and Bailey (1935) also related to predator/prey relationships and, in particular, to the interaction between a parasitoid and its host. The theory makes a number of assumptions: that the parasitoid lays only one egg per host; that each host is selected at random; that the host is distributed uniformly over an area which is itself uniform and constant; and that every host encountered is automatically utilised, i.e. that the appetite of the parasitoid (the capacity for oviposition) is insatiable. The authors introduced the idea of an 'area of discovery' which represents the searching efficiency of the female parasitoid in locating and ovipositing in a suitable host. If H_n is the host density in generation n and P_n is the density of the parasitoid, then the rate at which hosts of one generation are eliminated by parasitoids is:

$$\frac{1}{H_n} \cdot \frac{dH_n}{dt} = -aP_n$$

where a is the 'area of discovery', and the proportion of hosts remaining to reproduce after one generation is:

$$\frac{H_{n+1}}{H_n} = e^{-aP_n}$$

If F represents the host rate of increase per generation in the absence of the parasitoid, then:

$$H_{n+1} = F \cdot H_n e^{-aP_n}$$

The number of hosts attacked will be:

$$F \cdot H_n - H_{n+1} = F \cdot H_n (1 - e^{-aP_n})$$

and, assuming that each host parasitised gives rise to one adult parasitoid:

$$P_{n+1} = F \cdot H_n (1 - e^{-aP_n}).$$

The application of this model (by substituting values for H_0 and P_0 and then calculating H_1 and $P_1, \ldots, H_n$ and P_n by stages) predicts that numbers will oscillate with an amplitude which is dependent only on the capacity for increase and not on the area of discovery of the parasitoid. However, the amplitude of these oscillations continues to increase, a phenomenon which has not been observed in nature. Another part of this model predicts even more violent oscillations when two parasitoids attack the same host and this has not been shown to be true in nature or in experimental systems.

The assumptions made by Nicholson and Bailey clearly bear little resemblance to biological reality. The rate of reproduction in insects is not constant, and eggs are not deposited at random. In the case of the ichneumonid *Diadromus* sp., a parasite of the leek moth *Acrolepia assectella*, Labeyrie (1960) has demonstrated that the eggs are not laid at random but rather that females show a tendency to concentrate their eggs in the first host encountered. The rate of development of the eggs is faster in the presence of hosts who, in addition, stimulate the hatching of eggs. There is only limited egg production in the absence of suitable hosts, the host having a stimulating effect on fecundity.

Nicholson and Bailey's model was modified by Hassell and Varley (1969) who showed that the area of discovery for a parasitoid varied and was not constant. They introduced a new relationship:

$$a = \frac{Q}{P^m}$$

where a is the area of discovery, Q the quest constant and m the mutual interference constant. If the above modification is made to Nicholson and Bailey's model, replacing a, the revised model predicts oscillations which show a tendency to become more stable as the value of m increases. Furthermore, the coexistence of several parasitoids sharing the same host becomes theoretically possible, thus conforming more closely to reality. These theoretical ideas have important consequences especially in the field of biological control.

OTHER THEORIES

Other non-mathematical theories have been suggested to explain fluctuations in populations. Three examples are discussed below.

1 Feedback and cybernetic models

Many ecologists consider that populations show a degree of homeostasis, enabling them to regulate their numbers within limits compatible with the available resources. This control is achieved through feedback mechanisms which are familiar in cybernetics (figure 8.7). Some authors have put forward mechanisms of this kind to explain the regulation of some populations like, for example, the lemming (see figure 8.11). The operation of feedback mechanisms implies the existence of density dependent factors.

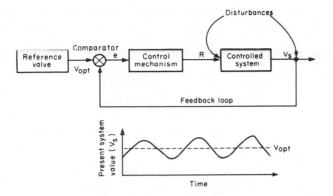

Figure 8.7   Above: theoretical plan for a feedback system.
  Below: example of regulation in a system controlled by feedback.
  The reference or required value, V_{opt} is compared with the present system value, V_s. The difference is an error, e, which is detected by the comparator and is used by the control mechanism to provide a modified input rate, R, to the controlled system.

2   The theory of Andrewartha and Birch (1954)

According to these authors, density independent factors have an important role in the regulation of animal numbers. The rate of increase of a species is continually being modified by its environment, especially weather. The rate of increase may be positive, negative or zero, according to the prevailing conditions. The observation that individuals of a species become increasingly rarer towards the edges of their geographical range is a good illustration of the effect of limiting climatic factors. An intermediate position has been taken by Milne (1957) who puts forward a comprehensive theory accepting the action of density-independent factors (climate, etc.), of imperfectly density-dependent factors (predation, parasitism) and of the one only perfectly density-dependent factor — intraspecific competition.

3   The action of genetic factors

According to density and environmental conditions, natural selection favours different genotypes, each having a different reproductive potential and this may have a considerable effect on the future development of a population.

III   NATURAL POPULATIONS

A   DENSITY-DEPENDENT FACTORS

The existence of density-dependent factors is well established. They can act on the rate of growth of a population in three different ways:

(i) The growth rate may decrease as the density increases. This mode of action is common, and explains the relative stability of animal populations. Decreases in birth rate as population density increases have been demonstrated in the laboratory and in the field. The clutch size for the great tit *Parus major* is fourteen when density is low, i.e. one pair per hectare, and it falls to eight at a density of eighteen pairs per hectare. In *Drosophila*, competition between flies for food and oviposition sites results in a reduction in fertility. Studies on the hymenopterous parasite *Nasonia vitripennis* and its host, the fly *Phaenicia sericata,* showed that, when the parasite density became too high, an increased proportion of its progeny was destroyed. The density-dependent factors acting in this case are of two kinds: one acts on the behaviour of the parasite, the other having a pathological action. Behavioural adaptations are, initially, a retention of eggs by the female when there are insufficient hosts, followed by a reduction in the proportion of females as a result of non-fertilisation of eggs. This effect occurs by virtue of the fact that *Nasonia* shows arrhenotokous parthenogenesis, where the sex of the progeny is determined by the female, who may or may not allow fertilisation of an egg by controlling the release of sperm from the spermatheca. An unfertilised egg gives rise to a male, and a fertilised egg to a female. Pathological action consists mainly of increased larval mortality as a result of the death of hosts harbouring more than one larva. These features of the life cycle of *Nasonia* that Walker recorded are applicable to the life cycles of many hymenopterous parasites.

Brereton (1962) has shown that an increase in population density in *Tribolium* cultures is followed by a decrease in the percentage of eggs hatching. He removed eggs by sieving periodically to prevent cannibalism, which had previously been considered the cause of this decrease in fecundity. Christian (1961) found that intra-uterine resorption of embryos took place when numbers of mice in breeding cages became too large, even when food was abundant. Normal reproduction was only possible when the number of mice was reduced to a value related to available space, and not to available food. Strecker and Emlen (1953) showed that the reproductive systems of female mice atrophy when they live in overcrowded cages.

An increase in mortality with density has been observed in other vertebrates. A population of bobwhite quail, *Colinus virginianus*, was counted twice a year, in April and November, for fifteen years by Errington (1945) in Wisconsin. His results (figure 8.8) show that, as the spring population increases, so the number of birds disappearing by the November count increases. Figure 8.8 also shows a direct relationship between mortality and litter size for the guinea pig and the alpine swift.

The age of mortality may be modified by population density. This has been shown in the vole, *Clethrionomys rutilus*, of the taiga of

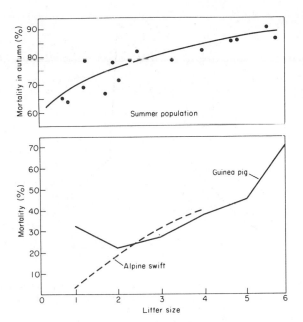

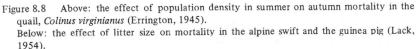

Figure 8.8   Above: the effect of population density in summer on autumn mortality in the quail, *Colinus virginianus* (Errington, 1945).
Below: the effect of litter size on mortality in the alpine swift and the guinea pig (Lack, 1954).

northern Russia. The proportion of individuals becoming sexually mature in the year of birth decreases when the density of this species is high in May. Similar results have been obtained for other mammals besides rodents. The African elephant, for example, matures at eleven to twelve years or at eighteen years depending on whether numbers are low or high and, in addition, births are more frequent (every four years) in the first instance than in the second (every seven years). The annual rate of increase varies, therefore, from 6.0–6.5 per cent at a density of 1.7 elephants per km² to 8–9 per cent at a density of 0.8 per km².

(ii) The growth rate may remain almost constant, even when the population density is high, and then suddenly fall. This is the case in species that show sharp fluctuations, like lemmings or some lepidopterous pests of forest trees.

(iii) Finally the growth rate may be highest at average densities. In some birds, gulls for example, the number of young per pair increases with the density of birds in the colony until it reaches a maximum, and then the number decreases. A similar phenomenon has been observed in the beetle *Callosobruchus sinensis*. This kind of effect of density on the rate of reproduction is characteristic of species showing group effects.

Density may have a retarding effect on the growth rate of a

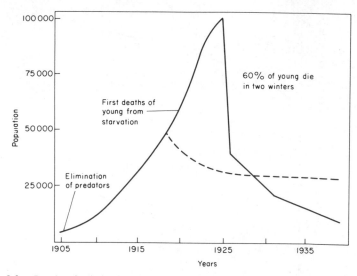

Figure 8.9   Result of eliminating predators (puma, coyote) on a population of the deer *Odocoileus* on the Kaibab plateau in Arizona. If the predators had not been eliminated, the curve would have followed the dotted line (Leopold, 1943).

population. Populations of the mule deer, *Odocoileus hemionus*, on the Kaibab Plateau in Arizona grew rapidly when hunting was banned and many of its predators destroyed. However, when the carrying capacity of the habitat was reached, many animals died of hunger. This sharp decline in the population was not followed by a rise in numbers, which were only ten per cent of the former maximum after fifteen years. This retarded density effect is due to interaction between the mule deer and vegetation, the latter being unable to regenerate as soon as numbers of deer were reduced (figure 8.9).

The important density-dependent factors are discussed in the following four sections.

1 Competition

In some species, the adults feed at the expense of the young, and in this way population size is limited. About eighty per cent of the food of larger perch (*Perca fluviatilis*) from lakes in western Siberia is composed of young individuals of this species. This adaptation allows the species to survive at the expense of the plankton when there are no other species of fish, as only the young can feed on plankton (Nikolsky, 1963).

Competition may be so intense that numbers are reduced below the carrying capacity of the habitat. This happens in *Lucilia cuprina*, a blow-fly living on carrion. Mortality is very high among the maggots, so that few survive. Fluctuations in *Lucilia* populations have been studied by Nicholson (1950).

In the Australian scarab beetle *Aphodius howitti*, the larva bites its companions when it comes into contact with them. The females also lay their eggs in batches, thus increasing the competition. This species is, therefore, maintained at a low level by its behaviour (Carne *in* Nicholson, 1958).

Intraspecific competition can produce physiological effects, known collectively as shock disease, especially in rodents. When density is too high, the disease causes a lowering of fecundity and increased mortality, thus restoring numbers to a normal level. Some authors have explained the cyclic oscillations in lemming populations in this way. Research at Vendée showed that, of the species studied, only *Microtus agrestis* is capable of rapid increase. This type of increase is especially common among species with restricted territories (as in *M. agrestis*), and is almost unknown in those with an extended territory like the wood mouse. When climatic conditions are favourable and there is ample food and space, the voles multiply rapidly, slowly occupying all the available space at a high density as the territory is restricted. Encounters between individuals become more frequent and, because they are intolerant of one another, they become involved in repeated fights and develop a physiological condition (state of shock) characterised by hypoglycaemia and by malfunctioning of the adreno-pituitary system. In addition, overcrowding favours the spread of parasites of the animals, who are more susceptible in their weakened state. The result is a fall in numbers, and an equilibrium is reached when the density is low enough for each pair to occupy its territory without having too much contact with other voles. The relative importance of shock disease in *Microtus* is further discussed in Andrewartha (1970) and Collier *et al.* (1973) and population cycles in small mammals are reviewed by Krebs and Myers (1974). Rodents with extensive territories like the wood mouse, *Apodemus sylvaticus*, do not show these rapid increases in numbers. In fact contacts between individuals will appear long before density has increased by a significant amount because of the size of the territories.

Cyclic fluctuations in lemming and other small mammal populations have been interpreted in various ways. Figure 8.10 shows an explanatory model in which feedback mechanisms operate. When numbers are low, the mutual interference between individuals is reduced and genetic mechanisms operate which select those individuals with a high reproductive potential. The population then increases in size. Mutual interference becomes increasingly important and this leads to a lower fecundity and a reduced population size. To this must be added the effect of emigration and the colonization of vacant areas. Some environmental factors also show similar cyclical changes to the lemmings. In the case of *Lemmus trimucronatus* investigated at Point Barrow in Alaska, this was especially true for the phosphate content of

215

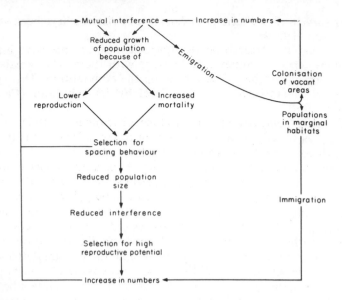

Figure 8.10   A hypothesis to explain the regulation of small mammal populations involving feedback systems (after Krebs et al., 1973).

the vegetation. This is included in the nutrient recovery hypothesis of Schultz (1964). The phosphate content of the vegetation, on which the lemmings feed, is related to the cycle of abundance of this rodent. After a year of rapid increase in lemming numbers, plant productivity is low, there is little litter and so mineral salts produced by bacterial decomposition of the litter are scarce. As the vegetation recovers, so its phosphate content is gradually restored and as the nutrient quality of the plants increases so the lemming population begins to increase. Cyclical changes in soil fertility can be related through this hypothesis, to the abundance of lemmings through a feedback mechanism.

The role of territorial behaviour and of competition in the regulation of bird populations has been demonstrated by many authors. Observations show that a varying number of individuals have no territory and are unable to breed. In this way the size of the population is restricted and prevented from exceeding the carrying capacity of the environment. Those birds without territories may migrate to a more favourable habitat and breed later. This behaviour results in a more even usage of the habitat. At a location studied in Poland, the relative stability of the number of species present, at about forty, suggests that interspecific competition plays a part in determining species diversity. The stability in overall density of numbers of bird species, as opposed to variability in numbers of individuals of each species, is a phenomenon which indicates homeostasis in the ecosystem, as opposed to the variability of

216

its components. During six years observations, changes in total density have not exceeded 11.8% while variations in numbers at the level of individual species have reached between 33 and 86% for the six dominant species of birds. From this it may be concluded that interspecific competition plays a fundamental part in the regulation of animal numbers.

2 Predation and parasitism

Predator-prey interactions are often dependent on density. The role of predation as a limiting factor is obvious. For the ruffed grouse *Bonasa umbellus*, it has been estimated that 39% of eggs and 63% of young are lost, mainly as a result of predation. The role of predators is just as obvious amongst mammals. Herds of *Ovis dalli* are regulated by wolves in Alaska, and lemmings by birds of prey and other carnivores. Many insect pests are held in check by predators. In Corsica, caterpillars of the oak procession moth have suddenly been destroyed by a rapid increase in numbers of the carabid *Calosoma sycophanta*.

The fecundity of predators increases as the prey becomes abundant, and this causes the population fluctuations that have been observed in the laboratory as well as in nature. While the effect of the prey on the predator is obvious, that of the predator on the prey is not always so. In fact the predator tends to remove diseased animals, thereby improving the quality of the surviving stock. A predator has no appreciable effect on the population size of the prey unless the two species have almost the same biotic potential. If this is not so, then the slower rate of reproduction of the predator does not permit it to limit the rapid increase of its prey. Insectivorous birds, for example, are not able to check the growth of insect populations by themselves. In fact when predators tend to have a much lower biotic potential than their prey, their effect tends to be independent of density. According to Tinbergen (1960), the tits make hardly any contribution to the control of phytophagous insects (table 8.1).

TABLE 8.1   Consumption of phytophagous insect larvae by tits in a forest

| | INSECT SPECIES | | | |
	Panolis flammea		*Acantholyda nemoralis*	
Mean density of larvae per square metre	1.26	0.46	0.06	0.80
Mean quantity of prey per day per tit	15.7	3.5	10.5	19.5

Holling (1961, 1968) studied the effect of predation on insects, and concluded that the number of prey taken usually increased with density up to a certain level, when the predators appeared to become satiated. Above this level the action of predators became almost independent of prey density. The aggregation of prey may afford them an effective

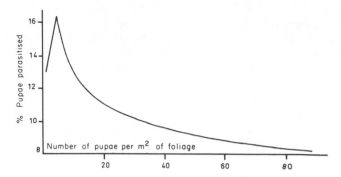

Figure 8.11   Relationship between the density of *Choristoneura fumiferana* pupae and the percentage parasitised by Hymenoptera (Miller, *in* Morris, 1963).

protection against enemies. This happens in the tenthredinid *Neo-piprion sertifer*, whose larvae live in fairly dense groups. The attack of birds or small mammals is checked by a resinous substance secreted from the mouth of the larva. When isolated larvae are attacked, they are unable to produce sufficient of this substance to protect themselves (see Southwood (1966) for a summary of Holling's work).

Miller (*in* Morris, 1963), by his work on the hymenopterous parasites of the caterpillar of *Choristoneura fumiferana* in Canada, has shown that the rate of parasitism is similarly increased until a certain host density is reached, and then becomes relatively lower (figure 8.11).

3   Food

Many species of birds become more abundant when the quantity of food increases. The example of *Loxia curvirostra* has been described already. Jesperson (1924) has shown that this is true for sea birds, whose density in the Atlantic is related to that of the macroplankton. In Holland, the tits are more numerous in deciduous woods than in coniferous woods, which offer poorer resources of food (Tinbergen, 1946). As Lack (1945) observed, two related species which live together usually have different food requirements. If food were not a limiting factor, it is difficult to understand why this differentiation should arise. Two examples of feeding in related species, the geospizids and the cormorants, have been given already (cf. p. 154 and 160).

The example of the deer from the Kaibab Plateau, described above, is another example of the role of food in the regulation of population density. Nikolsky (1963) has summarised the effects of increasing or decreasing the quantity of food available to fish populations. Increase in available food results in an acceleration of growth rate, a more rapid appearance of sexual maturity and the reproductive period, increased fecundity, a reduction of cannibalism of the young, an increase in the number of fertile eggs laid by females, a reduction in size variation in

218

any age class, and an increase in the fat content of the fish. A reduction in the amount of food available involves a retarding of growth, later appearance of sexual maturity, reduced fecundity, an increase in cannibalism and juvenile mortality, an increase in size variation in any age class, and a decrease in fat content.

Some authors consider that the quantity of available food does not act as a density-dependent factor, or at least not in some instances. They point out that numbers of insectivorous birds increase only slightly or not at all when insect numbers increase rapidly. According to Slobodkin et al (1967), studies on the transfer of energy in ecosystems (cf. chapter 13) show that animals use a relatively small proportion of the food available to them, at least in the case of herbivores. The percentage utilisation is greater, however, for seed-eaters and carnivores. Food is thus likely to act more frequently as a density-dependent limiting factor in carnivores and seed-eaters than in herbivores.

4 Disease

Disease may also act in a density-dependent way. In the case of myxomatosis, a virus disease of rabbits, the disease spreads more rapidly in dense populations, where most animals are killed, than in scattered populations. The disease is spread from animal to animal by insect vectors like the mosquito and the flea, which have a restricted range of activity. It can, therefore, spread only if healthy rabbits come into contact with infected rabbits. Lack (1966) considers that disease plays only a small part in the regulation of the population density of birds.

B GENETIC FACTORS

It has already been shown that genetic variability is a basic property of natural populations. This property, essential for the existence of natural selection, has important consequences for population ecology.

Two interbreeding forms of a sexually reproducing species are placed in a uniform environment to which they are both equally adapted. It is also assumed that both forms have the same fecundity and that they differ only by a simple allelomorphic character. One of these characters, the dominant, is represented by AA and the other, the recessive, by aa. If p is the frequency of gene A, and q that of gene a (clearly, by definition, $p + q = 1$), then according to Mendel's Law, the proportion of genotypes in the F_1 generation will be:

for genotype AA $= p^2$

for genotype aa $= q^2$

for genotype Aa $= 2pq$

and $p^2 + q^2 + 2pq = 1$, since $p + q = 1$.

If, for example, the frequencies at the beginning of the experiment are $p = 2/3$ and $q = 1/3$, then, in the F_1 generation:

AA = 4/9   aa = 1/9   Aa = 4/9

This is the basis of the Hardy—Weinberg law, which states that the relative frequencies of the two alleles A and a remain constant, provided there are no mutations or selection.

In practice one allele is nearly always more suited than the other to a particular set of environmental conditions. The *selection coefficient, s,* is the force acting on each genotype to reduce its *adaptive value* or *fitness, W,* so that $W = 1 - s$. If the allele A, which is also dominant, is more suited to a particular set of conditions than the allele a, which is also recessive, then after one generation subjected to selection, the frequency of each genotype will be equal to its initial frequency multiplied by its selection coefficient. If s_1, s_2, and s_3 are the selection coefficients of the genotypes AA, Aa and aa, then the frequencies of each genotype after one generation will be $s_1 p^2$, $2s_2 pq$ and $s_3 q^2$ respectively. The frequency of the gene A in the population will be:

$$p_1 = \frac{s_1 p^2 + s_2 pq}{s_1 p^2 + 2s_2 pq + s_3 q^2} = \frac{p(s_1 p + s_2 q)}{s_1 p^2 + 2s_2 pq + s_3 q^2}$$

similarly for the gene a:

$$q_1 = \frac{q(s_2 p + s_3 q)}{s_1 p^2 + 2s_2 pq + s_3 q^2}.$$

Clearly p_1 and q_1 differ from p and q since selection modifies gene frequency.

One question remains — Why, under these conditions, are certain genotypes not gradually eliminated by natural selection? It can be shown that the selection coefficient for any particular genotype varies with the frequency of that genotype. Thus in a mixed culture of *white* type and *wild* type *Drosophila melanogaster,* the female flies do not mate at random but, rather, show a preference for *white* type males when the frequency of the latter is less than 30%. The rarer type is therefore given some advantage. The existence of variable selection coefficients may explain how two species sharing the same ecological niche, or two genotypes of the same species, are able to coexist in apparent contradiction to Gause's hypothesis. The population cycles observed for predator and prey populations have been interpreted by some authors as the direct result of variable selection coefficients. The existence of polymorphism together with its associated mimicry ensures the protection of species against their predators. The predation of certain forms of the snail *Cepaea nemoralis* by birds is proportionally reduced when those forms are scarce. Batesian mimicry results in the protection of palatable insects through their dimorphism with an unpalatable one which they resemble. However the mimetic morph is protected only as long as its frequency is low in comparison with that of the model.

Natural selection acts as a function of population density in many instances, particularly in species showing population cycles like the rodents. According to Chitty (1967), genotypes showing increased fecundity are selected during the multiplication and rapid growth phase. When, however, the population reaches a high density, selection tends to favour the more aggressive genotypes and the others are lost. This conclusion is supported by research which has demonstrated that the frequency of particular genotypes varies during a population cycle. Tamarin and Krebs (1969) showed that for the vole *Microtus ochrogaster*, there is a distinct correlation between the abundance of the animal and the frequency of a specific gene, the allele transferrin-E. A genotype has been demonstrated corresponding to each phase in the cycle for this rodent. The relationship between genotype variability and environment has been demonstrated by Wellington in the moth *Malacosoma pluviale* in Canada and also by Morris for another moth, *Hyphantria cunea*. The variability of genotypes as a function of population density has been demonstrated for some insects including *Tribolium castaneum* and *Drosophila melanogaster*. Shorrocks (1970) found in the latter case that selection favoured those genotypes with a lower rate of reproduction at high population densities.

r-selection and K-selection

Two population parameters, r and K, can be derived from the equation for the logistic curve. Some species, or, within a species, particular genotypes, having a high rate of reproduction, do not survive in dense populations as the individual animals are inhibited by competition. The maximum population size is well below carrying capacity for these species which are subjected to r-selection. Conversely, species with a low rate of reproduction succeed in establishing themselves in dense populations where the value of K is high. These species are subjected to K-selection. Pianka (1970) listed some characteristics of these two categories. r-selection operates in regions with an unstable climate, where mortality is density independent (climatic factors) and competition is not important. Species submitted to r-selection have a short life span and are small in size. r-selection is most common among pioneer species. K-selection is characteristic of those regions with a relatively stable climate, mortality tends to be density dependent and competition is important. These species are long lived, develop slowly and reach a large size. K-selection predominates in climax communities.

When r-selection is strong, those genotypes which utilise the greater part of their energy resources for reproduction are favoured. This may be illustrated by the experiments of Gadgil and Solbrig (1972) on the

221

dandelion, *Taraxacum officinale*. Three stocks taken from three different habitats were cultured. The habitats were: (a) a mowed and trampled lawn where density-independent mortality was high; (b) grassland which was less trampled and infrequently cut; (c) relatively undamaged grassland where density-dependent mortality was important. Examination of isoenzymes by electrophoresis showed that the three populations contained four genotypes. The proportions of these genotypes were as follows:

HABITAT	PROPORTION OF EACH GENOTYPE %			
	A	B	C	D
Heavily mown lawn	75	13	14	0
Infrequently mown lawn	53	32	14	1
Uncut grassland	17	8	11	64

Genotype A was mainly present at those sites where density-independent mortality was operating and its frequency was reduced in competition with genotype D in the experimental cultures. However, it produced more leaves and flowers in pure culture. Genotype A is subject to *r*-selection and is eliminated by competition. On the other hand, genotype D is abundant at those sites where mortality is essentially density dependent and produces fewer flowers than A but more leaves. Thus it devotes its energy to increasing its biomass and is not inhibited by competiton. It is subject to *K*-selection. In conclusion, these experiments show that the adaptation of species to their environment, and their distribution, are determined not only by environmental factors (water, soil, food, etc.) but also by genetic factors which operate through natural selection.

C DENSITY-INDEPENDENT FACTORS

The best known examples of density-independent factors are the climatic factors responsible for population changes. There are numerous examples of these factors.

The effect of low temperatures on animal populations has frequently been recorded. For example, the cold spell in February, 1955 in the Camargue affected the numbers of many species (cf. p. 80).

In years when the snowfall is heavy in Kazakhstan, there is high ungulate mortality. The animals cannot feed, they are killed by wolves, and they are injured by the snow. When the snow is so deep that the ratio of snow depth/height of fore-leg is in the region of 0.9 to 1.0, the animal is unable to move about and it is rapidly 'drowned'. The critical depth of snow is 90–100 cm for the elk, 60–70 cm for the red deer, wild ass and Przewalskii's wild horse, 50–60 cm for the wild sheep, and about 40 cm for the Saïga antelope. The pressure exerted by the animal on the snow is also important. This is about 980 g/cm² for the wild ass, 480 g/cm² for the Saïga antelope, and 180 g/cm² for the reindeer, which is well adapted to life in snow by enlargement of its hooves. Wolves exert a pressure of between 140 and 160 g/cm², about three to four times less than that of their prey, and this enables them to move about more easily and to pursue their prey. During the winters of 1940/41, 1945/46 and 1948/49, as a result of the exceptional climatic conditions, the wolves were able to annihilate nearly all the roe deer and many of the Saïga antelope and young wild boar in Kazakhstan. In addition, the search for food underneath the snow is more difficult for herbivores, and they often injure their legs while walking over thin films of ice. This results in haemorrhages and, in many cases, gangrene. It has been estimated that, during the winter of 1826/27, the extremely harsh climatic conditions resulted in the death, between the Volga and the Urals, of 1 378 400 large mammals, including 10 500 camels, 280 500 horses, 75 400 head of cattle and 100 000 sheep. In 1953/54 between 10 000 and 15 000 gazelles were killed out of a total of 20 000 to 25 000 animals in the northern half of the Kyzyl-Kum mountain range.

Climatic factors can have a delayed action. Thus, in the coniferous forests of northern Europe, a favourable summer results in a good crop of cones and seed the following year. The slender-billed nutcracker, which then has an abundant supply of food at its disposal, lays a mean number of four eggs, instead of three, the following spring. There is, thus, a delay of two years between the appearance of the favourable climatic factor and the increase in numbers of birds.

The action of cold spells on the density of birds has been investigated quantitatively by Ferry and Frochot (1958) for tits in the forests of the Côte-d'Or (figure 8.12).

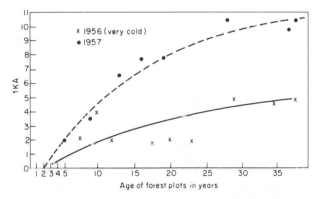

Figure 8.12   Effect of cold on the density of tits in a forest on the Côte-d'Or. The indices of kilometric abundance (IKA) are proportional to the absolute density (Ferry and Frochot, 1958).

Poikilotherms are far more sensitive to climatic changes. The annual decrease in the size of insect populations in the autumn is much greater than that of birds or mammals. The effect of climate is particularly marked in *Thrips imaginis*, which was studied by Andrewartha and Birch. This insect is able to reproduce throughout the year, but populations show regular fluctuations, with a rapid growth in spring. According to Andrewartha and Birch, 78% of the annual variation in populations can be related to changes in meteorological factors, and, allowing for sampling error, this leaves little room for other causes of variation. These authors reject the dogma of density-dependent factors, since it ignores the existence of variations in reproductive rates associated with weather (influence of seasonal variations and other modifications of the environment), as well as the uncertainties of the environment. The grasshopper *Austroicetes cruciata* also shows fluctuations in numbers that can be correlated with variations in the available moisture, according to Andrewartha and Birch.

The effect of prolonged drought was studied recently by French ecologists in the Sahel region of Africa, and was described in a special number of *La Terre et la Vie* (January, 1974). In this region there have been three successive years with virtually no rainfall. The cause of this is not known although some oceanographers have suggested that it results from large scale pollution of the oceans by oil, a thin layer of light oil spread over the surface of the sea having reduced evaporation. In a study of the bird fauna, G. and M. Y. Morel obtained results which showed close agreement with the theory of Andrewartha and Birch that animal numbers tend to be unstable, irregular changes in climatic factors being able to lead to the extinction of some species, at least in arid tropical environments. In the sahelienne savanna of Fété Olé in the north of Senegal, the bird fauna was reduced during 1972 and 1973 both in numbers of species and individuals. The number of resident species fell from 60 to 48 and of Ethiopian migrants from 17 to 7. The total bird fauna was reduced from 108 to 75 species. The mean annual density per hectare was reduced from 6.3 to 2.9 and the mean biomass (grammes fresh weight per hectare) fell from 402 to 186. Turtle doves, birds which need to drink every day, represented at least a third of the total bird population in the dry season during 1969 and 1970. This species had completely disappeared by 1972 and 1973.

There was a similar marked fall in numbers of mammals in this area at this time. Woody plants were also badly affected by the drought, many trees died and those surviving produced fewer fruits and seed than in previous years.

Numbers of the vole *Microtus arvalis* in Vendée show regular annual fluctuations (figure 8.13). The population density begins to rise during May, a maximum is reached between September and November according to the year and numbers fall to a minimum between January and April. It has been suggested that three factors are responsible for

224

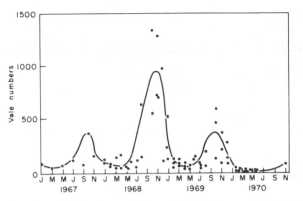

Figure 8.13   Variations in abundance of the vole *Microtus arvalis* at Vendée over four successive years.

this annual cycle. Photoperiod regulates growth and induces and maintains gametogenesis, the optimum photoperiod being between fifteen and twenty hours per day. Diet has an effect on growth and gametogenesis as well as on the survival of the young. For example, growth and fertility are reduced for voles fed on lucerne cut in the autumn rather than in spring. Temperature plays an important part in the survival of young voles. This is reduced from 80% at 21°C to only 25% at 5°C. The combined action of all three factors may explain the annual changes in numbers that have been observed.

The amount of available space is another example of a density-independent factor. In a wood the number of holes in trees suitable for nesting sites does not depend on the abundance of the birds, and yet this may determine the density of the birds.

The existence of a territory is also a limiting factor. This is well known in birds like the eagle *Aquila chysaetos*, grouse, *Lagopus scoticus*, tits, finches, in some mammals, and even in insects like dragonflies. Mortality due to predation is higher for muskrats that have not successfully established a territory than for those which have established one (Errington, 1956). Similarly, mortality is higher for grouse which have not located a suitable habitat (Jenkins *et. al.*, 1963).

Jourdheuil (1960) made a study of a relatively simple natural habitat. This was a field of rape together with its pests, phytophagous beetles and their parasites, entomophagous Hymenoptera and Diptera (cf. p. 163). Jourdheuil showed that in no instances were natural resources exhausted by either the phytophagous beetles or the entomophagous Hymenoptera. As the beetles and their parasites frequently failed to coincide in either time or space, numbers of parasites reached only a small fraction of those of the beetle population. The size of this fraction was independent of the density of

225

phytophagous insects, and showed irregular variations. In a host-parasite relationship the two components are affected in different and often opposite ways by changes in the environment. These observations led Jourdheuil to accept the hypothesis of Thompson (1929), who stated that 'the irregularity and variability of the natural environment are essential limiting factors to the increase of organisms'.

The importance of heterogeneity in both habitat and population has been stressed by several authors. A diverse habitat may increase population stability and so reduce the importance of density-dependent factors. The continued survival of scattered populations with low numbers, such as those of beetles studied by den Boer in Holland, may be explained by the action of density-independent factors. This is also true for populations of the oak moth pest *Tortrix viridana* for which the main mortality factor is the lack of coincidence between the hatching of the egg and bud burst. In a heterogeneous stand of oak, the buds burst over a longer period thus increasing the chances of survival of larvae and resulting in a lower mortality than occurs in a homogenous stand where the tree buds open at the same time.

IV CONCLUSIONS

What conclusions can be drawn from this discussion of the factors that determine population size? Homeotherms (birds and mammals) can be distinguished from poikilotherms (invertebrates, especially insects, which have been widely studied, and, among the vertebrates, fish) with regard to their responses to the different kinds of environmental factors. Homeotherms generally have a lower fecundity, longer life span, and slower replacement of numbers. The development of homeothermy renders them relatively independent of climatic conditions, and in every case more so than poikilotherms. The latter have a higher fecundity, shorter life span and, more frequently, several generations per year.

For homeotherms, biotic density-dependent factors are generally reponsible for the regulation of numbers. Intraspecific competition is often an important factor as, for example, in those birds that defend suitable nesting sites. The development by many homeotherms of complex social behaviour patterns provides a means of limiting population size by eliminating surplus individuals. Predator-prey interactions can also be interpreted in this way. Only predators with a biotic potential equal to that of their prey can keep numbers of the latter at a constant and fairly low level. This is the reason that insectivorous birds have been unable, by themselves, to control the rapid increases in phytophagous insect populations.

Poikilotherms are far more dependent on climatic factors. When climate is favourable, their rapid development allows them to increase quickly within a short time. At the end of the favourable period, the combined action of climate, predators and disease rapidly reduces their numbers.

It is clear that any distinction between density-dependent and density-independent factors is arbitrary. The rate of parasitism by the hymenopteran *Prospaltella berlesei* on the coccid *Pseudaulacapsis pentagona* depends on the location of the coccid on the tree branches, being higher on the shaded side than on the sunny side. Solomon (1957) concluded that, in regions where climate is relatively stable and favours the increase of organisms, biotic factors play an important role. In those regions where climate is less favourable and there is a marked winter season, climatic factors do play a decisive role. These statements justify the idea of Bodenheimer (1955) that none of the factors, such as food, predation, and competition, are independent of climate. 'To study the dynamics of a species without considering its climatic environment would be totally unscientific procedure.'

Finally, the complexity of the ecosystem is important. As ecosystems become more complex, so the number of species concerned increases, and the populations become more stable. Large and sudden fluctuations appear in simple ecosystems, such as the tundra, where few species are present or food chains are short. Every quantitative change in a trophic level is reflected violently in those levels above it, because those species are unable to choose alternative sources of food. Thus, the more complex an ecosystem, the more independent from external disturbing influences does it become. This tendency towards stability is called *homeostasis*. It may be compared with the gradual development in higher animals of an internal environment, the form of which remains relatively constant despite variation in the external environment.

The practical application of the theories of population regulation suggests three possible courses of action in planning a programme of pest control. Firstly, the pest population may be directly controlled by, for example, insecticides and other chemicals. Secondly, use may be made of biological control methods including predators and parasites. Thirdly, the physical environment may be changed, e.g. habitat modification, alteration of the microclimate, etc. If self-regulatory mechanisms of a density-dependent kind are operating for the species to be controlled, then it is likely that chemical control methods would have only a temporary effect as a decrease in density would interfere with the regulatory mechanism. In this case greater use could be made of natural regulatory mechanisms (parasites, predators, etc.) or the self-regulatory mechanisms could be augmented (genetic control). Beneficial species of this kind may be readily exploited as their numbers show rapid recovery. However, where a population is subject to the effects of density-independent factors like weather, economic control may be best applied through the various ecological control methods such as the alteration of microclimates by cultivation or the careful management of the habitat.

V KEY-FACTORS

In most instances a large number of factors are responsible in determining the size of a population. However, Morris (1957) concluded, as the result of his work on defoliating insects in forests, that only a limited number of factors have a significant effect on population trends. These factors are known as key-factors. Their identification has a practical value as they provide a relatively simple means for predicting changes in the abundance of pest and beneficial insects. There are a number of methods for the study of key-factors. These include the method of graphical analysis developed by Varley and Gradwell (1960) and a number of regression methods described by other authors, particularly for insects. The following example is given for the Californian quail, *Lophortyx californica.* Analysis of field data by multiple regression methods gave the following equation:

$$Q = 0.021A + 0.929M - 0.120P - 0.975$$

where Q represents productivity expressed as a ratio of the number of young birds (hatched in the previous year) to the number of adult birds during the shooting season; A is the number of adult females during the previous shooting season; M is the soil water content (in inches) at the end of April; P is the rainfall (in inches) between 1st September and 29th April. The close agreement between predicted and observed values may be illustrated by results for 1963. The predicted value for Q was 1.57 while the actual value recorded was 1.31.

Chapter 9

CONCEPTS OF COMMUNITY AND ECOSYSTEM

The zone of the earth occupied by living organisms is known as the *biosphere*. Since, in nature, living organisms are closely inter-related with one another, the biosphere should really be studied as a whole. This is obviously an impossible task, and the problem is usually solved by studying units of the biosphere of a more convenient size.

I DIFFERENT TYPES OF GROUP

The grouping of organisms within a clearly defined area is usually maintained by the action of well defined factors. By examining these factors, a distinction can be made between social groups and non-social groups.

1 SOCIAL GROUPS

According to the work of French authors like Rabaud and Grassé, a society may be defined as a group consisting of individuals with the following characteristics:

(i) They belong to the same species. There are some exceptions among the vertebrates, where groups of several mammal or bird species form true societies.

(ii) They are kept together by social attraction or by unilateral forces such as mutual feeding, called trophallaxis by Wheeler. Neither of these kinds of interaction are influenced by the physical environment. The binding force between members of a society 'is not due to an external influence, but rather from forces between each member' (Rabaud, 1937). This influence is maintained even when the physical environment is modified.

The grouping of individuals into a society suggests the presence of sensory bonds between them. This has been investigated for *Periplaneta americana*. When a cylinder of metal gauze containing two individuals was placed in a dish containing other cockroaches, the free insects collected against the side of the tube within an hour in ninety per cent of experiments. When a glass tube closed by gauze was used, a similar result was obtained; but if the tube was covered by cellophane, the free cockroaches showed no tendency to collect in the vicinity of the tube. The mutual bond in this instance is due to an olfactory stimulus.

Social appetitive behaviour plays an important part in the formation of vertebrate societies. Most primates live in societies and, when a chimpanzee is isolated from its group, its behaviour becomes very disturbed until it is returned to the group.

Behaviour plays an important part in the formation of animal societies. The importance of environmental factors must not, however, be overlooked. 'The ecologist has no less right to study societies than individuals, and they belong to his sphere of activity for two reasons. Firstly, a social animal is just as dependent on its physical environment as a solitary animal and, secondly, the social condition creates a particular environment where the action of individuals, taken separately, is as variable as it is unexpected' (Grassé, 1951, 1965).

2   NON-SOCIAL GROUPS

Four different categories may be distinguished here.

2.1   Aggregations

The idea of the aggregation was introduced by Rabaud (1929), who described it as a collection of individuals, belonging to one or more species, coming together through the attraction of one or more environmental factors, which act as the centre of attraction. An aggregation is accidental and temporary. There is no other relationship between the organisms that collect together, and they disperse as soon as the centre of attraction disappears. The assembly of various insects on summer evenings round electric light bulbs is an example of an aggregation, since they disperse as soon as the light is turned off.

2.2   Active parasite-like associations

Phoretic and commensal associations depend on the attraction of one species for another. This attraction is essentially unilateral and interspecific. In parasitic associations the parasite, attracted by thermotaxis or chemotaxis, lives at the expense of its host, which is harmed by the association. In phoretic associations, the animal transported (e.g. gamasid mites on the beetle *Geotrupes*) does not harm the host. Commensal associations represent an intermediate stage in the development of social phenomena. Mutual attraction tends to replace unilateral attraction, and associated species also show morphological and ethological modifications which are related to the association. The association between the hermit crab, *Eupagurus prideauxi*, and the

230

polychaete, *Nereis fucata* is a good example, although only the annelid appears to benefit from the association.

2.3 Pseudo-social groups

These groups are simply aggregations formed as a result of thigmotaxis. An example is that of the coccid *Orthezia urticae*. Hibernating females placed on a rough surface remain isolated from one another, while those on a smooth surface move together to form a group as the result of thigmotaxis.

2.4 Communities

The term biocénose (community) was first used by Mobius in 1877, during his studies on oyster beds and their associated organisms. His original definition of a community was a 'group of living organisms corresponding in composition (i.e. in numbers of species and individuals) with a particular set of environmental conditions. These organisms are bound together by mutual dependence upon one another and are able to maintain themselves, through reproduction, permanently in a particular locality . . . If any one of the conditions is changed for any length of time, the structure of the community will be changed . . . The community will also be changed if numbers of a particular species increase or decrease (e.g. through man's activities), or if a species disappears entirely from the community and another takes its place.' This definition of a community poses several problems:

(i) Mobius included in his community every organism, both microscopic and macroscopic, which occurred in the oyster beds. While clearly the correct approach, from a practical viewpoint it is impossible to study all members of a community, and this has probably never been attempted, partly because of the difficulty of identification. The most frequent solution has been to study one or more familiar groups in which the species 'are sufficiently numerous to be represented in each habitat by a different group' (Gisin, 1947).

(ii) The components of a community are kept together by bonds of mutual dependance. This distinguishes it from the aggregation (no relationship between organisms) and the society (social attraction). This interdependence of the members in a community is important, since a change in any one species may affect the entire community, even causing it to break up. Prenant (1934) has emphasised the importance of these interdependent bonds between the members of a community. 'It is difficult to accept the idea that these interactions are simply those of parasitism, commensalism and symbiosis and, even if predation is included, this list is unnecessarily restricted because there is also another type of interaction which may affect the structure of the community.'

The various species in a community can also affect one another indirectly, through the changes they make to the environment. During the nineteenth century, the North American buffalo prevented tree growth as the result of trampling and grazing, and the areas of grassland that were formed suited certain species of mammals, birds and insects. When the buffalo were exterminated, the grasslands were replaced by forest and a new fauna replaced the original one. A herbivore can, therefore, by its presence alone, modify the flora which, in turn, by its composition, determines which particular fauna develops.

(iii) A community is directly affected by the physical environment, while a society, since it is a closed group, is relatively independent of its environment. For example, temperature oscillations are largely eliminated in a beehive. The aggregation depends only on those environmental factors which form its centre of attraction.

(iv) According to Mobius, a community forms a group which is stable in time. This applies essentially to the human concept of time since, on the geological time scale, climatic and other changes are sufficient to cause the disappearance of existing communities in a particular locality and their replacement by new communities. Two types of community can be distinguished: *stable communities* which last many tens of years or even longer, and *cyclic communities* where succession may be very rapid, taking place over several days or even hours. The beechwood forms a stable community, while the many species of arthropod which succeed one another in mammal corpses, dead tree trunks or fungal fruiting bodies represent cyclic communities.

(v) According to Mobius, every member of a community reproduces within that community. This idea is no longer acceptable, Mobius's example being exceptional for a marine community, and unusual on land. The term 'community' has been used in a different context since Mobius, and the following definition is suggested:

A community is a group of living organisms brought together through the combined action of various environmental factors. It is characterized by a specific composition,· and by the existence of interdependent bonds, and it occupies a particular location known as a habitat.

This definition deliberately does not specify the extent or duration of the community, which is rather variable. The following terms are synonymous with the term community. The German *Lebensgemeinschaft* and *Abschluss*, the French *biocénose, communauté* and *association,* and the English *biotic community*.

The essential features of societies, aggregations and communities are shown in figure 9.1.

It will be seen that every community is based on a network of interactions between, on the one hand, the various members of the community, and, on the other hand, between the members and the physical environment. These interactions give a community a character-

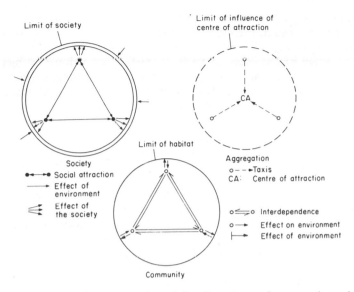

Figure 9.1   Diagrammatic representation of the three types of groups: the society, the aggregation and the community.

istic structure. It is for this reason that the community has been compared to a kind of superorganism which is relatively independent. Organisms and communities have several features in common: they both represent systems through which energy flows (cf. chapter 13); members of a community are closely dependent upon one another in a similar way to the various parts of a living organism; the community develops from a pioneer stage to a climax stage in a similar way to that by which an organism develops from egg to adult.

Despite these similarities, the concept of the community as a superorganism has been rightly criticised. Firstly, an organism cannot, by itself, consist of organisms. Secondly, the various parts of an organism always co-operate together, while this is not true in a community where phenomena like competition, predation, and parasitism are shown. Finally, a community is unable to reproduce itself, since it is made up of components from various sources. An organism, however, is fully capable of reproducing itself.

II   THE ECOSYSTEM CONCEPT

The concept of ecosystem is inseparable from that of habitat, which is the space occupied by a community. [Dajoz uses the term *biotope* instead of habitat. He considers that the use of habitat in French has too narrow a meaning, referring only to the space occupied by an individual, while the term biotope includes that of the community. The term habitat is used instead of biotope in this translation since it is the word normally used in the English language, and Odum (1971) makes it clear that, in English, habitat is synonymous with biotope.] The habitat is 'a geographical area of variable surface or volume, subject to conditions, the most important of which are homogeneous' (Pères and Deveze, 1961). According to Davis (1960), the habitat is a reasonably well defined space containing sufficient resources to ensure the continuity of life. It may be inorganic or organic by nature, as for example in the case-of parasites.

The community and the habitat it occupies form two inseparable components of a relatively stable system, the ecosystem (Tansley, 1935). An ecosystem is thus formed from two components: one, organic, the community of living organisms; and the other, inorganic or organic, the habitat which supports the community.

The idea of the ecosystem had already been suggested by Forbes (1887) with his idea of *microcosm,* and by Friederichs (1930) who introduced the term *holocénose.* The *biosystem* of Thienemann (1939) is equally synonymous with ecosystem. The ecosystem as a functional unit is fundamental to ecology because it includes living organisms and their environment, together with all those interactions that take place between organisms and environment. Soviet ecologists use the term *biogeocenosis*, which is also synonymous with ecosystem.

The ecosystem shows a certain uniformity in topography, climate, botany, zoology, hydrology and geochemistry. The exchange of materials and energy between its components takes place at a characteristic rate. From a thermodynamic point of view, an ecosystem is an open system which is relatively stable in time. Solar energy, water, nutrients and atmospheric gases enter the system, while heat, oxygen, carbon dioxide, other gases, humic compounds and living material are carried away from the system by water or other media.

Most ecosystems are formed as the result of a gradual process of evolution which involves the adaptation of species to their environment. Ecosystems are self-regulating and capable of resisting, to some extent, changes in environmental conditions and population density. The biosphere, the sum total of ecosystems, is distinguished by its mosaic structure and the great diversity of organisms found in it.

A complete ecosystem consists of non-living organic and inorganic materials, autotrophic organisms (producers) capable of manufacturing organic from inorganic materials, heterotrophic organisms (consumers) and, lastly, decomposers which degrade organic materials into inorganic compounds. The last three categories are not necessarily represented in all ecosystems. In particular, minor ecosystems lack producers and depend mainly on neighbouring ecosystems.

A lake provides a good example of an ecosystem. It forms a well defined unit in which the various components are closely linked with one another through many interactions.

III CLASSIFICATION OF COMMUNITIES

The term ecosystem may be applied to communities and habitats over a wide range of sizes. The following categories may be distinguished:

micro-ecosystems, e.g. a dead tree trunk;
meso-ecosystems, e.g. a wood or a pond;
macro-ecosystems, e.g. the ocean.

Plants generally form the major part of the biomass of living organisms, giving the landscape its characteristic appearance; the exceptions are in the aphotic zone of the oceans and lakes and in subterranean habitats. As a result, the terms used to define the different types of communities are, in general, borrowed from the vocabulary of the botanist.

The various categories used below are based on the ideas of American and European authors like Gisin (1949), Tischler (1949) and Wautier (1952). The word community is used here as a neutral term which does not prejudge the extent of the groupings studied. The categories, starting with the most extensive, are listed below.

(i) *Major communities* or *biota*, which are three in number. These are the terrestrial, freshwater and marine biota.

(ii) *Biomes*, also called *formations* or *complexes* by various authors. The biome is a unit of relatively uniform structure, and is independent of floristic composition, although based on the general type of plant cover, for example grassland. It covers a relatively large geographical area, and is largely determined by macroclimate. The American prairies of the last century provide a good example of a biome. They were covered by grass with no trees, and were occupied by herds of buffalo and groups of Indians. Another example of a biome is the African savanna, with its acacias, baobabs, large herbivores (giraffes, antelope, zebra) and lions. In the terrestrial biota, biomes correspond with the principal plant formations, the deciduous forests of temperate Europe providing a familiar example. According to Gisin (1947), the idea of the biome is a most satisfactory concept since it is relatively stable; the biome 'is perhaps the most natural ecosystem unit, and integrates well with biogeography.'

(iii) In practice regions as extensive as biomes are rarely uniform, and it is possible to distinguish more localised groupings of species which are *associations* or true *communities*. A beechwood, for example, with its flora and fauna forms a community, as also does a hedgerow or field.

(iv) Some very restricted communities are clearly delimited. A decomposing corpse, the trunk of a dead tree and the surface of a boulder all shelter *synusiae*. This term can be considered synonymous with the microassociation of Dice (1952). Synusiae are only relatively independent, and are of limited duration. They form, therefore, parts of communities rather than true communities. Botanists tend to use the term in a different context (cf. p. 245).

(v) Amongst smaller sub-divisions than the synusiae, the *layer* corresponds to a vertical division of the habitat. A forest, for example, will contain a tree layer, shrub layer, herb layer and field (moss) layer. In a stream, layers can be distinguished corresponding to organisms living on the surface (gerrids, gyrinids), and to benthic organisms. American authors use the term *consociation*, which corresponds to groups of organisms occupying restricted habitats which are not layered. A dead tree trunk and its fauna, or the riffle fauna of a stream may be described as consociations.

Tischler's classification employs subdivisions of lower rank than synusiae. He recognises units of horizontal and vertical distribution in a habitat. The unit of horizontal distribution is the *biochorion*, and examples include the trunk of a felled tree, a pile of stones or a mammal corpse. The group living in the biochorion is a *choriocénose*. The unit of vertical distribution is the *stratum*, and is occupied by the *stratocénose*. Finally, the *Strukturteil, structural element* and *mérotope* all correspond to a very small grouping, such as part of a plant (leaf, fruit) or a small part of a habitat (e.g. a submerged pebble). The mérotope is occupied by a *mérocénose*.

IV THE EDGE EFFECT AND THE ECOTONE CONCEPT

The transition from one community to another may be quite rapid. In every case there is a transition zone, and this may extend over tens of kilometres in the case of large biomes (the transition zone between the Canadian coniferous forests and the North American prairies, for example), or over a few metres for small communities. The transition zone is known as an ecotone. Examples include the marsh between a lake and the terrestrial zone, and the scrub which marks the limits between a wood and open fields. The fauna of an ecotone is more numerous and diverse than that of either adjacent community, since it includes species from both communities. This effect is also known as · the edge effect.

Chapter 10

DELIMITATION AND
DESCRIPTION OF COMMUNITIES

I DELIMITATION OF COMMUNITIES

The fauna of a habitat must first be examined, and this is done either by collecting the entire fauna, which is difficult or even impossible, or by sampling selected categories based on size or taxa. Where possible, samples should be quantitative, i.e. based on a known area or volume (for further information on sampling methods, see Chapter 6). The type of plant cover should be recorded, and samples from different woodland layers separated. The litter fauna of the soil can be separated from the soil fauna. Every sampling unit should be accompanied by a full description of environmental conditions, and this can be completed in the laboratory, using data obtained from other samples taken simultaneously, to measure salinity, particle size, water content, etc.

This kind of investigation provides quantitative data (numbers of individuals) and qualitative data (life cycles) about a particular habitat. Samples should be taken at different times of the year to show seasonal changes. In habitats subject to rapid change, these samples must be taken at close intervals. Each community presents its own particular problems, and these must be examined by the ecologist in the light of his previous experience.

The data obtained from samples can be used to describe the communities present in the area sampled. While the presence of several species in a sampling unit suggests that the environment satisfies their common needs, it is possible to talk in terms of a community only when the same species repeatedly occur together in a number of sampling units.

The comparison of samples from several habitats often shows that certain species are always present in some and absent from others. These are characteristic species, and may be used to define communities. It is more usual, however, to talk in terms of a characteristic group of species, because few species are restricted to a single community. While most members of a characteristic group occur separately in other communities, they only occur together in one

237

particular community, and the group can be regarded as typical of only that one community.

Quantitative techniques must be employed when there is no obvious characteristic group of species. The first step is to calculate indices of association for each species from pairs of samples. There are a number of such indices, and only a few simple examples are discussed here:

Sorensen's quotient of similarity (q) is found from the formula

$$q = \frac{2c}{a+b}$$

where a is the number of sampling units in a sample which contain species A, b the number of units containing B, and c the number of units containing both species together (Sorensen, 1948).

Jaccard's index (q) is

$$q = \frac{c}{a+b-c} \quad \text{(Jaccard, 1912)}.$$

Odum's index (q) is calculated from

$$q = \frac{a-b}{a+b}$$

and takes values from -1 to $+1$. It is based only on the numbers of units containing each species separately.

The extent of association between two species can also be measured using a χ^2 test. Using the above notation, the probability that any two species, distributed at random, will appear together in a sampling unit is

$$P = \frac{a \times b}{N}$$

where N represents the total number of sampling units. If P is greater than c, the two species are negatively associated; when P is equal to c there is no association; and if P is less than c, the two species show a positive association. In order to decide whether this association is a real one, i.e. not due simply to sampling error, the value of χ^2 is calculated from the equation

$$\chi^2 = \frac{N^3}{ab(N-a)(N-b)}(c-P)^2$$

If χ^2 is greater than 3.84, there is a 95% probability that the association is real and not due to chance. The probability is increased to 99% if the value exceeds 6.64.

When the association indices have been calculated, the communities can be delimited, using the technique of differential analysis, first used by Czekanowski in anthropological research, and subsequently by botanists (Guinochet and Casal, 1957) and zoologists (Kontkanen,

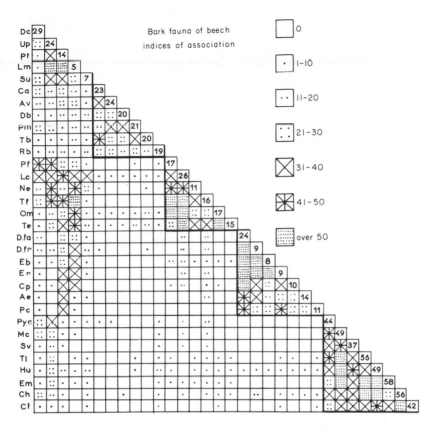

Figure 10.1   The use of the method of differential analysis for grouping beetles living under beech bark.

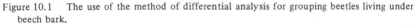

The indices of association $q = \dfrac{c}{a + b - c} \times 100$

are calculated for pairs of species. Four clearly delimited groups can be seen, each containing about five or six species. The initials on the left hand side represent the thirty-two species studied (Dajoz, 1966).

1957). The indices are grouped into classes, e.g. 1–20%, 21–30%, etc., and a trellis diagram (fig. 10.1) is constructed, each species being allocated to one row and one column in the same vertical and horizontal order. A different shading for each class is used at the intersections of rows and columns. Each species is arranged so that the highest indices are found along the diagonal of the figure. Several well defined groups of species can be seen in the completed diagram, each group separated by transition zones corresponding to those species which tend to occur in several communities. This arrangement will theoretically be unique for any particular set of data. The technique is, however, laborious when large numbers of species are concerned.

It is essential that samples are taken from areas which are relatively uniform with regard to species composition, so that only a single community is represented. The nature of the plant cover is a useful criterion for determining the uniformity of an area; for example, a sample should not be taken from the margin of a plant formation because of the edge effect. Uniformity can be tested more rigorously by plotting the number of species found per sampling unit against numbers of units, when the resulting graph should show an initial sharp increase in numbers of species and then levelling off and becoming almost horizontal. The minimal area is the smallest area that must be sampled to include all the species present. This depends partly on the animals being examined, rarely exceeding two square metres on grassland, but increasing to several square kilometres in desert. If the graph shows a further rise in numbers of species when the number of sampling units is again increased, it is likely that a second community is overlapping the first one.

A community can be described as soon as it has been delimited. Three different methods have been employed.

(i) The community may be named with reference to its habitat. For example, the association of beetles in coniferous forests of the subalpine zone, or the association from shaded banks of mountain torrents.

(ii) The community may be named using the name of one or more characteristic species. For example, *Amphioxus* sands.

(iii) The community may be named according to the botanical method, which uses the names of the most obvious species, the generic name being followed by the suffix *-etum* and the specific name used in the genitive case. This procedure was used by Bonnet (1964) for soil Testacea, Sacchi (1952) for terrestrial snails, and Amiet (1968) for soil-dwelling insects.

Communities may be arranged in a hierarchical order from smallest to largest. Bonnet (1964) adopted the botanical nomenclature, which distinguishes associations, alliances, orders and classes, going from smallest to largest. He distinguished twenty-six associations of soil testaceans belonging to eleven alliances, two orders and a single class. The class *Phryganelletea* (with the ubiquitous genus *Phryganella*) is subdivided into an order *Centropyxidetalia* (with the genus *Centropyxis*), from poorly developed soils, and an order *Plagiopyxidetalia* (with the genus *Plagiopyxis*) from well developed soils which are rich in organic matter. As a further example, the order *Centropyxidetalia* comprises five alliances of which the *Centropyxidion globulosae* alliance is subdivided into three associations: *Arcelletum arenariae*, *Diplochlamydetum timidae* and *Plagiopyxidetum intermediae*, all three of which are found in neutral, well aerated soils which are either freshly developed or continually renewed by erosion.

II DESCRIPTION OF COMMUNITIES

When a species list has been prepared for a community, it is possible to describe that community in terms based on either the component species or the community as a whole.

I ABUNDANCE (DENSITY)

Abundance, which can be defined as the number of individuals per unit area or volume, varies with time (seasonal, annual or accidental changes) and spatially (i.e. between one community and the next). It is difficult, in some instances, to calculate the abundance of a particular species and, where less precise estimates are acceptable, the following classes of abundance adopted by botanists may be used: 1: scarce; 2: infrequent; 3: frequent; 4: abundant; 5: very abundant.

2 FREQUENCY

The frequency of a particular species is the number of individuals present, expressed as a percentage of the total number of individuals present in a sample. (The use of the terms 'frequency' and 'constancy' (see below) varies widely, for further discussion see Kershaw (1973).) When frequencies have been calculated for each species present in a sample, a frequency histogram can be constructed showing the relationship between frequency classes and number of species falling within each of the classes. This histogram may be of three possible forms:
    (i) The frequency distribution may be very asymmetrical. In this case species diversity is high, although the fauna consists mainly of only one or two species.
    (ii) All the species are fairly equally represented, frequencies lying closer to one another. Species diversity is high, as in the previous example.
    (iii) Species diversity is low, and frequencies are rather variable.

3 CONSTANCY

Constancy (c) is the ratio p/P expressed as a percentage, where p is the number of units containing a particular species out of a total sample of p units. The following scale is based on values of c (Bodenheimer, 1955; Balogh, 1958).
    (i) *Constant* species, present in over 50% of units;
    (ii) *Accessory* species, present in 25–50% of units;
    (iii) *Accidental* species, present in less than 25% of units.
    Botanists classify species according to a scale of five constancy classes from I up to V. Class I corresponds to a constancy of 0–20%; Class II to 21–40%, etc. Raunkiaer (1928) showed that a frequency

diagram with the five frequency classes arranged along the abscissa and numbers of species in each class shown on the ordinate has two characteristic forms. The frequency distribution of species in classes I to V is either of the type $I > II > III \geqslant IV < V$ (typical of associations having numerous constant species, e.g. in temperate regions), or the type $I < II < III < IV < V$ (where constant species are less numerous, as in the dense forests of the Ivory Coast). When the observed distribution does not conform to these types, the group studied is not homogeneous but corresponds to a mixture of two or more communities. Raunkiaer's Law of Frequencies has, however, a number of disadvantages, and reference should be made to Greig-Smith (1964) for further discussion of this topic. Gounot (1970) considered that the type of frequency distribution simply reflects the particular sampling method applied to a uniform plant cover.

4 DOMINANCE

This is a difficult concept to discuss in quantitative terms. It describes the influence exerted by a species on a community. A particular species that is not abundant may have a greater influence on a community than one which is more abundant but less active. For example, a few large ruminants have a much greater effect on grassland than a very large number of phytophagous insects. In this instance the ruminants will be the dominant species. Dominance is often expressed in terms of a particular taxon, and not for the whole animal kingdom. With very few exceptions, plants may be regarded as the dominant group in most communities because of their abundance when compared with animals.

Braun Blanquet (1927) has proposed a combined scale of *abundance-dominance* which combines abundance (i.e. density) with dominance (i.e. area occupied). Species with the highest coefficient on the scale are the dominant species. The scale is;

+: sparse and occupying less than 1% of the area;
1: plentiful but occupying less than 5% of the area;
2: very numerous and covering at least 5% of the area;
3: any number of individuals, covering 25—50% of the area;
4: any number of individuals, covering 50—75% of the area;
5: covering more than 75% of the area.

5 FIDELITY

This concept is also difficult to describe in quantitative terms. Fidelity describes the degree of exclusiveness that a species shows towards a particular community. The following categories may be distinguished:

(i) *Characteristic species*, which are confined to one community or, more frequently, are much more abundant in one particular community than in others.

(ii) *Preferential species*, which are present in several neighbouring communities but which show a preference for one particular community.

(iii) *Accidental species*, which are rare and accidental intruders from another community.

(iv) *Indifferent* (or ubiquitous) *species*, which show no definite affinity for any community. These ubiquitous species have a high ecological valency, while that for characteristic species is low.

As a general rule, there are fewer characteristic species in a community than preferential or accidental species, and they are, therefore, the more abundant species. The following data have been given for insects of sand dunes in Finland:

	CHARACTERISTICS	PREFERENTIALS	ACCIDENTALS
Number of species (%)	8.8	31.3	59.3
Number of individuals (%)	48.8	33.1	18.5

6 DIVERSITY

Diversity is a measure of the number of species present in a community. Among the many indices of diversity that have been described, that of Fisher, Corbet and Williams (1943) is perhaps the most widely used. Their index of diversity (α) is given by the formula

$$S = \alpha \log_e \left(1 + \frac{N}{\alpha} \right)$$

where S is the number of species and N of individuals in a sample. If S and N are known, then α can be found from a nomogram which avoids laborious calculations.

The index of diversity is used to compare the faunal diversity of several communities, and is especially useful if numbers of individuals collected from these communities are widely different. The index of diversity forms a quantitative expression of Thienemann's first law of biocoenotics. When environmental conditions are favourable in a particular habitat, many species will be present, each being represented by a small number of individuals. The index of diversity will, therefore, be high. If conditions are more severe, then the number of species will be reduced, each one being represented by many individuals. The index of diversity will be low.

7 STRUCTURE

Every community has a characteristic structure depending on the vertical and horizontal distribution of animals and plants in relation to each other.

Vertical distribution (i.e. stratification) is usually the result of interspecific competition for light and water in plants and food in animals. It ensures the most efficient use of the available resources, and so results in increased productivity. It is best illustrated by reference to woodland, where there is a ground layer of cryptogams (mosses and lichens), a field layer of variable height (up to a metre or more), a shrub layer reaching up to eight metres in height, and the tree layer which includes the highest trees. Faunal stratification is superimposed upon this plant stratification, and includes species restricted to the soil, others to the tree layer and some, such as birds, which are found in various strata of the shrub layer. There is evidence that the roots of different plants are stratified in the soil. For example, in oak-hornbeam woods in the Paris region, the most superficial roots are those of *Galium aparine* and the root-tubers of *Ranunculus ficaria*. These are followed by the roots of *R. auricomus* and then, at greater depth by tubers of *Corydalis solida* and *Arum maculatum*. In lakes and ponds a distinction can be made between emergent, floating and submerged vegetation, and the fauna can be divided into surface (e.g. pond skaters), mid-water (e.g. backswimmers) and bottom-dwelling (e.g. water scorpions) species.

The horizontal structure of communities is well illustrated by the aggregated distribution of animals, and the alternation of open ground with patches of vegetation.

8 PERIODICITY

Communities show marked seasonal changes and even, in the case of animals, clear daily movements. Some open water animals, including fish and crustacea, show vertical migrations. The time of day plays an important part in regulating the activity of those species which are either diurnal or nocturnal. Seasonal periodicity is even more marked, affecting the physiological state of some species (flowering and leaf fall in plants. diapause and migration in animals), and the specific composition of communities, since some species are active for only a short period. Phenology (cf. p. 164) is concerned with the relationship between seasonal change (e.g. time of flowering) and climate. In the field layer of oak-hornbeam woods in the Paris region, the first plants to flower in early spring are geophytes (primroses, daffodils), while the larger herbaceous plants, mainly chamaephytes and hemicryptophytes like *Milium effusum* and *Circaea lutetiana*, flower later in May and during the summer. Botanists use the term synusia to describe a unit composed of species with similar life forms and having the same ecological requirements.

III EXAMPLES OF COMMUNITIES

1 THE *CAREX CURVULA* ASSOCIATION FROM THE UPPER ZONE OF THE ALPINE REGION

Guinochet (1938) has described a plant association called the *Caricetum curvulae alpinum* from the upper reaches of the Tinée (Alpes-Maritimes), which represents the climax community (cf. p. 254) of the upper zone of the alpine region. Table 10.1 has been constructed

TABLE 10.1   A table which compares samples taken from a plant association: the *Caricetum curvulae alpinum*, sub-association *typicum* from the Alpes-Maritimes (Simplified after Guinochet, 1938)

LIFE FORM	SAMPLE NO.	I	II	III	IV	V	CONSTANCY
	ALTITUDE (METRES)	2675	2550	2600	2350	2650	
	EXPOSURE	W	N	–	N	NE	
	SLOPE	2°	3°	0	5°	2°	
	COVER (%)	50	75	80	90	60	
	AREA OF SAMPLE (in m²)	10	50	2	50	100	
	Characteristic of association						
H	*Carex curvula*	2.2	3.3	4.4	3.4	4.4	V
H	*Hieracium glanduliferum*	+	+	+	1	1	V
Ch	*Silene exscapa*	1.4	–	+3	1.5	+3	IV
H	*Antennaria carpatica*	+	2.1	+1	–	1.1	IV
H	*Avena versicolor*	+	–	1.1	–	1.1	III
H	*Pedicularis kerneri*	–	+	+1	–	+	III
H	*Oreochloa seslerioides*	–	–	–	–	+	I
	Characteristic of alliance:						
Ch	*Minuartia sedoides*	1.3	1.3	2.3	2.5	2.3	V
H	*Phyteuma pedemontanum*	+	+	+	+1	+	V
Th	*Euphrasia minima*	–	–	+	3.1	1.1	III
H	*Veronica bellidioides*	–	–	–	1.1	–	I
H	*Hieracium glaciale*	–	–	–	2.1	–	I
H	*Senecio incanus*	–	+	–	+1	–	II
Ch	*Sempervivum montanum*	+1	–	–	–	+	II
H	*Luzula spicata*	–	+	–	1.1	–	II
H	*Agrostis rupestris*	–	–	+	–	–	I
H	*Luzula lutea*	–	–	–	1.1	–	I
H	*Minuartia recurva*	–	–	–	+1	–	I
	Characteristic of order:						
H	*Leontodon pyrenaicum*	–	1.1	–	–	+	II
H	*Statice montana*	–	–	–	1.1	+1	II
H	*Androsace carnea*	–	–	–	+	–	I
	Distinguishing the sub-association typicum:						
H	*Festuca ovina* sp. *laevis*	1.2	+2	3.2	2.2	2.3	V
H	*Juncus trifidus*	+1	+	+2	+	+1	V

from five samples taken from a sandy soil at an altitude between 2350 and 2675 m, and these samples have been used to describe the association. The first figure (or +) in each column corresponds to Braun-Blanquet's abundance-dominance index, while the second figure (1–5) is a measure of the way in which the plants are grouped (1: growing singly, isolated individuals; 5: in pure populations; +: scattered and not grouped.). The species are listed in the table in the following order: those characteristic of the association *Caricetum curvulae alpinum*; those characteristic of the alliance *Caricion curvulae*; those characteristic of the order *Caricetalia curvulae*; and, lastly, those species characteristic of the sub-association *typicum* which distinguish it from the sub-association *elyntosum*. The latter occurs on calcareous soils and includes *Elyna myosuroides*.

2   THE DAMP WOODLAND *PLATYNUS ASSIMILIS* ASSOCIATION OF THE UPPER VESUBIE VALLEY

This animal association has been described by Amiet (1967) from his studies on communities of terrestrial beetles in the upper Vésubie valley (Alpes-Maritimes). This group of moist woodland beetles occurs at altitudes between 750 and 1300 m in alderwoods on slopes and alder-sallow woods at the edges of streams. The absence of this association from moist localities at higher altitudes suggests that its distribution is determined by climatic factors. The figures in Table 10.2 show numbers of individuals collected in six samples, a plus sign indicating a species taken by non-quantitative sampling. The characteristic group formed by the species making up the association can be divided into four sections: species characteristic of streamside alderwoods, of temperate woodlands, of alpine woodlands, and of sloping alderwoods. The other species found (i.e. woodland, hygrophilous and other species) also occur in other associations. It will be seen that the characteristic species are the most numerous. Their large relative numbers suggest that they belong to a fairly specialised group, while the larger number of species with low frequency indicates that the association is not isolated from neighbouring communities since it contains species derived from these communities.

Communities defined in terms of animal communities often differ from those based on plants. For example, carabid beetles, living at the edges of pools in Languedoc form an association which moves from one plant formation to another at different times of the year, depending on the level of the water. In winter they are found amongst the osiers, in spring in the *Magnocaricion* (a formation of larger sedges), and in summer they occupy the *Scirpeto-Phragmitetum*. The specific composition of forest bird communities depends similarly on the nature of the plant cover (e.g. height of trees), rather than on the species of tree present.

TABLE 10.2   Samples taken from a damp woodland *Platynus assimilis* (Platynetum silva-ticum) association in the upper Vésubie valley (Alpes-Maritimes), after Amiet, 1967. Explanation in text.

| | SAMPLE NO. | | | | | % |
	922	714	859	861	533	ABUNDANCE
Characteristic groupings						
1. Alderwood and stream bank species						
Platynus assimilis	2	13	2		4	8.93
Platysma nigrita	1		10	1	3	6.38
Tachinus laticollis	+	+	+	+		
2. Temperate woodland species						
Philonthus decorus	18	6	4.	11	6	19.14
Trechus obtusus		+	+	+	+	
3. Mountain woodland species						
Pterostichus moestus	5	7	14	15	18	25.10
Trichotichnus nitens	6	13	2	3	8	13.61
4. Typical (?) of sloping alderwoods						
Nebria brevicollis			10	7		7.23
Platynidius peirolerii		1	6	11		7.65
Woodland species						
Cychrus attenuatus		1		2	5	3.40
Otiorrhynchus salicicola	+		+			
Chaetocarabus intricatus	2					0.85
Abax ater	1					0.42
Quedius lateralis	1					0.42
Leistus nitidus		1				0.42
Ceutosphodrus obtusus		1				0.42
Aptinus alpinus		1				0.42
Quedius solarii		1				0.42
Synuchus nivalis					2	0.85
Xantholinus jarrigei					2	0.85
Hygrophilous species						
Leistus fulvibarbis	1					0.42
Peryphus ustulatus	+					
Synechostictus decoratus	+					
Platysma nigrum	1					0.42
Other species	—					——
Amara ovata	2					0.85
Badister bipustulatus	1					0.42
Ocys harpaloides		+				
Gyrohypnus punctulatus	1					0.42
Aleochara curtula	1					0.42
Zyras humeralis		1				0.42

IV   CAUSES OF DIVERSITY IN COMMUNITIES

The numbers of species found in a particular area depend largely on geographical location. There is a marked increase in the number of species on moving from the arctic to the tropics (Table 10.3).

In a tropical forest it would be possible to find up to a hundred species of birds within a hectare, while a similar area of temperate forest would yield only about ten species; in both instances, however, the total number of individuals would remain about the same.

Simpson (1964) discovered from his work on North American mammals that species diversity increased along two main directions. The first was in a north-south direction, while the second showed an increase in species diversity with increased altitude.

Island faunas are often more restricted than those of neighbouring continents, and they tend to become less diverse if the island is smaller in size or further from the nearest mainland.

Species diversity is determined partly by past history, and partly by climatic and biotic environmental factors.

TABLE 10.3   The decrease in number of species with change of latitude in North America (Clarke, 1954)

ANIMAL OR PLANT GROUP	FLORIDA (27°N)	MASSACHUSETTS (42°N)	LABRADOR (54°N)	BAFFIN ISLAND (70°N)
Beetles	4000	2000	169	90
Terrestrial snails	250	100	25	0
Littoral molluscs	425	175	60	0
Reptiles	107	21	5	0
Amphibians	50	21	16	0
Fresh-water fish	–	75	20	1
Shore fish	650	225	75	–
Higher plants	2500	1650	390	218
Mosses and ferns	–	70	31	11

1   PAST HISTORY

Communities tend to become more diverse with time, the most ancient being richer in species than more recent communities. Simple, less stable ecosystems such as the tundra or arable land show a low species diversity. Diversity is high in stable and mature ecosystems, and this is especially true in the tropics. Temperate faunas, depleted by successive glaciations, are relatively recent. Lake Baikal, a very ancient water body, has a very diverse fauna with many relict species from the tertiary era. Over three hundred species of gammarid (Crustacea, Amphipoda) occur there; also seventy-nine flatworm species, representing nearly two-fifths of the known species, sixty-five of these being endemic. Cailleux (1953) used the data in table 10.4, for small islands in the Indian Ocean, to show that more ancient islands and those nearer the coast have richer faunas:

TABLE 10.4   Numbers of species of animals and plants found on various islands

ISLANDS	COCOS	CHRISTMAS	DURIAN
Distance from nearest land (km)	1 100	350	35
Area (km^2)	110	110	79
Total animal species	141	480	889
Total vascular cryptogams and mosses	1	43	?
Total flowering plants	22	126	?

The fauna of the island of Corsica includes about 2750 species of beetles, rather fewer than are found in a similar area of continental France. This latter number varies from 2900 in the department of Nord to at least 4000 for an area in the Alpes-Maritimes equal to that of Corsica.

2   CLIMATIC FACTORS

According to Klopfer (1962), regions of climatic stability (i.e. no frosts, average daily temperature range greater than average annual range, rainfall regular) encourage more specialisation and adaptation because food supplies remain stable. Animals of these regions are more conservative in their food requirements, which are more easily met, and their ecological niches become reduced in size, with a consequent increase in the number of species able to live together in the same habitat. Evidence to support this idea is provided by passerine birds, which are more abundant in temperate than tropical regions. Klopfer considered that these birds are more adaptable in their behaviour than other birds and so occupy broader niches, which enables them to survive in regions where the climate is less stable than in the tropics. This could explain their relatively greater abundance in temperate regions. Klopfer and Macarthur (1961) also suggested that there may be a partial overlapping between niches as faunal diversity increases, although this would conflict with 'Gause's principle'.

It has been suggested that the increased size of ecological niche might explain the paucity of island faunas. The niches of six species of bird were found to be much broader on islands than on the mainland, and this was correlated with a greater variation in beak length. It has also been suggested that the size of niche is reduced on the mainland by the action of environmental factors such as competition, which act less severely in island faunas.

249

3   EFFECT OF SPATIAL HETEROGENEITY

As a habitat becomes more complex, so the communities found there become more diverse. Topographic variability plays an important part in increasing habitat diversity and, thus, in the formation of new species (Mayr, 1963). However, tropical regions are no more varied topographically than temperate regions.

Floral diversity leads to spatial heterogeneity of animal habitats, especially in the tropics. (This idea relates essentially to faunal diversity. Evolutionary mechanisms are rather different in plants. Polyploidy is important, as plants are not mobile and there are also no interactions of the predator-type.) Pianka (1967) studied twelve genera of lizards which occur in deserts of western America extending over 2500 kilometres from north to south. The numbers of species present show a clear correlation with mean temperature, which also controls the length of the growth season for plants. Primary production increases with temperature increase, and the vegetation also becomes more stable. An increase in the number of plant forms, which is related to climatic change, further increases habitat diversity. Since the daily food requirement is less at higher temperatures, the diet becomes increasingly specialised. The net result is an increase in the number of available niches and, therefore, in species diversity.

Macarthur and Macarthur (1961) recorded the plant density in tropical forests by estimating the leaf area per unit volume for each vertical layer, i.e. field, shrub and tree layers. There was an almost linear relationship between plant density and species diversity of birds found in the forests. This is also true for temperate forests, except that species diversity at a particular plant density is lower than the corresponding value for tropical forests. These forests present a greater variety of habitats, the Bromeliaceae, for example, providing suitable habitats for many aquatic insects, including mosquito larvae, with the result that many of these insects which are absent from temperate forests are present in the canopy of tropical forests.

4   EFFECT OF COMPETITION AND PREDATION

Competition between species having similar ecological niches may be reduced by differences in the time of breeding that are made possible by the more stable tropical climate. This is important because competition for food and nesting sites is most intense during the breeding season. A displacement of the breeding season has been recorded for a number of tropical passerine birds (Klopfer, 1962).

According to Dobzhansky (1950), natural selection acts in a rather different way in the tropics, where density independent factors like cold or drought are rarely important. Paine (1966) considered that predators and parasites play a more important part in the tropics in maintaining populations at a level where competition is not important. The low intensity of competition has enabled new species of prey to appear and these, in their turn, support new predators. Paine compared three food webs in the intertidal zone, the first in southern California with forty-five species, the second on the Pacific coast of the U.S.A. with eleven species, and the third in Costa Rica with only eight species. The Californian community was the most diverse, and the top carnivores were starfish (genus *Heliaster*) and a carnivorous gastropod (*Muricanthus*). *Heliaster* consumed many of the carnivorous gastropods, allowing invasion by other species. Removal of starfish of the genus *Pisaster* from the Pacific coast community caused a decline in the number of species present. In Costa Rica the absence of second carnivores might explain the paucity of the fauna.

5   EFFECTS OF PRODUCTIVITY

According to Connell and Orias (1964), a habitat having a high productivity should also show high species diversity. In a stable environment, energy losses are reduced, and more energy is held in the form of living material. A higher level of productivity enables most species to develop larger populations, each with an increased potential for variation. These species may split into smaller semi-isolated groups due to increased available food, and these small populations can give rise to new species.

Chapter 11

THE DEVELOPMENT OF COMMUNITIES

'A community exists only as a function of the environment and the geological history (palaeography) of an area. It is continually developing, changing and then disappearing as it gives way to another community' (Grassé, 1929).

The idea of the community is essentially a dynamic concept. Observations made on abandoned arable land show that it is progressively invaded by perennial herbaceous plants, followed by scrub, and finally trees. The concept of succession has been known for many years, but its detailed study is due to Clements (1916).

I REASONS FOR SUCCESSION IN COMMUNITIES

The community is the product of its habitat and, conversely, this habitat is modified by the community. Since climatic, edaphic and biotic factors vary, communities are bound to change, the rate of change depending on the particular example studied.

1 ACTION, REACTION AND COACTION

These terms, employed by Clements, can be defined as in the following sections.

1.1 Action

This represents the influence exerted by the habitat on a community. This influence can take very different forms; for example, through the effects of climate and geological processes. The consequences are equally varied. Especially important are the appearance of morphological and physiological changes, the appearance of new species or the elimination of others (role of limiting factors), and the regulation of their numbers. The results of the action of a habitat on the components

252

of a community have been discussed in the section on autecology (Chapters 1 to 5).

1.2 Reaction

Reaction is the effect of the community on its habitat. It may destroy, improve or simply modify the habitat (Wautier, 1952).

There are numerous examples of the destructive effects of plants on their habitat. Algae, lichens and mosses are responsible for the breakdown of rocks in various ways. Roots of higher plants widen fissures in rocks, and also may exert a chemical effect through acid secretions. Animals are also destructive. Many marine invertebrates, including piddocks (bivalves), sea urchins and sponges, bore into rocks. Burrowing soil animals, such as earthworms and termites, play an important role in stirring the soil. It has been estimated that, in temperate regions, the entire upper soil, to a depth of ten centimetres in permanent pasture, has at some time passed through earthworms. In some parts of Africa, termitaria dominate the landscape of large stretches of savanna grassland.

The soil is improved by the accumulation of animal and plant debris (corpses, dead leaves, roots etc.), which are slowly converted into humus through a series of chemical changes known as bacterial decomposition. Guano deposits are built up through the ativity of large bird and bat colonies. Other deposits, like silt in ponds and peat, are also built up by the deposition of organic material. Familiar reactions in water are the formation of calcite deposits (fresh-water tufa, coral reefs and *Globigerina* oozes) and silica deposits (*radiolarian* oozes).

Communities may also modify local climates and create new microclimates. Beavers have completely upset the drainage pattern in British Columbia by the construction of barrages.

1.3 Coaction

Coaction is essentially the effect of one organism upon another. This subject has been discussed in chapter five. If a species suddenly becomes more abundant than previously, the resulting change in coaction may cause significant changes in the structure of the community.

This brief review of the various interactions which may take place between community and habitat shows that the factors which bring about succession in communities are mainly climatic, geological, edaphic and biotic.

2   CLIMATIC FACTORS

The effects of climatic factors are well illustrated by the changes that took place in Europe during the Quaternary glacial and interglacial periods. At the maximum extent of the glaciers, mid-Europe was covered by tundra, which included dwarf willows, *Dryas* and saxifrages, while the temperate flora was pushed down to the southern extremities of the continent. The 'cold fauna' included mammoths, woolly rhinoceros, musk ox and small rodents such as the lemming. The warming up during the interglacial periods allowed the return of the grapevine north of the Alps, and the development of a 'warm fauna' which included *Elephas antiquus* and the hippopotamus.

3   GEOLOGICAL AND EDAPHIC FACTORS

Geological processes like erosion, sedimentation, orogenesis and volcanic activity may sufficiently modify the habitat to cause a significant change in a community. The development of soils (edaphic factors) under the combined action of climate and living organisms results in a succession in parallel to that of the flora.

4   BIOTIC FACTORS

These are the most marked, and act rapidly. The role of the buffalo in the development of North American prairie communities has already been mentioned (p. 232). The combined action of lignicolous fungi, bacteria and xylophagous insects slowly changes the environment of a dead tree trunk, causing a series of successive changes in the fauna. One important environmental factor in the development of communities is interspecific competition, and this was discussed in chapter five (cf. p. 146).

The activities of man must not be overlooked since, without doubt, he represents the most important factor in the development of communities. Fire, deforestation and the introduction, intentional or otherwise, of new species are some of the human interventions which cause rapid changes in communities.

II   THE SERE AND THE CLIMAX

Clements (1916) developed the ideas of sere and climax.

Primary successions correspond to the introduction of living organisms into habitats which have never been colonised, i.e. 'bare areas'. The first colonisers are *pioneers*, and the communities which succeed them are *seres*. According to Clements, a stable community eventually develops at the end of a sere, and this is a *climax*. The climax community retains its identity over a period of several human life-spans. This stability is only relative, however, when considered on the geological time scale, each change in the habitat resulting in the replacement of one community by another. The climax, although appearing stable, is subjected to continuous renewal. In forest, trees are continually dying and falling, and new species of animals and plants appear in the glades that are formed, only to move to new clearings as fresh trees develop.

According to Clements, the climax is determined by the regional climate (i.e. it is a climatic climax), and only one climax can develop in any particular region. This theory is not accepted by many European ecologists, who believe that one of several climaxes may develop in the same region – the theory of the polyclimax as opposed to that of the monoclimax. The idea of the polyclimax appears more reasonable, as it allows for the existence of edaphic climaxes which are determined by soil type.

Secondary successions develop in habitats which have already been colonised but where living organisms have subsequently been eliminated by climatic changes (glaciations, fires), geological changes (erosion) or the intervention of man (land reclamation). A secondary succession often leads to the formation of a *disclimax*, different from the original climax.

A *destructive succession* is one which does not lead to a climax Modifications of the habitat, due to biotic factors, result in its gradual

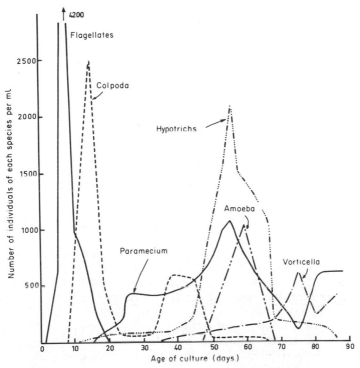

Figure 11.1   Succession of protozoa in hay infusions (after Woodruffe).

255

destruction by successive seral stages. Examples include the breakdown of carrion and dead wood.

III EXAMPLES OF SUCCESSION

1 WOODRUFF'S EXPERIMENTS

Woodruffe (1912) was responsible for the first experimental study of succession. He described the changes that took place in protozoan populations in a hay infusion (figure 11.1). First to appear were large numbers of flagellates, followed by ciliates of the genus *Colpoda* and then *Paramecium*. These were followed by hypotrich ciliates (genera *Stylonychia* and *Vorticella*), together with *Amoeba*.

2 SOME ILLUSTRATIONS OF PLANT SUCCESSION

There are numerous examples of plant succession. The climax formation which develops during the recolonisation of neglected arable land in the French Mediterranean region contains the oaks *Quercus pubescens* and *Q. ilex*. The seral stages are:

cultivation
↓
fallow with *Psoralea bituminosa*
↓
Brachypodium phoenicoides association
↓
Dorycnium suffruticosum
↓
Aphyllanthes monspeliensis
↓
juniper scrub
↓
pine forest (*Pinus halepensis*)
↓
oakwood (*Quercus pubescens* and *Q. ilex*)

The spontaneous development of vegetation in the Paris area leads, in the absence of human interference, to the formation of deciduous woodland. This can be seen in areas of the forest of Fontainebleau which have been left undisturbed for centuries as nature reserves. An example of this natural development of plants in the absence of competition was given in the section on competition (p. 159). Exposed chalk soils are slowly colonised by a thin cover of scattered vascular plants. As the soil develops, the flora gradually becomes more diverse, and a characteristic association, the *Mesobromion*, develops. This is a grassland formation which includes the grasses *Bromus erectus* and *Brachypodium pinnatum*, together with umbellifers, legumes and orchids. Small shrubs like dogrose and juniper become established, and these are followed by the development of open oakwood, a woodland formation with a dense field layer of herbaceous plants forming a meadow. If this formation is isolated, it is relatively stable and forms a-climax. If it is in contact with beechwoods, as it is at Fontainebleau, the oakwood is slowly replaced by beech, and the climax is chalk beechwood. This is the reason why those areas of the Fontainebleau forest where open oakwood is maintained, must be managed to prevent invasion by beech trees.

The stages in colonisation of moraines by plants have been described for the Aletsch glacier in Switzerland. The pioneer *Oxyria digyna* group is gradually replaced, after about ten years, by mosses, dwarf willow and some legumes (*Trifolium pallescens* and *T. badium*). Birch and larch begin to form scattered groups mixed with rhododendrons after about sixty years. At the end of the first hundred years, spruce and stone pine begin to form a forest, which takes several centuries to mature.

3   SUCCESSION OF BIRDS IN A FOREST

Ferry and Frochot (1958) studied bird populations of mature oak and hornbeam forest (*Querceto carpineto scilletosum* association) at an altitude of 500 metres on the Côte d'Or. The forest was divided into plots, which were felled on a forty year rotation. The number of species and the abundance of birds present in each plot were related to the age of the plot. The graph for total number of species showed a rapid rise during the scrub stage of regeneration, followed by a gradual decline; then a second increase, which would probably have continued if felling had not interrupted regeneration. This graph appears to be of general application, resembling that obtained by Johnston and Odum (1956) in their study of changes in bird faunas during the plant succession leading to the natural forest climax in North America.

Examination of results for individual species showed that some birds, such as the robin, are absent in the years immediately following felling, and that their relative importance increases gradually as the plots become older. This increase eventually reaches a maximum if felling does not occur. Other species appear at the begining of the rotation, soon after the trees have been felled, show a rapid increase in numbers, and then disappear. The yellowhammer is an example of this group. The tree pipit has a graph which begins above zero because this species is present in the plots as soon as the trees are felled (figure 11.2).

257

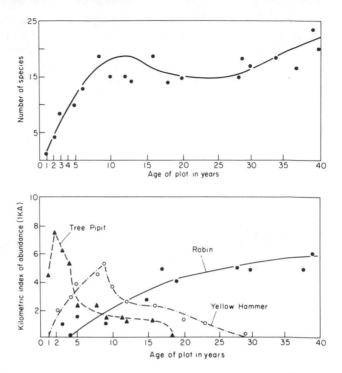

Figure 11.2   Above: Changes in numbers of bird species in a forest on the Côte-d'Or in relation to the age of the forest plots.
Below: Changes in density of three passerine bird species in the same forest (after Ferry and Frochot).

4   SUCCESSION IN MEADOWS

Boness (1953) studied the effect of seasonal changes and hay-making on numbers of insects in a meadow (figure 11.3). Maximum numbers were present at the end of May and early June, when weevils, flea-beetles, thrips and flies were at their peak. The first cut of hay reduced numbers of some flies, such as *Oscinella*, but these showed a rapid recovery. The cercopids *Philaenus* and *Macrosteles* and grass-hoppers were not affected by a second cut, and numbers of weevils and flea-beetles again recovered rapidly. There was a general decline in insect numbers during October.

5   FAUNAL SUCCESSION IN CARRION

The idea of destructive succession has already been introduced. The following stages have been described from a dead mammal corpse:
    (i) Flies belonging to the genera *Musca, Calliphora* and *Cyrtoneura*

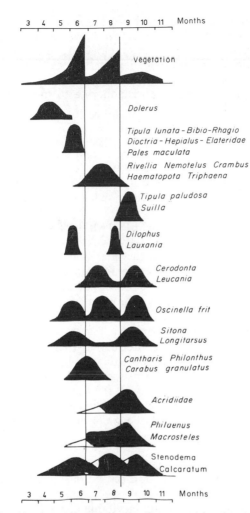

Figure 11.3   Monthly changes in the density of some meadow insect species in relation to hay-cutting, shown by the two vertical lines. Larval stages shown in white (Boness, 1953).

lay their eggs on the skin of the corpse, their larvae pupating within a week.

(ii) Other flies (genera *Lucilia* and *Sarcophaga*) lay eggs as soon as the corpse begins to evolve ammonia.

(iii) Larvae of the beetle *Dermestes* and the moth *Aglossa* are attracted to the carrion, and feed on the fats present.

(iv) These are followed by other beetles (genus *Necrobia*) and flies (*Piophila*), which are attracted by the ammoniacal decomposition of proteins.

(v) This stage includes additional flies (*Ophrya, Phora, Lonchaea* and *Tyreophora*) and beetles (*Hister, Saprinus, Silpha* and *Necrophorus*).

(vi) When the body has become mummified, the mites *Tyroglyphus* and *Uropoda* are abundant. Beetles of the genera *Attagenus* and *Anthrenus* also appear.

(vii) Finally, the remaining debris adhering to the bones is removed by beetles (*Ptinus, Tenebrio*).

6   SUCCESSION IN DEAD WOOD

The trunks of dead trees are attacked by insects, the most numerous of which are beetles. Three stages can be distinguished in the decay of dead oak wood, and the characteristic species are listed in Table 11.1.

7   SUCCESSION IN FRESH-WATER HABITATS

The progressive deposition of silt in water bodies, and the accompanying invasion by plants, represents a familar example of succession: the hydrosere. Lakes tend at first to be oligotrophic, i.e. deep, rich in oxygen, with a well developed hypolimnion, clear waters and low productivity. As the lake gradually fills by silting, by the deposition of organic debris on the bottom and by invasion of littoral plants, it becomes eutrophic and more productive. The process of silting continues, and the lake is converted to a marsh, then to grassland and finally into woodland. This progressive development takes place at varying rates depending on conditions. It has been estimated that Lake Constance in Switzerland has been in existence for about 12 000 years, and will become filled by the end of a similar period of time.

IV   CONCLUSION

The process of ecological succession can be described in the following way:

(i) it is an ordered, directed and inevitable process;

(ii) it occurs as the result of changes in the habitat brought about by the community present in that habitat;

(iii) it is terminated by a climax community in which biomass reaches a maximum value, species diversity is highest, and where, as a result, for a given flow of energy, there are a large number of interactions between the various organisms present. This climax community thus receives maximum protection against external influences, i.e. it shows a degree of homeostasis (cf. p. 227).

The study of ecological succession shows that communities possess certain features irrespective of the type of succession (for definition of terms, cf. Chap. 13):

(i) Food chains, whilst initially linear and dominated by herbivores,

TABLE 11.1. The main species of beetles characteristic of various stages in the decay of dead oak wood (Dajoz, 1966)

DURATION OF STAGE IN DECAY		BARK FAUNA	WOOD FAUNA
Stage 1: 1–3 years	Xylophagous beetles	Cerambycidae: *Pyrrhidium* sp. *Phymatodes* sp. Buprestidae: *Anthaxia* sp.	Cerambycidae: *Rhagium* sp.
	Predators	scarce	scarce
Stage 2: 2–6 years	Xylophagous beetles	Cerambycidae: *Rhagium* sp. *Morimus asper* Scolytidae: *Scolytus intricatus* *Platypus cylindrus*	Cerambycidae: *Rhagium* sp. Anobiidae: *Anobium* sp.
	Predators	Colydiidae: *Ditoma crenata* Histeridae: *Platysoma* sp.	Colydiidae: scarce
Stage 3: 5–10 years	Xylophagous lbeetles		Cerambycidae: *Morimus asper*
	Beetles feeding on rotting wood	*Pyrochroa coccinae*	Tenebrionidae: *Helops coeruleus* *Melasia culineris*
	Predators	Elateridae: *Ampedus* sp. *Melanotus* sp. Colydiidae: *Cerylon* sp. *Synchita* sp.	Elateridae: *Ampedus* sp *Melanotus* sp. Colydiidae: *Colydium elongetum*

develop into complex food webs, with detritivores becoming increasingly important.

(ii) Ecological niches become increasingly narrow and specialised. Organisms tend to increase in size, and life cycles become longer and more complex.

(iii) The total amount of organic material accumulating in the ecosystem is initially low, but gradually increases. Species diversity and biochemical diversity also increase.

(iv) The ratio: gross productivity/respiration, which is usually greater than unity in pioneer communities, tends towards unity for climax communities.

(v) The ratio: gross productivity/biomass, which is initially high, becomes smaller. This ratio represents the rate of renewal or turn over of the community (cf. p. 319).

(vi) Conversely, the ratio: biomass/energy flow, increases as succession tends towards the climax community.

These properties recall the concept of ecosystem maturation developed by Margalef (cf. p. 319).

Experimental aquatic micro-ecosystems have been used to verify observations made on natural ecosystems. An artificial system may consist, for example, of two or three species of unicellular non-flagellate algae, together with two or three species from each of the following categories: flagellates, ciliates, rotifers, nematodes and ostracods. These organisms are placed in a closed container which allows only gaseous exchange with the atmosphere. The temperature is maintained at 18−24°C, the photoperiod adjusted to 12 hours, and illumination to between 1000 and 10 000 lux. There is a rapid increase in daily net productivity for the first thirty days of the experiment and, since this is far greater than the energy lost through respiration, there is a rapid increase in biomass. This is followed by a decline in productivity and in respiration, and between the sixtieth and eightieth days these two values become almost equal, with the result that the biomass tends to remain stable. This stage represents the climax stage, where the ecosystem has reached maturity.

While it is not possible to draw too close a comparison between the natural ecosystem and the experimental system, the overall pattern of development is the same in both instances.

Chapter 12

THE MOVEMENT OF MATERIALS
IN ECOSYSTEMS

'The primary driving force of all animals is the necessity of finding the right kind of food and enough of it . . . The whole structure and activities of the community are dependent upon questions of food supply'
(Elton, 1927).

I. GENERAL PRINCIPLES

1 FOOD CHAINS

A food chain consists of a series of living organisms in which one organism feeds on the one which immediately precedes it in the chain and is, in turn, eaten by the organism which follows it. There are two types of food chain. The first begins with living plants which are eaten by herbivores, and the second begins with dead plant and animal material which is eaten by detritivores.

1.1 Food chains based on living plants

The following categories can be found in a food chain which begins with living plants:

(i) *Primary producers* are the green plants which synthesize organic materials, storing potential energy in the form of chemical energy. In terrestrial ecosystems, this synthesis is carried out mainly by higher plants, although pteridophytes and bryophytes make some contribution. In the sea it is mainly the work of microscopic planktonic algae (diatoms, dinoflagellates), with some contribution from littoral benthic algae and a few uncommon flowering plants. The primary producers in fresh water are algae, flowering plants or even both, depending on the type of water body.

(ii) *Primary consumers* feed on the primary producers and are, therefore, herbivores. Less common primary consumers are the animal and plant parasites of green plants which exploit their host without killing it. Herbivores of terrestrial habitats are mostly insects and, amongst the mammals, rodents and ungulates. The largest group of herbivores in the sea and fresh water are small crustacea and molluscs feeding on the phytoplankton.

(iii) *Secondary consumers* feed on herbivores and are, therefore, all carnivores. They are represented in many taxa.

(iv) *Tertiary consumers* feed on secondary consumers, i.e. they are carnivores feeding on other carnivores. Secondary and tertiary consumers are either predators, capturing and killing their prey in order to feed, or parasites, allowing their host to live, or carrion feeders. *Fourth* and *fifth consumers* are also known, but food chains rarely contain more than five or six links, for reasons which will be discussed below.

(v) *Decomposers* (biodegrading organisms) form the last stage in the food chain. These are mainly micro-organisms (bacteria, yeasts and saprophytic fungi) which attack dead animal and plant material and faeces, slowly breaking them down and releasing the energy and inorganic nutrients contained in them. Fungi are mainly responsible for the breakdown of plant cellulose, while bacteria attack dead animals. Decomposers have other roles, such as the production of inhibiting substances (e.g. antibiotics) or stimulating substances (e.g. certain vitamins), although the part played by these substances in the ecosystem is not properly understood.

A distinction can be made between predator chains and parasite chains. *Predator chains* lead from primary producers to herbivores, which are then eaten by smaller carnivores. the latter being eaten by large carnivores. If the predator chain is increased in length, individuals become progressively larger and less numerous in most instances. An example of a simple, relatively short, predator food chain is:

grass → rabbit → fox
*(primary   (primary   (secondary
producer)   consumer)   consumer)*

The longer, more complex chain below includes consumers of the fifth order:

Scot's pine → aphids → ladybirds → spiders → insectivorous birds → hawks

The following marine food chain shows the progression from a unicellular alga to a bird:

Chaetoceros → *Calanus* → *Ammodytes* → *Clupea* → *Phalacrocorax*

Parasite chains lead towards organisms which in contrast to the predator chain, become progressively smaller and more numerous. This is shown in the following example:

grass → herbivorous mammal → flea → *Leptomonas*

In this example a single mammal acts as host to many fleas, which are specific to the host. Each flea harbours many thousands of flagellates belonging to the genus *Leptomonas*.

The following example of a parasite chain is based on insects:

caterpillar → tachinid fly → chalcid (parasitic wasp)

2.2 Food chains based on dead organic material

In many instances food chains begin with dead organic material, the primary consumers being called detritivores. They may be found in a number of systematic groups, and are often small animals, for example the numerous insects which feed on decaying leaves in the soil, or bacteria and fungi breaking down the organic matter. The two groups are most frequently closely associated, the animals breaking decaying material into smaller particles, and so assisting the action of the micro-organisms.

Both herbivore and detritivore food chains are found together in most ecosystems, although one type is nearly always dominant. In some very specialised habitats, e.g. the abyssal region and caves where green plants are unable to survive in the absence of light, the food chains are based entirely on detritivores. This topic is discussed further in the next chapter.

2 TROPHIC LEVELS

Organisms can be regarded as belonging to the same trophic level when they are separated from the primary producers in a food chain by the same number of steps. Green plants, or decomposing organic matter in the case of a food chain based on detritivores, by definition form the first trophic level. It is useful to use the concept of trophic level and food chain even though the same animal may appear on several trophic levels. This is especially true for omnivores, which feed on both animal and plant material, and also for some carnivores which feed on different prey. For example, the mantis, a predatory orthopteran, may feed either on herbivorous acridiid grasshoppers (second trophic level) or carnivorous tettigoniid grasshoppers (third trophic level). The mantids occupy the third trophic level in the first instance, and the fourth level in the second.

The larva of the torymid wasp, *Monodontomerus aereus*, a primary parasite of caterpillars, may also develop as a secondary parasite of the hymenopterous larvae or tachinid fly larvae which are parasitic on the caterpillars.

A number of food chains usually interconnect to form a food web, since a single animal or plant often forms part of the food of several herbivores or carnivores. Grass, in the example given above, can be eaten by herbivores other than rabbits, which in turn are eaten by other carnivores like buzzards.

3 THE EXAMINATION OF DIETS

Information about the diets of animals is essential to an understanding of food webs and trophic levels. Little is known about the diets of many animals, and it is often difficult to obtain this information. There are several methods.

3.1 Direct observation

This is obviously the simplest method, but it is difficult to use with small animals and those which are difficult to observe. A good example is the work on the hare by Andersen (1950, *in* Macfadyen, 1963), who identified plants eaten by observation through a telescope.

3.2 The examination of gut contents

This method is especially suitable for birds. Ninety-two prey species were recorded in the food of the magpie in the Camargue. The most frequent prey, expressed as numbers of individuals per hundred gut contents examined, were as follows:

```
494 insects, including   302 beetles
                          42 ants
                          39 moths
                          35 orthoptera
                          16 bugs (Heteroptera)
                          14 flies
                          10 dragonflies
112 spiders
 35 crustaceans
  4 snails
  3 vertebrates
```

The size of prey varied, but over 80% of the animals eaten were between 0.6 and 1.5 cm in length. The choice of food indicated the feeding area, half of the prey being taken from grassland, which has a rich fauna in the Camargue.

The study of gut contents of carnivorous fish, such as the pike, also gives some indication of the type of prey taken. Analysis of pellets regurgitated by birds of prey is a classical method of determining which small mammals are eaten, and the diet of these birds has been widely studied. Studies on the food of goshawk in a game preserve near Hamburg may be taken as an example. The hunting territory of a single pair covered 3700 hectares in summer and up to 5000 in winter. The following prey, taken as the mean of ten years' observations, were recorded:

	%
Predators (domestic cat, stoat, weasel, small birds of prey)	1.02
Corvids (crow, magpie, jay, jackdaw)	7.04
Small game (hare, rabbit, partridge, pheasant, pigeon)	43.07
Water birds (teal, mallard, coot, moorhen)	1.01
Poultry (fowl, doves)	20.25
Various birds	27.20
Small mammals	0.24

The figures show that a wide range of prey was taken, including over sixty-five species of mammals and birds, although pigeons, partridges, starlings, crows and rabbits were the most frequent. By comparison, the kestrel on farmland in northern France has a diet of ninety per cent small rodents, the remaining ten per cent consisting of insects (mostly large beetles and grasshoppers). In southern France, on uncultivated land, the diet consists of between fifty and seventy per cent rodents and thirty to fifty per cent insects.

The remains of food in the gut are often difficult to identify, especially if the predator is a fluid feeder. The flatworm *Dendrocoelum lacteum* feeds by preference on the crustacean *Asellus*, removing only fluids and soft tissues which cannot be recognised. Examination of the gut contents reveals only oligochaete chaetae, although these worms are not an important food of the flatworm.

3.3   Serological techniques (the precipitin test)

Serological techniques have been used to identify gut contents by a number of research workers. Dempster (1960) investigated the predators of the chrysomelid beetle *Phytodecta olivacea*, which is common in England on the broom *Sarothamnus scoparius*. Anti-*Phytodecta* serum was prepared, using rabbits, and this was used to test the gut contents of predator species suspected of feeding on the chrysomelid. Ten regular predators of *Phytodecta* were found, including eight Heteroptera (five mirids, two anthocorids and a nabid), the earwig *Forficula auricularia* and a trombidiform mite. Dempster estimated that out of an estimated total of 200 000 eggs laid in an area of about 0.8 ha, between 100 000 and 110 000 *Phytodecta* were destroyed by predators, mainly as eggs or young larvae. Only the nabid and *Forficula* were able to attack *Phytodecta* larvae beyond the second instar.

Serological methods have also been used by Fox and Maclellan (1956) to study predators of the elaterid *Agriotes sputator* and by Hall *et al.* (1953) for several insects. The techniques are not always entirely specific, however. Anti-*Malacosoma disstria* serum will also react with *M. americana* (Hall *et al.*); anti-*Agriotes sputator* serum reacts with all elaterids, especially with *Athous affinis*, which is itself a predator of *Agriotes* (Fox and Maclellan).

3.4   Radio-isotope labelled food

These techniques are relatively recent. Odum and Kuenzler (1963) used radio-isotopes in their studies of food chains in old-field systems in the U.S.A. The three dominant plants were *Heterotheca subaxillaris, Sorghum halepense* (both grasses) and *Rumex acetosella* (Chenopodiaceae). The plants were watered with a solution containing ^{32}P, each plant being contained within a quadrat. Animals in the quadrats were sampled at regular intervals and their radioactivity measured. The cricket *Oecanthus nigricornis* and the ant *Dorymyrmex pyramicus*, two small herbivores feeding on the marked plants, showed signs of radioactivity after one or two weeks. Other herbivores, such as the grasshopper *Melanoplus* sp., became radioactive only after several weeks, indicating that they fed mainly on other plants. Among the predators, spiders showed the highest level of radioactivity after four weeks. Snails (*Succinea*) and carabid beetles became only slightly radioactive, and then only after a much longer interval, indicating that they were far less dependent on the plants studied or on the herbivores feeding on the plants (fig. 12.1).

The use of isotopes makes it possible to study those food chains for which other techniques are not applicable. It was also shown that the most abundant herbivores fed on a wide range of common plants, while less common species were often very selective in their choice of food. For example, no mordellid beetles, which live on flowers, were found on *Heterotheca* or *Rumex* during the experiment although the latter was in flower.

Radio-isotopes can also be used to study the quantity of food consumed, information which is useful but usually difficult to obtain. Crossley (1963-1966) made a study of insects living on the vegetation growing on the bed of a dried-up lake, which had been contaminated with a caesium isotope, ^{137}Cs. The caesium accumulated in the insects until an equilibrium was established between absorption and elimination. Measurements of caesium elimination by the beetle *Chrysomela knabi*, which feeds on sallow, showed that each larva consumed between seven and sixteen milligrammes of dry plant material a day. Laboratory measurements, for comparison, gave values between nine and ten milligrammes per larva per day. The close agreement of these estimates validates the use of the caesium method for measurements of food consumption of the whole community of phytophagous insects. For the association of 400 species found on the vegetation of the dry lake bed by Crossley, the use of caesium-137 gave an estimate of mean consumption of 0.77 mg of dry plant material per gramme of insect per day. Since there were about 300 mg of dry insect material present per metre^2, the total plant consumption was 0.23 g/m^2/day, and thus 23 g/m over a season lasting 100 days in a year. The plant biomass was estimated to be 600 g/m^2, and so the insects only consumed about four per cent of the available plant material.

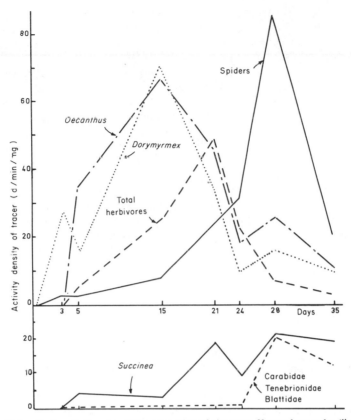

Figure 12.1   Radioactivity of animals on a plot of the grass *Heterotheca subaxillaris* after watering with 29 microcuries of ^{32}P per m². The graphs show activity density of the tracer in the cricket *Oecanthus pellucens*, the ant *Dorymyrmex pyramicus*, spiders (except Lycosids), the snail *Succinea*, carabid and tenebrionid beetles, and blattids (Odum and Kuenzler, 1963).

II   ECOLOGICAL PYRAMIDS

The trophic structure of an ecosystem may be described in terms of either numbers, biomass or energy. All these may be depicted graphically in the form of ecological pyramids.

1   PYRAMID OF NUMBERS

If the number of individuals present at each trophic level in a predator food chain is represented by horizontal rectangles having the same height but with widths proportional to numbers, the resulting figure is a pyramid of numbers. The pyramid becomes higher as the number of trophic levels in the chain increases. Since the numbers of individuals generally decrease from the first to the last trophic levels, the pyramid is triangular in shape with the point directed upwards. The concept of a pyramid of numbers was first described by Elton (1927). It is based on two observations:

(i) That, in an ecosystem, smaller animals are more numerous than larger ones and reproduce more rapidly;

(ii) That there is an upper and lower size limit to the prey taken by carnivorous animals. The upper limit arises because the predator is unable to catch and kill prey very much larger than itself. The lower limit is set for reasons of efficiency: very small prey would need to be taken in such large numbers that there would not be sufficient prey or time. Thus there is generally an optimum size of prey for each species of predator. Elton listed many examples: a lion requires fifty zebra per year, a pair of skuas (*Megalestris*) require the eggs or young of between fifty and a hundred pairs of Adélie penguins in order to survive. In the Falkland Islands, the oystercatcher *Haematopus quoyi* feeds on the limpet *Patella aenea*, which occurs on rocks at low tide. However, it can only dislodge medium sized individuals, and those larger than forty-five mm escape. The tse-tse fly, *Glossina palpalis*, found near Lake Victoria, feeds on mammals and birds having erythrocytes 7–18 μm in diameter, and is unable to feed on some lower vertebrates because their erythrocytes are too large to pass through the channel in the mouthparts of the fly.

There are some exceptions to the observation that size increases and numbers decrease on moving upwards from one trophic level to the next. These are in addition to the parasite and saprophyte chains referred to previously (cf. p. 264). Wolves in packs kill prey larger than themselves, deer for example. Spiders and snakes kill very large prey with the aid of their venom. In forest food chains, where the primary producer is a tree and the primary consumers are insects, there are more individuals at the primary consumer level than producers. Man is the only species capable of living on prey of any size.

The pyramid of numbers is of limited value, since it gives equal importance to all individuals irrespective of size or weight. This method of representation is rarely used, as data giving only numbers present in an ecosystem has little value.

2 PYRAMID OF BIOMASS

A second method of representation is by the pyramid of biomass, in which is shown the corresponding biomass of organisms at each trophic level. In a predator chain this pyramid usually takes the form of a triangle with the apex directed upwards. Some exceptions occur in aquatic ecosystems, where phytoplankton has a smaller biomass than zooplankton, but a much greater rate of growth. In a study of the first two trophic levels in mountain lakes of Colorado, Pennak (1955) found that the ratio:

Biomass of primary consumers (planktonic crustacea)

Biomass of primary producers (planktonic algae)

varied between 0.4 and 9.9, the biomass of zooplankton often being greater than that of the phytoplankton. Fleming and Laevastu (1956) found that the ratio zooplankton/phytoplankton at higher latitudes varies between one in winter and 0.04 in summer, showing that the shape of the pyramid may change with the seasons.

The pyramid of biomass is more useful than that of numbers since it shows the quantity of living material present in each trophic level. Two

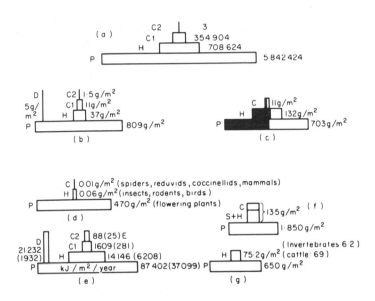

Figure 12.2   Examples of ecological pyramids.
  A: Pyramid of numbers (excluding decomposers) for *Poa pratensis* grassland in the U.S.A.
  B: Pyramid of biomass for Silver Springs, Florida (cf. p. 305).
  C: Pyramid of biomass for a coral reef at Eniwetok. Black: living material contained within
     the corals and representing three different trophic levels (cf. p. 301).
  D: Pyramid of biomass for a one-year abandoned field in Georgia, studied in September at
     the end of the season of plant growth.
  E: Pyramid of energy for Silver Springs expressed in $kJ/m^2/year$. The amount of energy
     available to the next trophic level is shown in brackets. The difference represents energy
     lost to the ecosystem through respiration (cf. p. 305).
  F: Pyramid of biomass in high altitude meadows on Mount Nimba, Africa, during the rainy
     season. An example of plants being exploited by animals.
  G: Pyramid of biomass in a Normandy meadow in August (fauna from surface and above
     ground vegetation only)
  Abbreviations: P: primary producers; H: herbivores; C: carnivores; C_1: primary carnivores;
  C_2: secondary carnivores; S: detritivores; D: decomposers (after various authors).

271

criticisms can be made of this method of representation, however. It gives equal importance to the tissues of all types of animals, although these may be of different chemical composition and so have different energy values. It also does not allow for changes in time. The actual biomass measured may have developed over some days (phytoplankton) or over several tens of years (forests). The role of bacteria in decomposition is particularly underestimated by the pyramid of biomass, because these organisms have very low biomass but a high rate of metabolism. This may be illustrated by comparing the pyramids of biomass and energy for Silver Springs (fig. 12.2).

3 PYRAMIDS OF ENERGY

The most satisfactory method of representation is the pyramid of energy, although the required data is often more difficult to obtain. Each trophic level is represented by a rectangle, the length of which is proportional to the amount of energy passing through that level, in unit time and per unit area or volume. Pyramids of energy always have the apex uppermost because of the energy loss which, as stated in the laws of thermodynamics, necessarily takes place between one trophic level and the next.

Figure 12.3 shows hypothetical examples of three types of pyramid given by Odum (1959). Using data from the literature, he built up an ideal ecosystem, consisting of a simple food chain. This was based on a lucerne field of four hectares as the primary producer. The lucerne formed the sole diet for calves which, in turn, would be eaten by a twelve year old child acting as secondary consumer. The figure shows

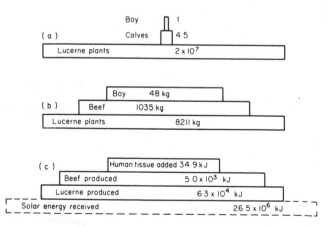

Figure 12.3 Ecological pyramids of numbers (A), biomass (B) and energy (C), for a hypothetical ecosystem based on a field of lucerne of 4 ha, eaten by 4.5 calves, which are themselves eaten by a boy over a period of one year (after Odum).

that 2×10^7 lucerne plants are required to support the 4.5 calves needed to feed a twelve year old child. This is clearly a hypothetical model, since man is omnivorous, but it serves as an illustration of energy flow in an ecosystem.

III EXAMPLES OF FOOD WEBS

Three examples of food webs are given here, two based on terrestrial habitats and the other from an aquatic habitat. Jones (1949) described the food web in a stream in South Wales. This functioned as an open ecosystem, receiving energy from outside in the form of dead leaves and detritus carried in by the water current. The trophic relationships are shown in figure 12.4. It will be seen that there are three trophic levels, with some caddis (*Hydropsyche* and *Rhyacophila*) making up an additional, intermediate trophic level.

Ricou (1967) described the trophic interrelationships which exist in the hypergaion and epigaion of a meadow in Normandy. It can be seen (figure 12.5) that, in the epigaion, fly larvae are primary consumers, acting as prey for spiders which are the secondary consumers.

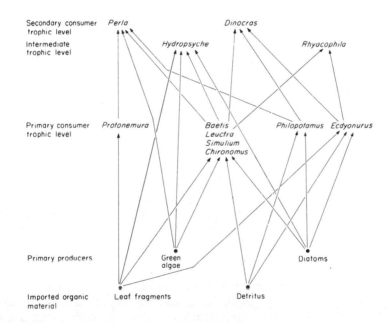

Figure 12.4   A portion of a food web for a small stream in South Wales. There are three trophic levels, although some species, such as *Hydropsyche*, occupy an intermediate level. The genera listed are stoneflies (*Protonemura, Perla, Dinocras*), caddis (*Rhyacophila, Philopotamus, Hydropsyche*), a mayfly (Baetis) and two flies (*Simulium, Chironomus*) (after Jones, 1949).

273

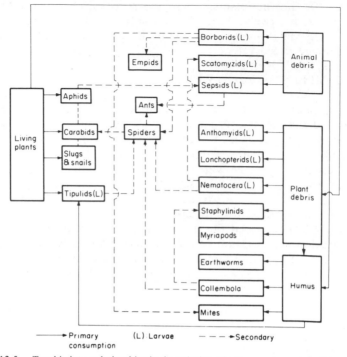

Figure 12.5   Trophic inter-relationships in the epigaion of a Normandy pasture (Ricou, 1967). In a plant formation such as a pasture, the hypogaion corresponds to the fauna living in the soil, the epigaion to the animals living on the soil surface, and the hypergaion to the fauna of the above ground vegetation.

Figure 12.6 shows the main food chains for a beach on an Atlantic shore.

IV   BIOGEOCHEMICAL CYCLES

Nutrients tend to circulate along characteristic pathways in eco-systems, in marked contrast to energy, which flows through the ecosystem, being finally degraded and lost to the system in the form of heat. Energy is never re-cycled.

The synthesis of protoplasm requires about forty elements, the most important of which are carbon, nitrogen, hydrogen, oxygen, phosphorus and sulphur. Other elements are required in smaller quantities, including calcium, iron, potassium, magnesium and sodium. These elements circulate in the biosphere, alternately between living organisms and the environment, through cycles known as biogeochemical cycles. These can be divided into two groups: (i) *gaseous cycles*, in which the atmosphere is the reservoir for the element (e.g. carbon, nitrogen and water); (ii) *sedimentary cycles*, in which the element is

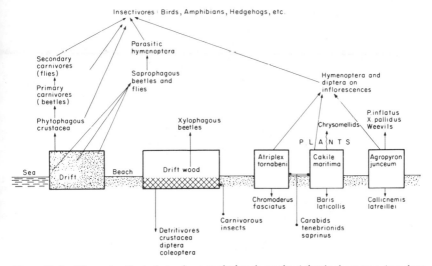

Figure 12.6   The major food chains for a sandy beach on the Atlantic shore near Arcachon (after Amanieu, 1969). There are essentially three habitats on the beach:

(i) *Drift* formed from fragments of *Zostera marina* and various animal debris deposited by the waves. These are inhabited by phytophagous crustaceans like *Talitrus saltator, Orchestia microphthalma* and *Tylos latreillei*, and by numerous saprophagous insects including beetles (staphylinids like *Phytosus nigriventris* and *Cafius xantholoma*, and tenebrionids like *Phaleria cadaverina* and *Xanthomus pallidus*), and flies. As the drift decomposes, the fauna gradually moves towards the surface (decomposing drift is shown in the diagram by dots which become more dense towards the right). Primary predators are beetles like the carabids *Cicindela trisignata* and *Nebria complanata*, which form the second link in the trophic chain. The asilid fly *Philonicus albiceps* is a secondary predator, and attacks the tiger beetles. Numerous hymenopterous parasites develop at the expense of the saprophages.

(ii) *Driftwood* of pine transported by the waves contains xylophagous beetles characteristic of pine, and those that depend on wood saturated with sea water such as the weevil *Mesites aquitanus* and the oedemerid *Nacerda melanura*. Carnivorous insects (*Nebria complanata* and the earwig *Labidura riparia*) hide during the day under the dead wood. In wood that has more or less completely decomposed (shown by crosses), are found detritivores like isopods and amphipods (Crustacea), Diptera and staphylinid and tenebrionid beetles.

(iii) *Vegetation* poor in species and consisting essentially of *Atriplex tornabeni, Cakile maritima* and *Agropyron junceiforme*. The weevil *Chromoderes fasciatus* attacks the roots of *Atriplex*, and *Baris laticollis* those of *Cakile*. The scarab *Callicnemis latreillei* feeds on roots of *Agropyron*. The chrysomellids *Psylliodes marcida* and *Phyllotreta poeciloceras* feed on *Cakile*, while weevils, the tenebrionid *Xanthomus pallidus* and the carabid *Pelor inflatus* eat the spikes of *Agropyron junceiforme*.

stored in the sediments of the earth's crust (e.g. phosphorus and sulphur).

1   THE CARBON CYCLE

The only source of carbon available to autotrophic plants for the synthesis of organic material is carbon dioxide in the atmosphere or

275

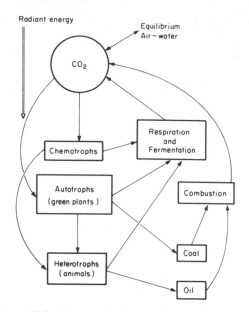

Figure 12.7   Diagram of main stages in the carbon cycle.

dissolved in water. The carbon present in rocks (mainly as carbonates) is not available. The main pathways of the carbon cycle are shown in figure 12.7. Carbon dioxide is converted, through photosynthesis, into organic compounds, which then form the food of animals. The gas is returned to the atmosphere through respiration, fermentation and combustion.

Some numerical data may be used to indicate the importance of the cycle. It is estimated that the carbon dioxide reserves in the atmosphere total 7×10^{11} tonnes, and those in the hydrosphere total 5×10^{13} tonnes. The total annual requirement for photosynthesis has been estimated, variously, as 3×10^{10} and 15×10^{10} tonnes, the corresponding lengths of the cycle being between three or four centuries, and a thousand years. It is clear that the carbon dioxide content of the atmosphere is not decreasing, and will remain steady as long as carbon dioxide is replaced by respiration, fermentation and combustion.

2   THE NITROGEN CYCLE

The main reservoir of nitrogen is the atmosphere. Atmospheric nitrogen is made available through the action of nitrogen-fixing organisms, such as bacteria (*Azotobacter, Clostridium* and bacteria associated with legumes) and some blue-green algae, and by conversion to nitrous and nitric compounds by electrical discharges during storms. The proteins present in carrion are gradually converted by bacteria to ammoniacal compounds and then to nitrates or nitrites, the former forming an important source of nitrogen for green plants.

3 THE WATER CYCLE

Water is distributed over the earth as shown in table 12.1.

Up to twenty-five per cent of rain water may be intercepted by the plant cover in temperate regions. The water which reaches the soil either percolates through or flows over the surface. This water is then returned to the atmosphere through evapotranspiration. A balance sheet has been produced for the Federal Republic of Germany. It is estimated that the water reaching the sea, through rivers, streams etc., is equivalent to a rainfall of 367 mm, less than half the annual rainfall (771 mm). The remainder, about 404 mm (fifty-two per cent), returns to the atmosphere through evapotranspiration. About thirty-eight per cent of the rainfall is taken up by plants and then returned to the atmosphere. Only one per cent of the rainfall is retained in the form of living protoplasm. Evapotranspiration is even greater in equatorial regions, and, in the Congo basin for example, two-thirds of the rainfall is returned through evapotranspiration (Duvigneaud, 1967).

TABLE 12.1   The distribution of water over the earth

FORMS OF WATER	VOLUME (km^3)	PERCENTAGE OF SUM TOTAL
Surface waters:		
Freshwater lakes	94 000	0.009
Saline lakes and inland seas	75 000	0.008
Watercourse beds	900	0.000 1
Ground water:		
Humidity of the soil and vadose water	50 000	0.005
Ground water to a depth of 500 m	3 000 000	0.31
Deep ground water	3 000 000	0.31
Glaciers and ice caps	21 000 000	2.15
Atmosphere	9 600	0.001
Oceans	980 000 000	97.2
Total	1 007 000 000	100.0

4 THE PHOSPHORUS CYCLE

This is a relatively simple cycle, the main reservoir being rocks which are slowly eroded, releasing phosphates to ecosystems. These phosphates are taken up by plants and used in the synthesis of organic compounds, phosphorus being an essential component of living protoplasm. The bacterial breakdown of carrion also releases phosphates into the soil, where they are available to plants. Some phosphates are carried into the sea through leaching, and are available to the phytoplankton and the food chains derived from this. A portion of the phosphorus in the sea may be returned to the land in the form of guano.

TABLE 12.2   Above ground biomass and mineral content for evergreen oak forest, aged between 130 and 150 years, on heathland in the Montpellier region (Rapp, 1970)

	LEAVES	BRANCHES AND SHOOTS	WOOD	TOTAL
Biomass (tonnes/ha)	7	27	230	264
Mineral content (kg/ha)				
Na	1	8	23	32
K	43	90	493	626
Ca	70	493	3290	3853
Mg	9	25	117	151
P	10	40	174	224
N	93	153	517	763
Fe	1.2	2.6	14.1	18
Mn	2.4	2.3	14.1	19
Zn	0.4	1.3	4.7	6
Cu	0.1	0.5	4.7	5
Total mineral content	230.1	815.7	4 651.6	5 698

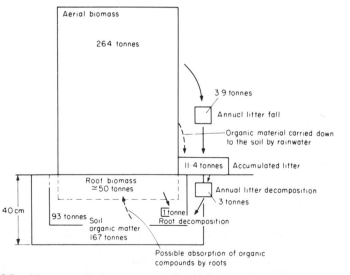

Figure 12.8   Distribution and movement of organic materials in evergreen oak forest near Montpellier (Rapp, 1970).

5 BIOGEOCHEMICAL CYCLES AND PRODUCTIVITY

Terrestrial plants contain, on average, five to eight per cent of their dry weight as mineral salts, this figure increasing to forty per cent in halophytes. Each year, therefore, between 10^8 and 10^9 tonnes of mineral salts are held by plants. The quantity of mineral salts entering the cycles shows wide variation between different plant formations. It may be as high as 500 to 700 kg/hectare/year for steppe grassland and other herbaceous plant formations, but only 70 to 200 kg for coniferous forests. The demand for nutrients and nitrogen is higher, for similar levels of productivity, in herbaceous plant formations than in deciduous forests, and very much higher than in coniferous forests. The ratio

Litter dry weight

Dry weight of dead plant material produced annually

gives some idea of the speed of decomposition of the litter (including all dead plant material reaching the soil) and of the return of nutrients to the soil. It is, therefore, a measure of biological activity.

The ratio exceeds fifty in swamp forests, and varies between twenty and fifty for the tundra. It is between ten and seventeen for coniferous forests of the taiga, and between one and one and a half for steppe. In subtropical forests it is estimated at 0.7, falling to only 0.2 for savanna and 0.1 for tropical rain forest. Biological activity is very high in those regions where little or no plant material accumulates.

Biogeochemical cycles of evergreen oak (*Quercus ilex*) forest in the Montpellier region have been studied by Rapp (1970). The biomass of the aerial parts of the trees was 264 tonnes of dry matter per hectare, and the root biomass was fifty tonnes. The mineral salt content of this material was estimated at 5698 kg per hectare for the aerial parts, and 1547 kg for the roots. The organic matter content of the soil to a depth of forty centimetres, was estimated at between 93 and 167 tonnes per hectare. Table 12.2 shows the dominant mineral salt to be calcium, concentrated mainly in the wood. This is true for temperate forests generally. Figure 12.8 shows how organic materials are distributed over the year in this type of forest.

During a single year, it has been estimated that 3.9 tonnes of dead leaves and branches, per hectare, fall to the ground and form litter. In addition, organic material is added through rainwater falling on the leaves and running down the trunks. This material contains about 53.5 kg of mineral salts. The biomass of the litter is about 11.4 tonnes per hectare, and three tonnes of this, together with 1 tonne of dead roots, decompose annually to form the organic material (humus) of the soil. Figure 12.9 shows the main routes in the biogeochemical cycle for the more important mineral salts (Na, K, Ca, Mg, P, N).

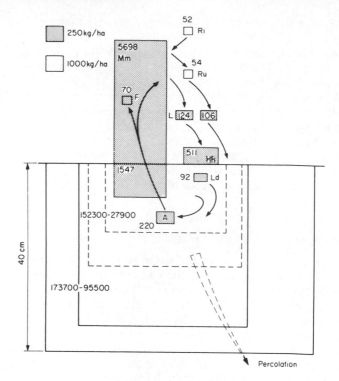

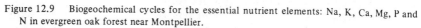

Figure 12.9   Biogeochemical cycles for the essential nutrient elements: Na, K, Ca, Mg, P and N in evergreen oak forest near Montpellier.

•Mm: mineral mass;   Ld: litter decomposition;
Hh: holorganic horizon;   A: absorption;
L: litter;   F: fixation (after Rapp, 1970).
Ri: incident rain;   Grey: essential elements in plants and litter.
Ru: runoff;   White: essential elements in soil.

Chapter 13

ENERGY FLOW AND
PRODUCTIVITY IN ECOSYSTEMS

'Probably the most important ultimate objective of ecology is an understanding of community structure and function from the viewpoint of its metabolism and energy relationships'
(Park, 1946).

The study of energy flow and productivity in ecosystems is a rapidly expanding field. Investigations can be carried out on populations, trophic levels and entire ecosystems. Comparisons can, therefore, be made of different ecosystems and different populations based on a common unit – the joule.

This type of investigation cannot, however, replace or dispense with traditional descriptive studies, but is, instead, complementary to them, adding to existing data on numbers, biomass and distribution. Thus the study of ecological energetics provides a better understanding of the biology of the various species in an ecosystem.

I DEFINITIONS

A Every living organism requires a certain amount of energy for growth and reproduction. This energy is used in the following ways:

(i) To provide for basal metabolism, including the energy required for the continual renewal of tissues throughout life.

(ii) To allow for the movement of mobile animals, i.e. energy expended through activity. The energy required for basal metabolism and for activity together represent the total energy required for the maintenance of an organism.

(iii) To provide for growth and formation of new protoplasm.

(iv) To supply the necessary structures for reproduction (eggs, embryos, seeds, etc.) and the formation of starch (plants) or fat reserves (animals).

B The amount of living material produced in unit time (usually per year) at a particular trophic level, or by one of its components, is the *gross productivity. Net productivity* corresponds to gross productivity less that amount of organic material broken down through respiration.

The organic material produced by autotrophs (green plants) is *primary productivity*, while that produced by other trophic levels, i.e. consumers and decomposers, is *secondary productivity*.

C Energy transfer between different trophic levels in a food chain can be outlined in the following way:

(i) Solar radiation supplies the energy for primary producers. Only a small part of the total radiation (*TR*) received is available to the plants, the remainder (NU_1) is not used. The absorbed radiation (*AR*) is partly lost from the plants in the form of heat (*H*). The remainder is used in the synthesis of organic materials, and corresponds to the *total photosynthesis* or *gross primary productivity* (*GP*). The *net primary productivity* (*NP*) or *apparent photosynthesis* corresponds to the gross productivity less that quantity of organic material lost through respiration (R_1).

Table 13.1 gives some indication of the difference between *GP* and *NP* for three ecosystems which have been studied. In natural ecosystems respiratory loss accounts for almost fifty per cent of gross productivity. Even under the most favourable experimental conditions, about thirty-eight per cent of gross production is lost through respiration, although this loss may fall to about twelve and a half per cent during the growth period. Because natural populations usually consist of individuals of all ages, it is hardly surprising that more energy

TABLE 13.1 Comparison of gross and net primary productivity expressed as grammes dry weight per metre^2 per day

	ECOSYSTEMS			
	SILVER SPRINGS, ANNUAL MEAN (after H. T. Odum, 1957)	EXPERIMENTAL LUCERNE FIELD (after Thomas and Hill, 1949)		SARGASSO SEA, ANNUAL MEAN (after Riley, 1957)
		GROWTH PERIOD	MEAN FOR 6 MONTHS	
Gross productivity	17.5	56.0	30.1	0.55
Net productivity	7.4	49.0	18.7	0.26
Proportion of gross productivity used in respiration (%)	57.5	12.5	38.0	53.0

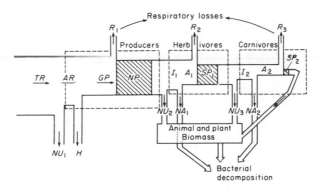

Figure 13.1   Energy flow through a simple food chain consisting of only three trophic levels: producer, herbivore and carnivore.

is lost from these than from populations of young plants which have been artificially selected for an experiment.

The *energy flow* across a particular trophic level corresponds to the total energy assimilated by that level, i.e. the sum of net productivity and respiratory energy loss. In the case of primary producers, the flow of energy across this trophic level is: $GP = NP + R_1$.

(ii) Part of net primary productivity (NP) serves as food for herbivores, which assimilate an amount of energy I_1. Another portion of net primary productivity (NU_2) remains unused and passes, as dead plant material, to bacteria and other decomposers. The energy I_1 may be divided into two components. one of which, A_1, is assimilated by herbivores while the other, NA_1, is discarded in the form of faeces and other wastes. Part of the assimilated energy (A_1) becomes secondary productivity (SP_1) and the remainder is expended in respiration (R_2), so that

$$SP_1 = A_1 - R_2.$$

The energy flow acoss the herbivore trophic level is thus

$$A_1 = SP_1 + R_2.$$

Similarly, energy flow across the carnivore trophic level is

$$A_2 = SP_2 + R_3.$$

Figure 13.1 shows these principles through a simple food chain having only three trophic levels.

D   The *ecological efficiency*, or *trophic level assimilation efficiency*, is the ratio between energy assimilated by trophic level n and that assimilated by trophic level $n - 1$. Using the above notation, ecological efficiency is thus $A_1/GP \times 100\%$ and $A_2/A_1 \times 100\%$. This ratio is

TABLE 13.2   Efficiency of energy transfer in some ecosystems

| | | ECOSYSTEMS | | | | | | | | |
TROPHIC LEVELS AND EFFICIENCIES		OPEN SEA (after Clarke, 1954)	POND WITH DENSE VEGETATION (after Clarke, 1954)	CHLORELLA CULTURES (2% in strong light, 30% in poor lighting) (after Tamiya, 1957)	LAKES (after Clarke, 1954)	CATTLE PASTURES (after Clarke, 1954)	TROUT FISHERIES (after Clarke, 1954)	LAKE CEDAR BOG, MINNESOTA (after Lindeman, 1942)	LAKE MENDOTA, MINNESOTA (after Juday)	SILVER SPRINGS FLORIDA (after H.T. Odum, 1957)
PRODUCERS	Photosynthetic efficiency AR/TR	1%	14%							
	Assimilation efficiency GP/AR			2 to 30%	0.043 to 0.38%					5%
	Ecological growth efficiency AP/GP							0.10%	0.40%	
CONSUMERS	Assimilation efficiency of 2nd trophic level (*herbivores*)					2.5%		13.3%	8.7%	11%
	Assimilation efficiency of 3rd trophic level (*primary carnivores*)						20%	22.3%	5.5%	16%
	Assimilation efficiency of 4th trophic level (*secondary carnivores*)								13%	6%

always low. For primary producers, the ratio NP/TR is the *photosynthetic efficiency*.

Several other ratios are sometimes used:

ecological growth efficiency (SP_1/I_1), which is of interest to farmers because it expresses the maximum amount of meat produced by a species from a given quantity of food.

the *production efficiency* or *tissue growth efficiency* (SP/A). This does not allow for energy lost in the form of faecal material.

assimilation efficiency (A/I) expresses the ability of a species to use the chemical energy contained in natural food.

It will be shown later that green plants can only convert between one and five per cent of the solar radiation received into chemical energy through photosynthesis. Between eighty and ninety per cent of carbohydrates produced are used by the plant in respiration, and so the photosynthetic efficiency is only of the order of 0.1 to 0.5 per cent. Duvigneaud (1967) obtained a value of only 0.1 per cent for the whole biosphere. Herbivores use only about one per cent of the plant energy consumed for growth, although this efficiency may increase to ten per cent for some other trophic levels. Thus, for a mean of 4200 kJ/day/m² of radiant energy fixed by plants, only 42 kJ is assimilated by herbivores, 4.2 kJ by primary carnivores and 0.42 kJ by secondary carnivores. The quantity of energy becomes so small that the number of animals at the secondary carnivore trophic level is very low. For this reason food chains rarely include more than four trophic levels.

All animals do not have the same capacity to assimilate food in a given time. For animals of similar weight, homeotherms (birds and mammals) are less efficient than poikilotherms because they must use up a large proportion of their food in the maintenance of body temperature. This is why more protein can be produced from a given area by fish farming than by grazing cattle.

The pig is the best 'converter' of energy known at present, twenty per cent of the energy consumed by this animal being converted into food available for man. Table 13.2 compares the efficiencies of energy transfer in some ecosystems.

II   PROBLEMS OF MEASURING PRODUCTIVITY

1   METHODS FOR MEASURING PRIMARY PRODUCTIVITY

The measurement of primary productivity presents difficulties. Some of the more important methods are outlined below.

A simplified equation for photosynthesis is:

$$CO_2 + H_2O + \text{radiant energy} + \text{enzymes} + \text{chlorophyll} \rightarrow$$
$$\text{protoplasm} + O_2$$

Sverdrup *et al.* (1942) estimated experimentally values for this equation from studies on marine phytoplankton, and obtained the following results:

5447 kJ radiant energy + 106 CO_2 + 90 H_2O + 16 NO_3 + PO_4 + mineral elements → 54 kJ potential energy contained in 3258 g protoplasm (106 C, 180 H, 46 O, 16 N, 1 P, 815 g mineral ash) + 154 O_2 + 5393 kJ heat energy dispersed.

The yield or ecological efficiency is only about one per cent, and while these figures do not relate directly to terrestrial habitats, they do give some indication of the size of the processes involved.

1.1 The sampling (harvest) method

This involves the collection, at monthly intervals, of all living plant material (aerial parts and roots) trom quadrats. The material is then dried in an oven at 100°C and its energy content measured using a bomb calorimeter. The advantage of this method is its simplicity, but it has some disadvantages. It is not possible to differentiate between the energy contained in living and dead plant material in the samples. It is also only applicable to annual plants, since it does not distinguish between the production of material for successive years. No allowance is made for plant material which has been eaten by herbivores, and a correction must be made for this.

1.2 Oxygen measurement

Since there is a definite relationship between oxygen produced, carbon dioxide used and organic material produced, oxygen production or carbon dioxide absorption can be used to measure productivity. However, organisms in an ecosystem respire while photosynthesis is taking place, and it is necessary to isolate the two processes. The problem was solved by Gaarder and Gran (1927) in marine and fresh water environments by their 'light-and-dark bottle method'. They used two similar bottles, one transparent and one 'dark', i.e. covered by black tape or foil. Water samples containing phytoplankton are placed in the bottles, which are then suspended in the water. After a fixed period, the oxygen concentration in each bottle is determined and compared with the concentration at the begining. Oxygen is used up in the dark bottle by respiration while, in the light bottle, oxygen has been produced by photosynthesis and used up by respiration. The sum of respiration (oxygen consumed in dark bottle) and net productivity (oxygen produced in transparent bottle) gives an estimate of gross primary productivity.

This method assumes that the rate of respiration is the same in the light as in the dark, and this has been demonstrated by studies on radioactive oxygen. It is not possible to discover, however, whether a closed bottle has any effect on photosynthesis.

1.3   Carbon dioxide method

This method is more suitable for terrestrial habitats. Two identical quadrats are selected, one being covered by a transparent container (e.g. plastic box) and one by an opaque container. In the first, the decrease in carbon dioxide is a measurement of photosynthesis less respiration, while in the second the amount of carbon dioxide produced gives a measure of respiration. Gross productivity can be calculated from the carbon dioxide measurements obtained from the two containers. The method is suitable only for vegetation which is sufficiently short to be enclosed in air-tight containers.

1.4   Methods using radioactive materials

This is the most precise method as it does not interfere with the functioning of the ecosystem. The rate of transfer of a radioactive isotope is measured from the intensity of the radiation emanating from the organism which has assimilated the isotope. Steeman-Nielsen (1952) used carbon-14, in the form of sodium carbonate, to measure productivity in tropical waters during the Galathea expedition. Samples of phytoplankton were removed from the water a short time after it had been in contact with the isotope. They were dried, and their radioactivity measured to give an estimate of net productivity. The method assumes that carbon-14 is assimilated at the same rate as carbon-12, and only through photosynthesis. This is not entirely true, however, because some organisms are known to take up small amounts of carbon dioxide in the dark at rates which are not related to photosynthesis. This phenomenon can, however, be ignored in sea water, which contains relatively few bacteria, since it is equivalent only to one or two per cent of the rate of photosynthetic uptake. Where the photosynthetic rate is low, e.g. under low intensity illumination or in polluted water containing many bacteria, an appreciable quantity of carbon dioxide may be absorbed during darkness. Carbon-14 is also assimilated more slowly than carbon-12. When the rate of photosynthesis is high, it can be demonstrated that carbon-14 is fixed ten per cent more slowly than carbon-12 if the respiratory rate is ten per cent of that of photosynthesis. If respiration is five per cent of the rate of photosynthesis, the value is reduced to seven per cent, and it is increased to fifty per cent if the rate of respiration rises to fifty per cent of the photosynthetic rate. A correction must be applied, therefore, to measurements made with carbon-14 to allow for light intensity. According to Steeman-Nielsen and Jensen (1957), the oxygen method gives values for productivity which are about thirty-three per cent higher than those obtained with carbon-14.

2 METHODS FOR THE MEASUREMENT OF SECONDARY PRODUCTIVITY

It is even more difficult to obtain measurements of secondary productivity.

2.1 At the level of the individual

The type and quantity of food injested must be carefully determined, and its energy content measured with a bomb calorimeter. It is more convenient to prepare balance sheets of food taken for animals in cultures maintained under conditions which are as natural as possible. Some attempts have been made to estimate the energy requirements of animals under natural conditions rather than in the laboratory. The different efficiencies listed on page 283 can then be calculated (table 13.3)

Table 13.3 shows that, while the A/I ratio is high for the pig, it is very low for the slow growing myriapod *Glomeris*, which uses only a small proportion of its food, rejecting most of it in the form of faeces. The values of SP/A and SP/I are generally lower in mammals than in invertebrates as the result of the higher energy requirements of homeotherms for maintaining body temperature.

Table 13.3   Ecological efficiency ratios for a variety of species (Macfadyen, 1966)

SPECIES	RATIOS A/I	SP/A	SP/I
Pig (mainly vegetarian, but tending towards an omnivorous diet)	76%	13%	9%
Glomeris (saprophagous, invertebrate detritivore)	10%	3 to 50%	0.5 to 5%
Mitopus (carnivorous invertebrate)	46%	55%	20%
Caterpillar of *Hyphantria* (phytophagous invertebrate)	29%	57%	17%

2.2 At the level of the population

The difficulties here are even greater, since information must be available about the number and biomass of individuals in different age classes as well as the rates of emigration and immigration. Productivity has been estimated for some populations, and the size of the animals appears to be important. If populations of three herbivorous mammals – the elephant, the vole (*Microtus pennsylvanicus*) and the deer (*Odocoileus virginianus*) are compared, it can be shown that the vole population produces one and a half times its own weight in a year, compared to only a quarter in the deer and a twentieth in the elephant (Petrides and Swank, 1965). The environment also plays an important part. Wiegert (1965) showed that the ratio SP/I for the bug *Philaenus spumarius* varied between 4.8% in neglected pasture and 16% in a lucerne field, the latter providing a better food source.

2.3 At the level of the ecosystem

It is necessary to determine the diets of the different species present and to define the trophic levels. In order to measure energy change from one trophic level to another, data relating to the different populations may be added together, if the ecosystem contains relatively few species, or a method of synthesis may be used. This involves making a single measurement of secondary productivity using radio-active isotopes such as caesium-137. The primary producers are marked with the isotope, and the rate of assimilation of this substance can then be followed through the different trophic levels.

III ENERGY TRANSFER AND PRODUCTIVITY IN THE BIOSPHERE

1 PRIMARY PRODUCTIVITY

1.1 Introduction

It has been estimated that 21×10^{20} kJ of radiant energy enter the earth's atmosphere each year, i.e. about 64×10^5 kJ/m²/year. Most of this energy is reflected away by the atmosphere or used by the evaporation of water. The average amount of energy available to plants varies with geographical location (cf. chap. 2). In Great Britain it has been estimated at 10.5×10^5 kJ/m²/yr; in Michigan (U.S.A.) at 19.7×10^5 kJ/m²/yr; and in Georgia (U.S.A.) at 25.1×10^5 kJ/m²/yr.

Odum (1959) summarised the world distribution of primary production (figure 13.2). There are four distinct zones:

(i) The deep oceans and deserts have a very low productivity, measured in grammes of dry organic matter, which is usually about 0.1 g/m²/day, and always less than 0.5 g.

(ii) Semi-arid grasslands, areas of temporary cultivation, deep lakes, alpine forests and waters of the continental shelf all have a productivity varying between 0.5 and 3.0 g/m²/day.

(iii) Rain forests, shallow lakes and permanent arable land have an average productivity of between 3 and 10 g/m²/day.

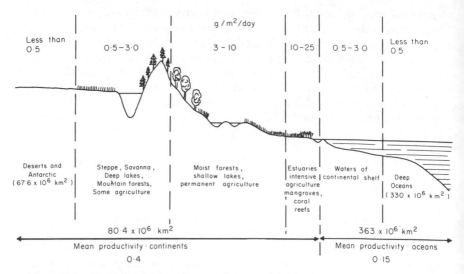

Figure 13.2   World distribution of gross primary productivity in grammes of dry matter per m² and per day (after Odum).

(iv) There are a few specialised ecosystems where productivity exceeds ten grammes and may reach 20 g/m²/day. These include estuaries, coral reefs, plant associations on alluvial plains, and some intensive agriculture such as sugar cane. Productivities higher than 20 g/m²/day have been recorded only for short periods in sugar cane crops, some polluted waters and in a few natural ecosystems.

The figures show that primary productivity is mainly limited by the availability of water and nutrients, light intensity and the ability of the ecosystem to use the available resources. For example, the limiting factors for deserts and the deeper regions of the oceans are very different. In one case available water is the limiting factor, and in the other it is light intensity and the availability of nutrients. These two regions represent true biological deserts, with very low productivities. There is no significant difference, however, between other terrestrial or aquatic ecosystems with regard to productivity, providing that environmental conditions are favourable. The structure of an ecosystem is important. Primary productivity in a deciduous forest is normally greater than that in a cornfield because the forest shows vertical layering and each layer (tree, scrub, field and ground) fixes a proportion of radiant energy. This is also true for aquatic ecosystems, where high productivity is only possible in shallow waters which are able to warm up quickly. These water bodies must, however, be large enough for an optimum density of phytoplankton to develop so that the maximum amount of radiant energy may be fixed.

Much of the earth's surface is covered by oceans or deserts having a low productivity, while high productivity is limited geographically to a few areas of the globe. An increase in primary production can be achieved in only three ways: by the irrigation and cultivation of deserts (possible in some instances), by increasing the productivity of the seas through increased circulation of nutrients from deeper waters, and by increasing crop yields. Primary productivity is especially low in regions with Mediterranean type climates and in warmer regions, where the human population is increasing most rapidly. The figures given in figure 13.2 also emphasise the need for the conservation of those natural ecosystems with high productivities, such as wetlands. This will be discussed later.

Examination of variations in primary productivity shows that in the arctic, terrestrial productivity is low due to the very short period of photosynthetic activity. Yet the Arctic and Antarctic Oceans are among the most productive in the world. In the tropics, much of the land is covered by desert, and the seas are not very productive. However, there are some very productive ecosystems in equatorial regions; coral reefs, mangroves, wetland communities of estuaries and marshes and, especially, tropical rain forests, where optimum conditions of temperature, humidity and light occur throughout the year. Attempts to clear and cultivate areas of rain forests have met with little success: crop yields are low and there is rapid erosion of the soil. This type of forest functions as a closed system, and includes many symbiotic organisms which reduce the loss of nutrients. The slightest interference upsets the delicate balance of this ecosystem.

1.2   Primary productivity of the continents

The oceans cover 363×10^6 km^2, which represents seventy-one per cent of the surface of the earth, while the land surface is only 148×10^6 km^2 in area. The 21×10^{20} kJ of solar energy reaching the earth every year can be divided into 5.9×10^{20} kJ falling on the land masses and 15.1×10^{20} kJ on the oceans.

The terrestrial ecosystems may be divided into four groups:

(i) *Silva*, representing the different types of forest which cover 40.7×10^6 km^2, i.e. twenty-eight per cent of the land surface.

(ii) *Saltus*, representing the steppes, prairies and permanent grasslands, both natural and artificial, covering 25.7×10^6 km^2, i.e. seventeen per cent of the land surface.

(iii) *Ager*, including all cultivated land, covering 14×10^6 km^2, i.e. ten per cent of the land surface.

(iv) *Desertus*, including natural and artificial deserts (including areas occupied by man, e.g. towns, roads, etc.), high mountains with permanent snow cover and the Antarctic continent. This group occupies 67.6×10^6 km^2, i.e. forty-five per cent of the land surface, of which 13×10^6 km^2 is in the Antarctic.

Table 13.4 Total annual primary productivity of the biosphere (after|Duvigneaud)

	AREA (10^6 km^2)	PHOTOSYNTHETIC EFFICIENCY	CARBON FIXED IN TONNES/ha/year	ORGANIC MATERIAL PRODUCED IN TONNES/ha/year	TOTAL ORGANIC MATERIAL PRODUCED IN 10^9 TONNES	ENERGY EQUIVALENT IN 10^{16} kJ
Silva	40.7	0.33%	2.50	5.0	20.4	34.36
Ager	14.0	0.25%	2.00	4.0	5.6	9.64
Saltus	25.7	0.10%	0.75	|1.5	3.8	6.29
Desertus	54.9	0.01%	0.10	0.2	1.1	2.10
Antarctic°	12.7	0.00	0.00	0.0	0.0	0.00
Total	148.0				30.9	52.39

Duvigneaud (1967) estimated the total annual primary productivity of the biosphere from data obtained by various authors on the annual productivity of various ecosystems (table 13.4).

This table shows that all terrestrial ecosystems together produce about thirty-one thousand million tonnes of organic materials each year, using fifteen thousand million tonnes of carbon dioxide gas taken from the atmosphere, which contains reserves estimated at nearly 700 thousand million tonnes. The table emphasises the importance of forests, since they, alone, account for twenty thousand million tonnes, i.e. two thirds of the annual primary productivity of terrestrial ecosystems.

There is considerable variation between ecosystems in the biomass of terrestrial plants. The following figures are given: for subtropical and tropical deserts, the biomass is less than 2.5 tonnes/ha; for polar and sub-boreal deserts and on saline soils, it lies between 2.5 and 5.0 tonnes/ha; it varies between 12.5 and 25.0 tonnes/ha for tundra rising to fifty tonnes/ha for forest-tundra; for steppe, pampas, mangrove and savanna, it can vary between 12.5 and 150 tonnes/ha; in the taiga, the value is about 400 tonnes/ha; the biomass reaches a maximum in forests, deciduous forests producing up to 500 tonnes/ha and tropical rain forests from 500 tonnes/ha to 1 700 tonnes/ha in Brazil. Plant formations which differ in structure may have similar biomasses and productivity, although their chemical composition varies considerably (cf. p. 279).

1.3   Primary productivity of the oceans

The important marine primary producers are phytoplanktonic organisms, littoral benthic macroscopic algae and angiosperms being relatively unimportant. There is a high rate of turnover of phytoplankton and, as a result, productivity can only be estimated by direct measurements of photosynthetic rates. The carbon-14 method of Steeman-Nielsen (1952) produced much lower estimates than Riley (1944) obtained using the method of oxygen measurement. The mean productivity of the oceans has been estimated at 0.15 g/m^2/day, equivalent to 15×10^9 tonnes of carbon or 3.0×10^{10} tonnes of organic matter per year, with a corresponding energy value of 50.3×10^{16} kJ.

The total annual primary productivity of the biosphere is, therefore, in the region of 61×10^9 tonnes of organic matter, equivalent to 10.5×10^{17} kJ. The estimates of some American authors are more optimistic, varying about a mean of 140×10^9 tonnes. Taking a value of 100×10^9 tonnes (i.e. 17×10^{17} kJ) and the radiant energy available to plants as 21×10^{20} kJ, then the photosynthetic efficiency at a global level is

$$\frac{17 \times 10^{17}}{21 \times 10^{20}} \times 100$$

i.e. approximately 0.1%. Despite this low photosynthetic efficiency, the quantity of organic matter produced each year is high if compared with the total production of the chemical and mining industries, which is only 10^9 tonnes, or one hundred times less.

1.4 Primary productivity of specific ecosystems

Having considered primary production at a global level, some data are now given for the productivity of specific ecosystems, where this has been measured.

TABLE 13.5   Net primary productivity in tonnes of dry matter per hectare per year for some terrestrial habitats (after various authors)

a) *productivity of some crops, estimated for the whole plant*

CROP	WORLD AVERAGE	AVERAGE FOR COUNTRIES WITH HIGH PRODUCTIVITY	COUNTRY
Wheat	3.44	12.5	Netherlands
Oats	3.59	9.26	Denmark
Maize	4.12	7.90	Canada
Rice	4.97	14.40	Japan, Italy
Hay	4.20 (USA)	9.40	California
Potatoes	3.85	8.45	Netherlands
Sugar beet	7.65	14.70	Netherlands
Oil palm	–	19.50	Nigeria
Sugar cane	17.25	67.00	Hawaii
Algal cultures, best yields outdoors:		45.30	Japan

b) *productivity of some uncultivated ecosystems determined by cropping methods*

Pine plantations, England, average during years of most rapid growth (20–35 years old)	31.80
Deciduous plantations, England, similar to the above pine plantations	15.60
Tall grass prairies, Oklahoma and Nebraska	4.46
Short grass prairies, Wyoming, with low rainfall (32.5 cm per year)	0.69
Desert vegetation, Nevada (12.5 cm of rain per year)	0.40
Salt marshes with *Spartina*, Georgia	33.00
Giant ragweed (*Ambrosia trifida*) stand, Oklahoma	14.40
Rain forest with *Tectona, Altingia* and *Ochroma*, Java	54 to 69
Papyrus beds in tropics	72.00
Sphagnum bogs, Germany	2 to 10
Rhododendron maximum community	7.3
Alpine meadow (excluding roots)	0.0007 to 0.004
Non-deciduous forest canopy, Thailand	91.00

c) *productivity of agricultural grasslands in Germany*

Festuca grassland	10.5 to 15.5
Lolium grassland	5.5 to 7.9
Arrhenatrerum grassland	9.1
Deschampsia with clover, three cuts per year	22.4

a Terrestrial habitats

Table 13.5 gives estimates of primary productivity for some important crops and natural ecosystems. Table 13.6 shows the plant biomass for several ecosystems. A hectare of temperate forest produces about ten tonnes of dry matter per year, consisting of six tonnes of wood and four tonnes of leaves (in the case of beech), and this, on combustion, yields about 193 million kJ. The solar energy available is about 38 thousand million kJ per hectare per year, and so the photosynthetic efficiency is about 0.5%.

Productivity is highest in the tropics, since the vegetation remains green and photosynthesis can take place throughout the year. Secondary equatorial forest at Yangambi produced twenty tonnes/ha/yr. Under suitable conditions (fertile soil and adequate water supply), a tropical forest will produce abut twice as much material as a temperate forest. The productivity of the field layer is also higher in the tropical forest: African *Pennisetum purpureum* savanna may produce up to thirty tonnes/ha, and sugar cane up to sixty-seven tonnes. This demonstrates the important role of climatic factors in primary production. The South African steppe has a productivity above ground which varies between one and six tonnes/ha and shows an almost linear relationship with rainfall, which varies between 100 and 600 mm depending on the region. In both alpine and arctic regions, where temperatures are low, productivity is reduced to between 0.5 and five tonnes/ha according to the ecosystem.

In temperate climates, where photosynthesis can occur throughout the year, productivity is higher. In England coniferous forests produce

TABLE 13.6 Biomass of total vegetation (except underground parts) estimated in most cases at the end of the season of maximum growth and expressed in kilogrammes of dry weight per hectare (Bourlière and Lamotte, 1962)

Sedge-meadow above an iron pan, Guinea	200 to 500
Sahel savanna, field layer only, Senegal	590
Themeda and *Heteropogon* veld, Rhodesia, field layer only, mean for 9 years	1 450
Field uncultivated for 15 years, Michigan	510 to 3 850
Carex marsh, south Michigan	4 650
Festuca grassland, Montana	5 250
Themeda savanna, fieldlayer only	6 300
Andropogon grassland, North America	6 550
Andropogon savanna, Guinea	5 000 to 10 000
Masai steppe, field layer only	7 000 to 14 000
Rhizophora mangle mangrove, Puerto Rico	63 000
Quercus robur woodland, Great Britain	128 000
Secondary forest 20 years old, Ghana	123 750
Pinus sylvestris plantation 35 years old, England	163 000
Secondary forest, Zaire	150 000 to 200 000
Oak/beech wood, 120 years old	275 000
Secondary forest 40 to 50 years old, Ghana	362 360
High primary forest of *Scorodophloeus zenkeri*, Zaire	1 003 600

31.80 tonnes/ha, while deciduous forests, on the same soils and under the same climatic conditions, produce only 15.60 tonnes.

Most of the important crops in temperate countries have a productivity which averages about ten tonnes/ha, at least in those countries where modern agricultural methods are employed and conditions are most favourable. This value also applies to managed grasslands.

b Marine habitats

The productivity of the oceans varies with the region (Steeman-Nielsen and Aabye Jensen, 1957). The productivity of coastal waters varies between 0.8 and 3.7 grammes of carbon/m^2/day. These waters are green in colour and rich in nutrients, derived either from water flowing from the land masses or from deeper offshore waters coming to the surface. The productivity of the North Sea is about 800 kg/ha/year (Steele, 1958 in Duvigneaud). Benthic algae have an important role, but little is known about their productivity. The biomass of algae along the Californian coast varies between sixty and one hundred tonnes per hectare, and annual productivity between ten and fifteen tonnes. On the Scottish coast the biomass is between twenty and forty-five tonnes/ha, while the corresponding value for Nova Scotia lies between twenty and 130 tonnes with a productivity of 3.58 tonnes/ha/year of dry matter. In those areas where the deep waters of the Benguela current surface off the South African coast, productivity varies between 1.8 and 11.0 tonnes of carbon per hectare per year.

Where surface waters mix with deeper waters in the open sea as the result of convection currents, productivity varies between 0.25 and 0.52 g of carbon/m^2/day. This happens, for example, in temperate waters of the North Atlantic. In contrast, the surface waters of tropical oceans are warm and, since there is little mixing with deeper water, they are poor in nutrients. Their productivity ranges from 0.1 to

TABLE 13.7   Gross primary productivity in grammes of organic matter per $metre^2$ per day for some marine ecosystems (after Kohn and Helfrich, 1957 and Odum 1959)

1. **Annual mean**

Open sea, Pacific Ocean	0.1 to 0.2
Sargasso Sea	0.5
Shallow inshore waters, Long Island Sound	3.2
Benthic algae, Nova Scotia coast	3.58
Laguna Madre, a Texas estuary	4.4
Benguela current	0.9 to 5.2
Rhizophora mangle mangrove, Puerto Rico	16.0
Coral reefs, Pacific	10 to 23

2. **Mean for shorter periods**

A Texas estuary	23
Marine turtle grass flats, Long Key, Florida	34

0:2 g/m²/day. The blue surface waters of the oceans are practically deserts, with very low productivity, estimated at 0.05 g/m²/day. These are typical of equatorial waters.

Coral reefs and mangroves are habitats marked by a high productivity (Table 13.7).

c Fresh water

There is a wide variation in gross productivity in freshwater. Table 13.8 gives values for some ecosystems.

TABLE 13.8   Gross productivity in grammes of dry organic matter per metre² per day for some fresh water ecosystems (Odum and Odum, 1959).

1. **Annual mean**

Bog lake, Cedar Bog Lake, Minnesota	0.3
Lake Weber, low calcium, Wisconsin	0.7
Lake Mendota, high calcium, Wisconsin	1.3
Shallow eutrophic lake, Japan	2.1
Silver Springs, Florida	17.5

2. **Mean over shorter periods**

Lake Erie, winter	1.0
Turbid river with suspended clay, North Carolina, summer	1.7
Fertilised pond, North Carolina, May	5.0
Lake Erie, summer	9.0
Pond with treated sewage wastes, Denmark, July	9.0
Pond with untreated wastes, South Dakota, summer	27.0
Algal cultures with extra carbon dioxide added	43.0
Polluted stream, Indiana, summer	57.0

2   PRODUCTIVITY OF OTHER TROPHIC LEVELS

There are relatively few detailed studies on productivity and energy flow from one trophic level to another in ecosystems. It is possible to obtain estimates of secondary and even tertiary productivity by estimating the animal biomass obtained either by cattle rearing or by fishing in selected habitats.

2.1   Terrestrial habitats

The biomass of terrestrial animals is generally less than one per cent of the corresponding plant biomass, ninety to ninety-five per cent of this biomass consisting of invertebrates. The ratio plant biomass/herbivore biomass varies from 10^2 for steppe and deserts, to 10^3 for wooded steppe, 10^4 for tundra and 10^5 in the taïga and coniferous and deciduous forests. The herbivore/carnivore biomass ratio usually has a value of about 10^2.

TABLE 13.9 Biomass of wild herbivorous mammals in tonnes per km² for some natural ecosystems (after various authors)

Ruanda-Rutshuru plains, Albert National Park, Zaire	24.4
Forest edge savanna, north of Kiwu	23.5
Queen Elizabeth National Park, Uganda	19.5
Bush savanna, Western Uganda	18.8
Nairobi National Park, Kenya	15.6
Masai steppe, Kenya	13.2
Serengeti plains, Tanzania	5.2
Henderson ranch, Rhodesia	4.9
Open savanna, Tanzania	4.7
Scrub savanna, Rhodesia	4.4
North American prairie, Montana	3.5
Kruger National Park, Transvaal	1.8
Scottish deer forest (*Cervus elaphus*)	1.0
Canadian tundra with caribou	0.8
Primitive forest, Mount Polana, Czechoslovakia	0.5
Steppe, southern Russia with saïga antelope	0.35
Sahel-type steppe, Chad	0.08
Tropical rain forest, Ghana	0.075
Sahara sand hills, Mauritania	0.005 to 0.02

Tables 13.9 and 13.10 give some examples of productivity. They show (assuming that productivity is proportional to biomass) that productivity for herbivorous mammals is fairly high in tropical grassland formations, and much lower in temperate and arctic regions and also in forests. The pastures of Western Europe can support up to 125 000 kg/km² of cattle, while the pampas of Argentina and the African steppe support only 5000 kg/km². Bird biomasses vary between 0.5 and 130 kg/km². Values for invertebrate biomasses in some ecosystems are given in Table 13.11.

TABLE 13.10 Biomass of birds in kg/km² for some ecosystems (after several authors in Bourlière and Lamotte, 1962)

Heathland, NW Germany	0.5
Dry forest, NW Germany	8
Coniferous forest, Finland	22.5
Sahel savanna, Bas Senegal	20 to 50
Mixed forest, Finland	58
Primitive forest, Mount Polana, Czechoslovakia	116
Lakes and shores, NW Germany	130

TABLE 13.11 Total biomass of insects in the litter and field layer in kg fresh weight per km² (after several authors in Bourlière and Lamotte, 1962)

Sedge-meadow over iron pan ('cuirasse ferrugineuse') at the end of the rainy season, Guinea	700 to 2 500
Sansouire, Camargue	5 500
High meadows, Mount Nimba, Guinea, in rainy season	18 700
Andropogon grassland, Guinea, in rainy season	25 000

2.2   Marine habitats

F.A.O. statistics show that 35.6×10^6 tonnes of fish and invertebrates were taken from the sea in 1959 by commercial fisheries. Epipelagic fish accounted for only seven per cent of this total, indicating that most fishing takes place over the continental shelf in the rich neritic zone. Although invertebrates are very numerous in the sea, they made up only fifteen per cent of the catch (5×10^6 tonnes) and, of this, sixty-eight per cent were molluscs and twenty-two per cent crustaceans. The cephalopod catch was 704 900 tonnes in 1960, of which 600 000 tonnes were taken by Japan (Clarke, 1963). In France, 60 000 tonnes of oysters and between 100 000 and 400 000 hectolitres of mussels are collected each year. The Far East takes about half the world catch of fish and invertebrates. Man obtains only six per cent of his protein and less than one per cent of his total food from the sea.

When the 35.6×10^6 tonnes of fish and invertebrates taken annually from the sea is expressed in terms of carbon assimilated, it can be seen that only 0.015 per cent of primary production in the sea is used by man as food. Much of the ocean is not exploited, and can probably never be economically used. In sheltered waters at higher latitudes, such as the North Sea, as much as 0.2 to 0.3 per cent of the carbon assimilated by plants is removed by fisheries. Such a high value can only be attained in areas where a proportion of the catch is of fish, like the herring, which is partly herbivore and partly primary carnivore. This increases the yield because there is only one trophic level between the phytoplankton and fish (Steeman-Nielsen and Jensen, 1957). A similar thing is seen in fresh water, where the yield of herbivorous fish is greater than that of carnivores.

Some figures have been obtained for the biomass of benthic animals in some localities, and Table 13.12 (after Pères and Deveze, 1961) shows a correlation between biomass and the number of fish caught.

Examples of the biomass of marine micro- and mesoplankton are given below in kg per 10 000 m^3 (after Pères, 1961)

Barents Sea between 0 and 25 metres	0.5 to 10
Sea of Japan	up to 12
Sea of Azov	1060 in October
	2700 in August

The biomass of the benthos rapidly decreases with increasing depth, reaching values of less than 0.1 g/m^2 at 4000 metres. It is difficult to obtain a precise estimate of the density of the abyssal benthos. Sparck in Pères (1961) gives a mean value of one gramme per metre^2 in the Pacific. For abyssal zones in the Atlantic, the mean is about 0.7 g/m^2, while in the tropics the biomass is much lower, being about 0.08 g/m^2 in the Pacific Ocean and 0.03 g/m^2 in the Indian Ocean. However, local conditions may have a considerable effect on these figures. The

TABLE 13.12   Mean biomass of marine benthos (mainly invertebrates) on loose substrates of the continental shelf in kg/ha, and the amount of fish taken in kg/ha (after Pères and Deveze, 1961)

LOCALITIES	BENTHOS BIOMASS	FISH CAUGHT
Small estuaries, Ghana	72	–
Mediterranean	100	1.5
Gulf of Guinea, Ghana	117	–
White Sea	200	1.2
Baltic Sea	330	6
Mouths of Congo and of Volta Rivers	300 to 400	–
English Channel	400	–
Barents Sea	1 000	4.5
Bering Sea, NW region	1 650	–
Sea of Japan	1 750	28.8
Sea of Azov	3 210	80
North Sea	3460	24.5
100 to 200 m depth, Antarctic	13 470	–

Galathea recorded an invertebrate biomass of 12.5 g/m^2 at a depth of 6580 metres in the Banda Sea, Indonesia. In the Bering Sea the biomass of sublittoral animals may reach 1000 g/m^2, while it is only 0.1 or 0.2 g/m^2 at depths of 3500 to 4000 metres.

2.3   Fresh water habitats

The total quantity of freshwater fish taken annually has been estimated at 3×10^6 tonnes, of which over half is in the U.S.S.R. This represents only a small proportion of the secondary, tertiary and quaternary productivity of fresh waters. The figures in Table 13.13 show fish production in some fresh water ecosystems. The carp, a herbivore, gives higher yields than carnivorous fish but, where lakes are managed as fisheries, they must contain sporting fish because these interest the angler. Some biomasses of freshwater fish, in kg fresh weight per hectare, are given below (after Bourlière and Lamotte, 1962):

Trout lakes, U.S.A.	62.5
Heated water lakes in Canada and U.S.A.	155 to 187
Mean of 132 rivers in the U.S.A.	187
Lakes of the Mid-West, U.S.A.	250 to 375
Streams in U.S.A.	625

The yield from heated fish ponds in China may reach 5000 kg per hectare (cf. p. 161).

IV   ENERGY FLOW AND PRODUCTIVITY IN SPECIFIC ECOSYSTEMS

The autotrophic and heterotrophic components of an ecosystem may be spatially separated. For example, the plants and herbivorous animals

TABLE 13.13   Fish production in some fresh water ecosystems (Odum, 1959)

ECOSYSTEM AND TROPHIC LEVEL	KILOGRAMMES OF FISH HARVESTED PER HECTARE PER YEAR		
1. Unfertilised waters			
a) Herbivores and carnivores mixed:			
American Great Lakes	0.9	to	8
African lakes	1.5	to	247
Small lakes, U.S.A.	2.0	to	178
b) Carnivores:			
Fish ponds, U.S.A.	44	to	165
c) Herbivores:			
German carp fish ponds	110	to	383
2. Artificially fertilised waters			
a) Carnivores:			
Small fish ponds, U.S.A.	220	to	550
b) Herbivores:			
Marine ponds, Philippines	495	to	990
Carp fish ponds, Germany	980	to	1540
3. Fertilised waters with additional food added			
a) Carnivores:			
Small fish ponds (0.5 ha), U.S.A.	2 409		
b) Herbivores:			
Hong Kong	2 200	to	4 400
South China	1 100	to	14 850
Malaya	3 850		

living above the soil are partly separated from the detritivore primary consumers living in the soil. In the oceans, the phytoplankton and herbivores are found above the benthos, which includes many detritus feeders. There is also, sometimes, a delay before the organic material produced by the autotrophs is available to the heterotrophs. There is, as a result, a dichotomy between the first and second trophic levels which corresponds to the two types of primary consumer. On the one hand there are herbivores and plant parasites feeding directly on living plant material; and, on the other, detritivores feeding, after some delay, on dead animal and plant materials. In most ecosystems one or other of these branches predominates, and ecosystems may be divided into two groups depending on whether herbivores or detritivores are dominant.

1   ECOSYSTEMS WHERE HERBIVORES ARE DOMINANT

1.1   Coral reefs

There is a rich and diverse fauna of coral reefs, and biomass and productivity is high. A general study of the Eniwetok Atoll in the Pacific Ocean was made by Odum and Odum (1955).

a   Structure of the ecosystem

Six horizontal zones can be distinguished across the coral reef as one moves from the open sea to the lagoon:

(i) *The outer slope of the reef*, about sixty-six metres wide. This zone was not studied.

(ii) *The outer ridge of the reef* (Zone A in figure 13.3), about seventeen metres wide, and covered by red algae and corals, the most abundant being *Millepora platyphylla, Acropora* sp. and *Pocillopora danae.*

(iii) *The zone of encrusting corals* (Zone B), about sixty-six metres wide, covered at low tide by only fifteen centimetres of water. This zone contains encrusting algae and the corals of zone A, together with *Favia pallida, Favites halicora, Pocillopora verrucosa* and *Plesiastrea versipora.* Sea anemones also occur in this zone.

(iv) *The zone of erect corals* (Zone C), containing small blocks of coral about thirty centimetres in diameter and height. These corals are mainly *Cyphastrea* sp. *Favia pallida* invades this zone as the encrusting *Acropora* of the outer zones disappears. Numerous species of fish occur in this zone.

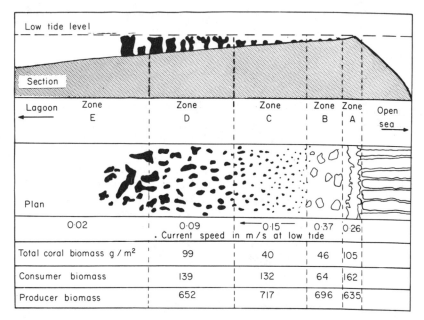

Figure 13.3   Diagram of the structure of a coral reef.
   Above: Section of reef showing zones A, B, C, D and E from the open sea through to the lagoon.
   Centre: Plan of reef.
   Below: Current speed at low tide in m/s in the various zones, and biomasses of some trophic levels (Odum, 1955).

(v) *The zone of larger, branched, erect corals* (Zone D), consisting of blocks 0.6 to 1.2 m or more high, and from 0.6 to 6.0 m in diameter. The space between these blocks is filled with white sand. Branched corals like *Acropora gemmifera* and *A. cymbicyathus* are dominant here. The water depth increases and current speed decreases in zone D.

(vi) *The zone of sand and gravel formed from dead corals* (Zone E), about 150 to 200 m wide, and sloping gently towards the centre of the lagoon. Corals found here include *Turbinaria, Heliopora* and *Millepora.*

Figure 13.3 shows a cross section of the reef from zone A to zone E. The biomasses in g/m² of corals, consumers and producers are shown for each zone. It can be seen that, although the species of coral vary from one zone to another, their total biomass varies only slightly (less than by a factor of 2.5), and this is also true for the total biomass of producers.

b Trophic structure of the community

(i) *Producers.* Madreporarian corals and many other reef-dwelling organisms harbour unicellular algae (zooxanthellae) in their protoplasm, and these utilise carbon dioxide and nitrogenous waste produced by the animals, providing in exchange oxygen and organic materials. Odum (1955) also found that madreporarians and alcyonarians harbour, additionally, filamentous green algae in the pores of their calcareous skeleton. These algae are sixteen times more abundant than zooxanthellae (table 13.14).

Corals thus contain three times more plant than animal protoplasm. Although the algal filaments are located up to three centimetres deep in the coral, they are able to obtain light because the aragonite skeleton is translucent, and because of the clarity of tropical waters. The reef as a whole has a biomass of dry plant material of between 500 and

TABLE 13.14. Percentage biomasses of plant and animal tissues in a coral reef, estimated from sections and by chlorophyll extraction

		PLANT TISSUE (%)	ANIMAL TISSUE (%)
Polyp zone	in polyps	4.5 (zooxanthellae)	25
	between polyps	26 (green filaments)	–
Skeletal zone below polyps		44 (green filaments)	–
Total (approximate)		75	25

1000 g/m², the total biomass (dry weight) for plants and animals being between 665 and 1335 g/m², including the extensive calcareous skeleton.

Symbiotic algae occur in the following groups of animals and substrates:

— in living corals (twelve species) belonging to the order Zoantharia (genera *Leptastrea, Favia, Porites, Acropora, Turbinaria, Pocillopora* and *Lobophyllia*), the order Alcyonaria (genus *Heliopora*) and the order Hydrocorallina (genus *Millepora*);

— in animals other than corals, such as sea anemones and the giant bivalve *Tridacna*;

— in the sand which covers the floor of channels running across the reef;

— in broken fragments of colonies resting on the floor of the reef;

— in dead portions of reef remaining *in situ* and standing above the level of the reef;

(ii) *Consumers.* These include the corals, which act as primary consumers because they use the symbiotic algae. Corals also act as secondary consumers, capturing zooplankton at night. Herbivores, other than corals, include echinoderms (urchins, holothurians and brittlestars), molluscs, crabs and fish. Carnivores include polychaetes, crustaceans, a few molluscs and some fish. The biomass (in grammes dry weight per square metre) was estimated at 703 for producers, 132 for herbivores and 11 for carnivores (figure 12.2).

(iii) *Decomposers and detritivores.* These include bacteria and foraminifera. Blennies (fish) are also included in this group because they use organic detritus suspended in the water. The decomposer biomass was not estimated.

c   Productivity

In addition to a high biomass, another feature of the coral reef was its high rate of productivity, equivalent to twelve and a half times the biomass. The gross primary productivity was more than 24 g/m²/day of dry organic matter. The rate is higher than for most other communities, the corresponding value for the open sea in the tropics being only 0.2 g/m²/day, i.e. less than 1%. The energy balance for that part of Eniwetok Atoll studied was:

$$
\text{GAINS} \begin{cases} \text{organic matter derived from outside} & 2.0 \text{ g/m}^2/\text{day} \\ \text{gross productivity} & 24.0 \text{ g/m}^2/\text{day} \end{cases} 26.0
$$

$$
\text{LOSSES} \begin{cases} \text{organic matter lost into lagoon} & 0.4 \text{ g/m}^2/\text{day} \\ \text{respiration} & 24.0 \text{ g/m}^2/\text{day} \end{cases} 24.4
$$

The inaccuracy of the measurements make it difficult to decide whether this ecosystem shows a gain in materials or is in a steady state. The efficiency of primary producers has been estimated. The available radiant energy during August is about 6900 kJ/m^2/day, and gross primary productivity is 400 kJ/m^2/day, giving an efficiency of about 5.8%. This figure is less than for some experimental crops, but higher than for most agricultural crops.

The complex trophic structure of the coral reef has enabled this type of ecosystem to survive in an almost steady state for millions of years. The symbiotic relationship between algae and coelenterates allows the development of a closed nutrient cycle which is essential for rapid growth in tropical water 'deserts', which are very poor in plankton. Corals, themselves, may be regarded as ecosystems since they are, at the same time, producers, herbivores and carnivores. A considerable quantity of calcium carbonate is assimilated, Odum and Odum (1955) estimating an annual increase in thickness of reef of 1.6 cm. However, erosion appears to remove almost as much calcium carbonate from the reef. It still seems likely that coral reefs, covering 190 million km^2, represent the main point of accumulation of the calcium that is carried by rainwater into the oceans from the continents.

1.2 Silver Springs

These springs have a water temperature that is almost constant, varying only between 22.2 and 23.3°C over the year. The springs are located in central Florida at latitude 29°N. The water from these springs enters the river Oklawaha after flowing eight kilometres along a stream. The mean rate of flow is about 0.21 m^3/second, and the water is almost pure with only traces of calcium, sodium, magnesium, potassium, sulphates and chlorides. The water is very clear, and plants growing there receive intense illumination under sub-tropical conditions.

a Trophic structure of the community

The dominant plants of Silver Springs are one angiosperm (*Sagittaria lorata*) and several algae, mainly diatoms and filamentous algae epiphytic on *Cocconeis placentula*, *Melosira granulata* and *Navicula minima*. Other less common plants include the angiosperms *Najas guadalupensis*, *Ceratophyllum demersum*, *Vallisneria neotropicalis* and *Potamogeton illinoiensis*, and numerous unicellular algae.

Herbivores include terrapins (*Pseudemys nelsoni* and *P. floridana*), fish (*Mugil cephalus*, *Lepomis microlophus* and *L. punctata*), crustaceans (*Paleomonetes paludosus*, *Gammarus* sp.), gastropods (*Pomacea paludosa*, *Viviparus georgianus*) and insect larvae (*Elophila* sp., *Calopsectra* sp., *Tendipes* sp., *Hyaroptila* sp.).

Primary carnivores include fish of the genus *Lepomis* (an omnivorous fish) and *Gambusia affinis*, amphibians (*Amphiuma* sp. and three species of *Rana*), birds (*Ardea, Fulica, Gallinula*), coelenterates (*Hydra* sp., *Craspedacusta sowerbyi*), insects (gyrinid beetles, veliid bugs), mites and leeches. Secondary carnivores are fish (*Lepisosteus* sp., *Amia calva*, *Micropterus salmoides*) and reptiles (*Alligator mississippiensis*).

Decomposers are mainly bacteria, and the crayfish *Procambarus fallax* is a detritivore, feeding on animal and plant debris.

The pyramid of biomass for this ecosystem (figure 12.2) shows a total biomass of 863 g/m^2, including 809 g of producers, 37 g of herbivores, 11 g of primary carnivores, 1.5 g of secondary carnivores and 4.6 g of decomposers.

b Productivity

The autotrophic living plants are the main source of energy in Silver Springs, and these are eaten directly by herbivores. Gross primary productivity is 6390 g (dry weight) of organic matter per square metre per year, including thirteen per cent protein and twenty-three per cent mineral ash. Additionally, 120 g/m^2 of plant material is carried into the springs annually, and is largely consumed by herbivores. Respiratory losses amount to 600 g/m^2/year, organic matter carried away in the water or lost through the flight of adult insects amounts to 766 g/m^2/year. Productivity in spring is two or three times greater than in winter, as the result of changes in light intensity. However, increased productivity is accompanied by increased consumption, with the result that the biomass hardly changes.

The ratio : net productivity/biomass is approximately eight. The present community has probably remained stable for at least one hundred years and represents a climax. *Sagittaria* represents the bulk of producer biomass, but only a third of the productivity, algae playing an important part in productivity of 4490 g/m^2/year. Figure 13.4 shows energy flow in Silver Springs as kJ/m^2/year.

2 ECOSYSTEMS WHERE DETRITIVORES ARE DOMINANT

2.1 Root Spring, Massachusetts

a Structure of the ecosystem

This simple natural ecosystem studied by Teal (1957), consisted of a small hollow, two metres in diameter and between ten and twenty centimetres deep, fed by water at a relatively constant temperature, varying only between eight and eleven degrees centigrade throughout the year. The producers were filamentous algae (*Spirogyra, Stigeoclonium*), diatoms (*Nitzchia*) and the duckweed *Lemna minor.* There was no true plankton, and the fauna was fairly simple, consisting of about forty species. Many of these were detritivores, including an oligochaete *Limnodrilus*, an isopod *Asellus*, an amphipod *Crangonyx*, a bivalve *Pisidum*, a snail *Physa* and the chironomid *Calopsectra.* Dead leaves and other plant material falling into the spring provided a substantial part of the food resources each year.

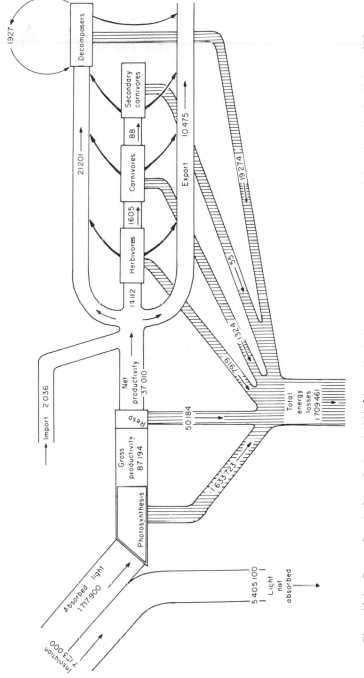

Figure 13.4 Energy flow in Silver Springs, in kJ/m²/year. Hatched areas: heat losses from plants and through respiration (Odum, 1957).

b Productivity

The number and biomass of consumers were measured each month, together with their respiration rates and speeds of development. Falling dead leaves contributed 9847 kJ/m² each year, while only 2975 kJ were supplied through photosynthesis (gross productivity). With an immigration of 75 kJ/m² in the form of living animals, there was an annual input into the ecosystem of 12 897 kJ/m². Annual losses were 9293 kJ/m², of which 9155 kJ/m² were due to respiration and 138 kJ/m² due to the loss of emerging adult insects. Energy flow in the ecosystem may be summarised in the following way:

(i) Much of the energy entering the system (seventy-six per cent) was provided by plant debris, and served as food for detritivores.

(ii) Part of the energy entering the system remained there in the form of organic debris slowly accumulating at the bottom of the pool.

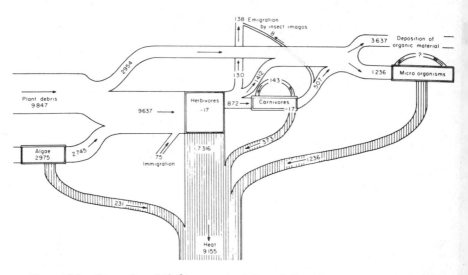

Figure 13.5   Energy flow (kJ/m²/year) through Root Spring, Massachusetts. Hatched areas: heat losses from plants and through respiration (Teal, 1957).

If the latter is not regularly cleared of these sediments, it will gradually fill up. About twenty per cent of the energy contained in dead herbivores is not used, and one per cent of the energy entering the ecosystem is lost in the form of adult insects.

(iii) The amount of energy held in the tissues of herbivores decreased by 17 kJ/m²/year, and this was also true for carnivores. This ecosystem is not, therefore, in a steady state, productivity tending to decrease. The ecological efficiency for the herbivore trophic level is 2300/208, i.e. eleven per cent, and this is a typical value (figure 13.5).

2.2 A Georgia salt marsh

a Structure of the ecosystem

The dominant plant and main producer in a salt march studied by Teal on the Georgia coast was the grass *Spartina alterniflora*. Several algae played a secondary role in primary production. Several herbivorous insects, including the grasshopper *Orchelinum* and the fulgorid bug *Prokelisia*, lived on the *Spartina*. These insects served as food for carnivores, which included spiders, passerine birds and dragonflies. Other animals which lived on the mud surface fed on dead and decaying *Spartina* and on algae. These included crabs (*Uca, Sesarma*), gastropods (*Littorina*), annelids and the bivalve *Modiolus*. Carnivores which preyed on these mud-dwelling animals included other crabs, the raccoon (*Procyon lotor*) and a bird *Rallus longirostris*. Few animals are adapted for life in this environment with its wide variations in salinity (12 to 30‰).

In this salt marsh ecosystem, there are two components, one deriving its energy from living *Spartina* and the other depending on algae and detritus.

b Productivity

Gross productivity has been estimated at 6.1 per cent of the total energy received, a much higher value than for most other ecosystems, where values vary between 0.1 and 3.0 per cent. This higher productivity is due to an abundance of nutrients, which are continually renewed through the circulation of the tides. Net productivity has been estimated at 1.4 per cent of the energy received by the ecosystem. Bacteria and detritivores feeding on dead *Spartina* play an important

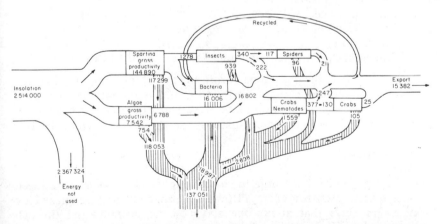

Figure 13.6   Energy flow (kJ/m²/year) through a *Spartina* salt marsh in Georgia. Hatched areas: heat losses from plants and through respiration (Teal, 1962).

role. This ecosystem is very stable, despite the small number of species present. This conflicts with the view that ecosystems with low species diversity are subject to the large fluctuations which occur, for example, in arctic and desert regions. The stability of the present system may be due to its antiquity, since it has been in existence for a very long time, in contrast to arctic ecosystems which are of recent appearance. The omnivorous diet of many species, such as the crabs *Uca* and *Sesarma*, also contributes to the stability of the system (Macarthur, 1955).

Figure 13.6 shows that forty-five per cent of net primary productivity is exported via the waters of the estuary. The latter are very muddy and primary production is almost nil, although crustaceans (crabs and shrimps) are abundant. These animals depend on the productivity of the salt marshes.

2.3 Temperate deciduous forests

The forest forms a typical example of an ecosystem where the detritivore food chain is far more important than that of the herbivore. Dead wood forms a significant part of the biomass, and it is broken down mainly by fungi and insects and, subsequently, by oligochaetes and some arthropods. Leaves are mostly used as food after they fall on the soil. Bray and Gorham (1964) showed that, on average, only seven per cent of the leaf biomass is consumed in the form of living tissue, representing 0.85 per cent of primary production. Some data is available from Duvigneaud for 120 year old oak/beech forests in central Europe. He gives the following biomass per hectare of the important producers and consumers:

Woody plants: 274 tonnes
Herbaceous plants: 1 tonne
Large mammals (wild boar, red deer and roe deer): 2 kg
Small mammals (rodents, carnivores and insectivores): 5 kg
Birds: 1.3 kg
Soil animals: 1 tonne, including 600 kg of earthworms.

It is clear that the soil fauna forms an important component of the ecosystem, includes mainly detritivores, and represents an important part of secondary production in a forest. Unfortunately little is known about energy flow in forest ecosystems. (See however Reichle, (1970).)

2.4 Grasslands

Macfadyen (1963), from data in several published works, has produced a generalised plan for energy flow in permanent pasture grazed by cattle. He finds that about one sixth of the energy fixed through photosynthesis is lost through respiration, and that two thirds of the energy passes in the form of dead organic material to decomposer organisms in the soil. Nearly half the material eaten by herbivores is returned to the soil in the form of un-assimilated faecal material. Thus, for grazed grassland, less than one per cent of the total production is cropped by man, and about seventy-five per cent is returned to the soil. In the soil, over eighty per cent of respiratory loss is of bacterial origin, eight per cent by protozoa, and the remainder by metazoans (myriapods, collembola, mites, etc.).

2.5 Soil

The last section leads to a study of metabolism in the soil. In terrestrial ecosystems most of the net primary production is used, not by herbivores, but by detritivores and decomposers in the soil. These organisms have a much more important role than is indicated by their biomass or even their metabolism.

The data given by Macfadyen (Table 13.15) give some idea of the relative importance of the main groups of soil organisms. The importance of the role of soil animals in the breakdown of organic matter and re-cycling of nutrients has often been underestimated, recent research having shown that soil and litter fauna greatly assist the bacteria in breaking down dead leaves which accumulate in woodland. Edwards and Heath (1963) showed that the decomposition of dead leaves is retarded by the absence of earthworms, leaves showing little trace of breakdown after nine months when micro-organisms are left to act by themselves. Mechanical breakdown by animals probably assists

TABLE 13.15   Relative importance of the major groups of soil organisms

SYSTEMATIC GROUPS	NUMBER per m^2	BIOMASS (g/m^2)	METABOLISM IN kJ/m^2 per year
Bacteria	10^{15}	1000	–
Fungi	–	400	–
Protozoa	5×10^8	38	474
Nematodes	10^7	12	1488
Lumbricids	10^3	120	754
Enchytraeids	10^5	12	670
Molluscs	50	10	260
Isopods	500	5	402
Opiliones	40	0.4	159
Parasitiform mites	5×10^3	1.0	21
Oribatid mites	2×10^5	2.0	268
Spiders	600	6.0	126
Beetles	100	1.0	143
Diptera	200	1.0	34
Collembola	5×10^4	5.0	25
Myriapods	500	12.5	641

chemical breakdown (Nef, 1957), but little is known about this subject. The importance of soil animals lies, not in their efficiency of energy transfer between different trophic levels, but in the speed at which decomposition liberates essential plant nutrients.

According to Weis Fogh (1948), the amount of organic matter present per hectare of soil is about 115 500 kg dry weight, including 6368 kg of animals and micro-organisms, 11 550 kg of living roots, and 97 582 kg of decaying organic material. The two important groups of soil animals are the annelids and arthropods.

The Lumbricidae and Enchytraeidae are the main annelid groups in temperate regions, but other families, such as the Acanthodrilidae, Megascolecidae and Eudrilidae, become important in the tropics. Enchytraeids play an important part in mor and moder soils, where they are more abundant than earthworms. The work of Nielsen (1955) and of O'Connor (1957) gives an indication of the extent of enchytraied activity. Nielsen found on pasture a mean density of between 30 000 and 70 000 worms/m^2, having a biomass of $3-10.5$ g and an oxygen consumption of $7-32$ litres m^2/year. O'Connor gives figures of a similar order of magnitude from studies on litter fauna of a *Pseudotsuga douglasi* plantation. There was a mean density of 134 000 enchytraeids/m^2 and a biomass of 10 794 g. The rate of oxygen consumption was 31.4 litres/m^2/year, corresponding to a metabolic rate of 626.4 kJ/m^2/year. Enchytraeids are essentially detritivores and, since their populations appear to be renewed annually, their productivity is at least equal to their biomass.

Earthworms form the largest component of the soil fauna in terms of biomass and activity. Darwin drew attention to the importance of this group as early as 1881. Their biomass usually varies between 100 kg and two tonnes per hectare, but may reach four tonnes/ha on pasture. It is lower on arable soils which are poor in organic matter, and on acid and saline soils. Earthworms increase the productivity of soils mechanically, through aeration and by bringing soil from deeper horizons up to the surface. The quantity of soil stirred by earthworms varies from six to eighty tonnes/ha/year in Europe, to 210 tonnes in the Cameroons. Earthworms also have a chemical effect on the soil through the action of their digestive enzymes.

Soil arthropods can be divided into macroarthropods, larger than two millimetres (woodlice, millipedes, insects like beetles and fly larvae), and microarthropods, less than two millimetres (mainly mites and collembola). Macroarthropods include predators (mainly beetles) and many detritivores which contribute to the breakdown of organic matter in the soil. According to Dudich *et al.* (1952), they consume annually about forty per cent of forest leaf litter. Among the microarthropods, mites occur at densities of 100 000 and 400 000/m^2 in forests, and from 50 000 to 250 000/m^2 in grassland. Oribatids are the dominant group, and they may consume up to fifty per cent by weight of the litter in an oak forest.

Most collembola are found in moister habitats than oribatids. They are about as abundant as mites, but their role is not fully understood. Data given by Macfadyen suggest that the part played by animals in soil metabolism is not proportional to their biomass, and that their metabolic rate increases as size decreases (Table 13.15).

Possibly the most important feature of soil animals is that most (about eighty per cent) are detritivores and, in conjunction with the bacteria that are also present, feed on dead organic material derived from animals and plants. Phytophagous animals are uncommon in the soil community (ten per cent), and they feed on plant roots. They are, however, more abundant in drier soils which lack humus, for example in arid areas, and here they may cause considerable damage. Predators are also scarce (ten per cent), and include some mites and several mammals such as the mole. There are few, if any, producers in the soil, since soil animals are dependent on plant debris from living plants. Macfadyen made a quantitative study of the rate of energy flow through the different groups of organisms in grassland (figure 13.7). The results show that eighty-seven per cent of dead organic material is broken down by bacteria, eight per cent by protozoa and only five per cent by other invertebrates. The final products of decomposition are carbon dioxide, nitrates, ammonia, phosphates, etc., all of which are available to living plants.

Plant debris falling on the ground forms litter, which is then slowly converted to humus in the soil. Soil generally consists of three fairly distinct horizontal layers or soil horizons, superimposed upon one another. The upper, or A, horizon contains organic matter, but may be deficient in colloids and iron since these are leached out and carried away by rain water. The intermediate horizon, horizon B, is enriched by materials such as clay and iron and is a zone of deposition. The lower horizon, horizon C, includes unchanged parent material from the rock beneath. Each horizon may be sub-divided. Horizon A, for example, comprises a layer A_0 which is entirely organic, and a layer A_1 containing a mixture of organic and mineral materials (figure 13.8).

Mull or mild humus is generally slightly acid (pH 5.5–7.0) and well mixed with mineral material. It is formed when the soil contains many earthworms and enchytraeids. Decomposition of plant debris is rapid and, as a result, the horizon A is not well developed. Mull humus has a stable, spongy structure which is due to the crumb structure of the soil and the pellets of soil voided by earthworms. It is formed mainly in deciduous forests. *Mor* or raw humus forms in localities where earthworms are scarce. Litter decomposition is slow and a thick layer of dead plant material accumulates, forming a well marked horizon A_0. Mor humus is acid (pH about 4) and is characteristic of coniferous

313

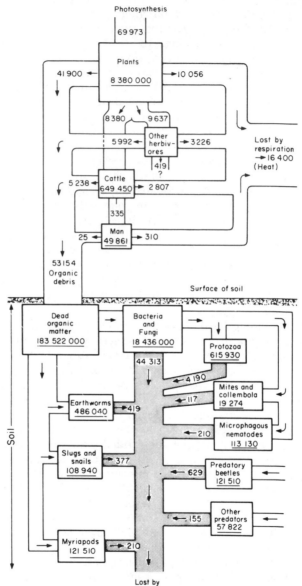

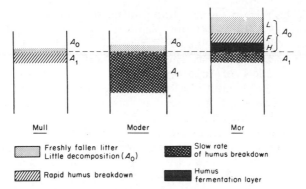

Figure 13.8   The three main types of humus.

Mor type humus shows an upper layer, A_0, which is dense, black and dominated by undecomposed material and intermediate compounds; fungal mycelia are abundant. At the surface there is a litter layer, L, then a fermentation layer, F, where substances like lignin are liberated, and, finally, a humus layer, H. The A_1 horizon is clearly marked, less dense, black and is a zone where mineral and organic particles are mixed.

Mull type humus has a very thin A_0 horizon as litter decomposition is very rapid. The A_0 horizon is brown in colour and formed from an intimate mixture of clay particles and humus rich in calcium.

Moder humus has an A_0 horizon between 2 and 3 cm thick, and the boundary between the A_0 and A_1 horizons is not distinct. The A_1 horizon is about 10 cm thick and consists of a simple mixture of clay and humus, i.e. these components are not formed into organo-mineral complexes as in mull soils. The faecal pellets of arthropods are abundant in moder soils.

forests and heathlands. Faunal differences between mor and mull soils can be clearly seen in the following table, which relates to soil invertebrates of Belgian forests (Nef, 1957).

mull		*mor*	
Biomass	1000 kg/ha	Biomass	300 kg/ha
Earthworms	70%	Enchytraeids	30%
Arthropods	20%	Arthropods	50%
Nematodes	10%	Nematodes	20%

Figure 13.7   Diagram showing energy flow through grassland, separating flow above ground from that below ground. The biomasses of the various components are shown in the rectangles (underlined figures) as kJ of organic matter present. The figures which are not underlined show the energy assimilated (through photosynthesis by plants, as food by other living organisms) and the energy lost in the form of heat as $kJ/m^2/day$. It can be seen that photosynthesis (i.e. primary productivity) corresponds to 69973 $kJ/m^2/day$; cattle use only 8380 kJ, which is about twelve per cent, and man only 335 kJ, less than 0.5 per cent. The 'other herbivore' category includes animals other than cattle, i.e. phytophagous insects, rodents, etc. A total of 53154 kJ reaches the soil, i.e. seventy-six per cent of primary production. Bacteria and fungi liberate eighty-seven per cent of this energy (44313 kJ out of a total 50740 kJ), Protozoa eight per cent (4190 kJ) and other invertebrates only five per cent (after MacFadyen).

315

Moder is the form of humus found in nutrient-poor soils. It consists of a horizon, A_0, formed from plant debris, and a poorly marked horizon, A_1, where the humus and clay mix but do not form stable organo-mineral complexes as they do in mull soils. Moder contains many faecal pellets of arthropods. The fauna consists mainly of oribatid mites, which multiply rapidly (several hundreds of thousands per square metre) and are able to resist drought, together with collembola which prefer moister parts of the soil. Other arthropods of the moder include woodlice, ants, beetles, fly larvae and myriapods. The faeces of these animals form a brown powder, rich in cellulose and lignin, which is not attacked by digestive enzymes. There is intense fungal and bacterial activity in moder soils.

V   ENERGY FLOW AND PRODUCTIVITY IN FOOD CHAINS

Some studies limited to specific food chains have been made in ecosystems.

a Golley (1960) studied energy flow in an old-field community in Michigan. The vegetation consisted of *Poa compressa, Daucus carota, Cirsium arvense* and *Linaria vulgaris.* The rodent *Microtus pennsylvanicus* was the main herbivore, and the carnivore *Mustela rixosa* was the predator on *Microtus.* Golley showed that, for this relatively simple food chain, the ecological efficiency from the radiant energy reaching the plants to the carnivores was only $5 \times 10^{-5}\%$ because of wastage along the food chain. The largest proportion of the energy lost was through respiration, primary producers losing fifteen per cent of the energy assimilated in this way, *Microtus* sixty-eight per cent and *Mustela* ninety-three per cent (figure 13.9).

b   Buechner and Golley (1967) made a study of the kob (*Adenota kob thomasini*) in the toro reserve (Uganda). The kob is the most abundant ungulate in this reserve, which covers 400 km^2, and the population has been estimated at 15 000 individuals with a biomass of 2 174 kg/km^2. Energy flow studies have produced the following data (in kJ/m^2/year) for the population:

> food assimilated: 310.5
> energy flow: 261.5;
> metabolism: 258.1;
> population growth: 3.39.

The growth efficiency is only about one per cent, corresponding to a production of 577 kg of living material per year per km^2.

Comparison of energy flow through populations of three herbivores

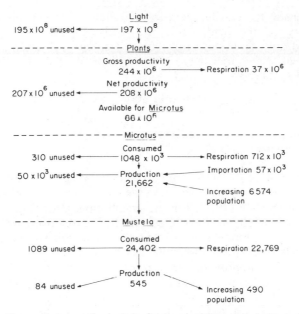

Figure 13.9   Energy flow in a food chain from an old-field community in the south of Michigan. Values in kJ/ha/year (Golley, 1960).

(the African elephant, white-tailed deer *Odocoileus* and the Ugandan kob) and a rodent *Microtus pennsylvanicus* gives the following results in kJ/m² /year:

> *Microtus*: 73.3
> elephant: 97.6
> *Odocoileus*: 180.6
> Ugandan kob: 261.5

The Ugandan kob population shows the highest rate of energy flow known for mammals. This is due to the large size of this species and the density of the population. The highest value known for any animal is for the homopteran bug *Prokelesia* of Georgian salt marshes, being about 1152 kJ/m² /year.

It was estimated that the Ugandan kob consumes ten per cent of primary production, which is a higher proportion than for other mammals: two per cent for *Microtus* and 4.5 per cent for *Odocoileus*.
c   Odum *et al* (1962) studied the energy flow in populations of two seed-eating vertebrates, the finch *Passerculus sandwichensis* and the rodent *Peromyscus polionotus*, and three herbivorous grasshoppers (*Melanoplus femur rubrum, M. bilineatus* and *Oecanthus nigricornis*). These studies were carried out in old-field ecosystems in Georgia. The

317

leaves eaten by the grasshoppers formed eighty-five per cent of the net productivity of the field, while the seeds eaten by the vertebrates represented only seven per cent. The percentage utilisation of available food showed wide variations, however, depending on the diet, as the following figures show:

	AVAILABLE FOOD (kJ/m²/year)	ENERGY USED (kJ/m²/year)	% UTILISATION
Seed-eating vertebrates	210– 420	42–109	10–50
Herbivorous grasshoppers	3350–5030	92–231	2– 7

It is interesting to note that the low percentage utilisation for herbivorous grasshoppers (two to seven per cent) is similar to that obtained by Smalley (1960) for the grasshopper *Orchelinum fidicinum* from a *Spartina* salt marsh in Georgia, the insect using only two per cent of available food. The percentage utilisation is higher for seed eating animals, i.e. between ten and fifty per cent.

These results support the theory of Hairston *et al.* (1960), who considered that herbivore populations were rarely limited by food shortages, unlike those of predators, detritivores and seed-eating animals. As food becomes scarce in winter, it may act as a limiting factor for seed-eaters, since these, although primary consumers, occupy an intermediate position in the ecosystem between carnivores and herbivores.

VI CONCLUSION

Although much research is taking place into the productivity of ecosystems there are many questions still to be answered. They include the following:

(i) Is productivity in ecosystems regulated by man different from that of natural ecosystems? There does not appear to be any difference.

(ii) Can the productivity of ecosystems be increased by improving the efficiency of photosynthesis? Experiments with cultures of the algae *Chlorella* and *Scenedesmus* seem to provide an answer to this question. The highest level of efficiency (ten to fifty per cent) has been obtained in small cultures with low intensity light, when the production of carbohydrates is obviously low. When direct solar illumination and larger culture vessels are used, efficiency falls to between two and six per cent, i.e. hardly greater than in natural ecosystems or intensive agriculture. Also, as the size of cultures increases, it becomes difficult to prevent the development of zooplankton, which then feed on the crop. It would appear that, at present, it is not possible to improve the efficiency of photosynthesis. However, some countries such as Japan will continue to be interested in algal culture, since they are short of arable land.

(iii) Is it possible to increase the efficiency of energy flow from one trophic level to the next?

(iv) Are the trophic structure and energy relationships in recently established ecosystems similar to those in ancient ecosystems, where species have become more closely adapted?

(v) Is it possible to compare ecosystems with high species diversity with those showing a low diversity?

The ideas of Margalef (1963) partly answer these questions. Study of the ratio Q which is equal to

$$\frac{\text{gross primary productivity}}{\text{total biomass}}$$

shows that this ratio becomes smaller as the ecosystem increases in efficiency and as species diversity is increased. The quotient Q represents the rate of renewal or turnover of the ecosystem. For marine plankton

$$Q = \frac{250 \text{ g of carbon/m}^2/\text{year}}{25 \text{ to } 50 \text{ g dry weight/m}^2} = 5 \text{ to } 10.$$

For coral reefs

$$Q = \frac{2500}{800 \text{ to } 900} = 2.6 \text{ to } 3.0$$

The degree of organisation and species diversity in the plankton is considerably less than on the coral reef. A similar comparison shows that Q is less for tropical rain forest than for terrestrial ecosystems in temperate regions.

The seral stages of a succession show a tendency towards progressive reduction in Q; and, in community succession, the better adapted and more efficient species eliminate the more primitive species (cf. p. 000).

The maturity of an ecosystem is a function of its organisation. Mature ecosystems are more complex, better organised and have a low value for Q. Less mature ecosystems may be more easily exploited than others, although exploitation always brings about changes that are equivalent to a reduction in maturity of the system. In the sea, the following series represents increased maturity:

littoral plankton → oceanic plankton → deep-water plankton →
benthos on loose substrate → benthos on solid substrate →
coral reefs.

In the terrestrial environment, soil represents the most mature ecosystem, and the ratio (plant biomass)/(animal biomass) is often

319

higher in terrestrial ecosystems than marine ecosystems. This predominance of plants may be related to the predominance of detritivore over herbivore food chains. In the sea, the circulation of nutrients takes place in the water, while on land it is in the soil. It can be stated in conclusion that, although there are general principles valid for all ecosystems, there also appear to be specific laws relating to each of the major groups of habitats.

Chapter 14

THE MAJOR ECOSYSTEMS OF
THE WORLD

*'In the eyes of the ecologist, nature can be compared to a vast laboratory.
Each location there is an example of a particular equilibrium reached
between a type of vegetation and its environment. These represent experiments that have already been carried out, and this is just as true for
crops as for natural forests. However, the components of the equilibrium
between natural environment and endemic vegetation are complex, and
change from location to location in an uncontrolled manner. In order to
study these components it is necessary to use methods suited to this
variability and capable of explaining it. These presuppose a careful choice
of locations, the description of all the important elements in each location
and, finally, the interpretation of the mass of observations collected. The
relationships established between environment and vegetation will
represent the true laws of ecology for the region investigated'*
(Emberger, 1967).

The communities that are found over the surface of the globe are almost
infinite in number. This chapter will be restricted to a brief description
of those major communities known as biomes.

Techniques for the description of communities are much more
advanced for the plant kingdom. Plants also impose their basic
characteristics on communities and, for this reason, boundaries set up
by botanists are also used by botanists are also used by zoologists.
There is a spatial coincidence between animal and plant associations at
the level of the biome, although this is not always true for smaller
communities, where an animal community may encroach on several
plant communities and *vice versa*.

The map (figure 14.1) shows the geographical distribution of major
plant formations. It will be seen that, with the exception of tundra, the
biomes extend over several bio-geographical regions. In passing from
one region to another, examples are often found of species not
systematically related, but showing close convergence with one another
as a result of almost identical environmental conditions.

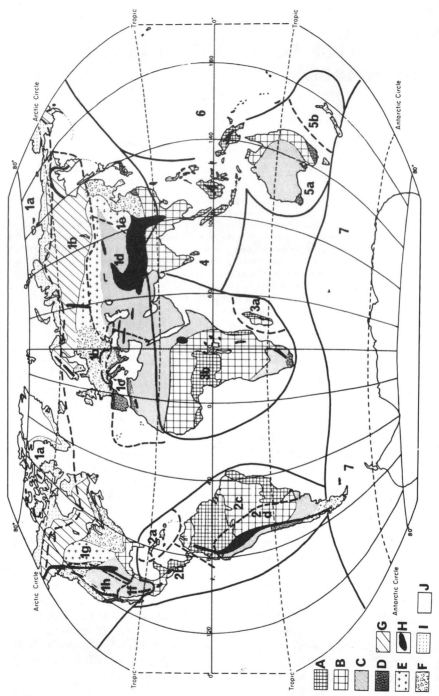

Figure 14.1   Map of the major world biomes and bio-geographical regions.

A: tropical forests   E: steppe and prairie   H: high mountain
B: savanna and open forest   F: temperate forest   I: tundra
C: deserts   G: boreal forest (taiga)   J: snow
D: maquis and chaparral

1. Holarctic region (1a. Arctic sub-region; 1b. Euro-Siberian sub-region; 1c. Mediterranean sub-region; 1d. West and Central Asiatic sub-region; 1e. Siro-Japanese sub-region; 1f. Sonoran sub-region; 1g. Atlantic North American sub-region; 1h. Pacific North American sub-region).
2. Neotropical region (2a. Caribbean sub-region; 2b. Mexican sub-region; 2c. Brazilian sub-region; 2d. Andean sub-region).
3. African region (3a. Madagascar sub-region; 3b. Ethiopian sub-region).
4. Oriental region.
5. Australian region (5a. Australian sub-region; 5b. New Zealand sub-region).
6. Polynesian region.
7. Antarctic region.

323

I TERRESTRIAL BIOMES

1 POLAR REGIONS – TUNDRA

Tundra can be defined as a zone of vegetation situated beyond the natural tree limit. This limit almost coincides with the Arctic Circle (latitude $66°33'$), reaching $72°$ in central Siberia and down to $53°$ in Labrador and Alaska.

The climate is characterized by a frost-free period of less than three months, and a mean temperature of the warmest month not exceeding ten degrees Centigrade, thus limiting the season of plant growth. The ground remains frozen during the brief summer, apart from the upper few centimetres, and the permanently frozen deeper layer, the permafrost, hinders drainage, causing the formation of extensive marshes.

The vegetation along the southern limit of the tundra at the edge of the forests consists of dwarf scrub (*Betula nan* and various Ericaceae) mingled with *Sphagnum* bogs. Further north this is replaced by grassland and moorland with *Carex* and *Eriophorum*, while at the northern limit the plant cover consists of mosses and lichens. In the southern hemisphere the colder and more humid climate results in a tree limit at about $45°$ latitude in Chile and $53°$ in New Zealand. There is no true tundra, but only a plant formation characterized by endemic umbellifers of the genus *Azorella*.

The fauna of the tundra comprises sixty-one species of mammals including eight insectivores (shrews), thirty-four rodents and lagomorphs (squirrels, marmots, beaver, rats, mice, voles, lemmings, hares and muskrat), thirteen carnivores (martens, stoats, weasels, otters, mink, wolverines, brown and polar bear, wolf, coyote, red and arctic fox and lynx) and five ungulates (elk, reindeer, wild sheep, musk ox and caribou). However, only about ten species, including the Arctic fox (*Alopex lagopus*), the reindeer (*Rangifer tarandus*), the caribou (*Rangifer caribou*), the musk ox (*Ovibos moschatus*), three voles and two lemmings, are indigenous tundra animals.

Mammals are adapted in many different ways to tundra life. These are, first, feeding adaptations. The diet is often more specialised than in other regions, and food rich in vitamin A plays a large part in maintaining normal activity. Mammals of the tundra have a remarkable capacity for using unusual foods such as lichens, fungi and even, in the case of the reindeer, small animals. The reindeer is, therefore, an ungulate with a partly carnivorous diet. Many species are able to remain active even when the amount of available food is reduced.

Reproduction similarly shows modifications related to environmental conditions. Arctic mammals generally reproduce before the spring and, therefore, before the air begins to warm up and days begin to lengthen perceptibly. Lemmings can breed throughout the year. Fecundity is increased, the young grow more quickly and these factors affect the population dynamics of rodents, small insectivores and their predators. In May, for example, voles born during the previous winter form up to seventy per cent of the population, while in June ninety per cent of the population consists of young born in the winter and spring. Since some years are more favourable to the survival of the young, there are cyclical oscillations in numbers of rodents, and then of carnivores like the Arctic fox *Alopex lagopus*.

Temperature regulation mechanisms are equally evident in Arctic mammals. The fur provides increased insulation (cf. fig 2.39), and in winter the density of hair increases. The development of thermo-regulatory mechanisms is rapid (at nineteen days in the lemming *Lemnus obensis*). The relative decrease of body surface area (Bergmann's rule) is also important. Hibernation usually occurs in holes dug into the snow, as the frozen soil prevents mammals from making burrows. The reindeer and caribou migrate in winter as far as the coniferous forests, although they can survive temperatures of the order of $-50°C$ to $-60°C$ and can feed solely on lichens (*Cladonia rangiferina*), which are collected by scraping the snow with enlarged hooves.

Birds, which are fairly numerous (about forty species), are mainly migratory species which come to nest in the summer. The most characteristic are resident species like the snow bunting (*Plectrophenax nivalis*), ptarmigan (*Lagopus albus* and *Lagopus lagopus*, the American ptarmigan), snow goose (*Anser caerulescens*), eider (*Somateria mollissima*) and snowy owl (*Nyctea scandiaca*). Less than a third of the species of birds feed on land, the others feeding in lakes or on their shores.

Arctic mammals and birds frequently become white in winter. This has been interpreted as protective coloration against predators, white species blending with their habitat. This seems to be caused by a decrease in pigment formation similar to that found in desert animals.

Reptiles and amphibians are rare, and none are characteristic of polar regions. Insects are numerous and able to tolerate very low temperatures (cf. p. 91). They become active as soon as the thaw sets in, and some species are able to multiply rapidly, especially among Collembola and Diptera, which are the dominant orders. It has been shown that these two orders increase in relative importance at higher latitudes (table 14.1).

Mosquitos (particularly *Aedes*), simulids and tabanids are abundant in July. Bumblebees (*Bombus*) are common, as they can easily feed and build up reserves by taking advantage of the long summer days for collecting food. Hemiptera, Orthoptera, Odonata. Neuroptera and ants are rare or absent. Amongst Lepidoptera, species of the genera *Colias* and *Erebia* are particularly well represented.

TABLE 14.1   Insect populations as a function of latitude

REGIONS	PERCENTAGE OF TOTAL NUMBER OF INSECT SPECIES	
	Diptera	Collembola
Sweden (55–70°N)	22	1
Iceland (64–66°N)	34	5
Greenland (60–80°N)	52	8
Nouvelle-Zemble (70–77°N)	51	3
Spitsbergen (78°N)	56	19
Jan Mayen Island (71°N)	64	24

The large population fluctuations that take place in the tundra can be explained by the simplicity of the ecosystem (cf. Fig. 7.1).

The fauna is even more sparse in the Antarctic, where mammals are absent and birds are mainly represented by penguins, which are more dependent on the water, especially for their food, than on the land.

2   MOUNTAINS

High mountains present very specialised conditions for life, and these have been described on p. 43. A succession of vegetation zones can be seen on mountains, and the altitudinal limits of these zones vary with geographical distribution (fig. 2.16).

The following zones are found in France in the eastern Pyrenees:

0 to 300 m:	lower Mediterranean stage, with cork oak, *Arbutus*-type maquis, tree heathers and *Cistus*.
300 to 500 m:	upper Mediterranean zone, with holm oak.
500 to 1000 m:	sub-Mediterranean zone, with *Quercus pubescens*
1000 to 1600 m:	mountain zone, with beech and fir.
1600 to 2400 m	subalpine zone, with pitch pine.
2400 to 3000 m:	alpine zone above the tree limit.

The formations situated above the tree limit can be regarded as alpine tundra. Above this tundra is the permanent snow zone.

The vegetation shows considerable variation, depending on exposure, soil type and the length of snow cover. Many of the plants have an arctic-alpine distribution, and are found both in the alpine zone and in the arctic tundra. This kind of distribution is also found among animals, especially in insects (beetles *Amara erratica* and *Hypnoidus hyperboreus*, and the butterfly *Colias palaeno*), in birds (ptarmigan) and mammals (blue hare).

Mountain vertebrates are not numerous. Mammals include chamois, yak, marmot and blue hare, and are nearly all herbivores. The birds include ptarmigan and alpine accentor. The few reptiles are viviparous (cf. p. 75). Amongst invertebrates, insects are abundant. Beetles include many species associated with snow, mainly from the family Carabidae. Numbers of collembola are very high (*Isotomurus saltans*, the 'glacier flea'), and butterflies of the genera *Erebia* and *Parnassius* are also numerous. There is a marked frequency of melanic forms among the insects and some vertebrates like the black salamander (cf. p. 113). Many insects are apterous, and this is probably a response to high winds. Adaptations to cold are numerous, and include a marked tolerance to low temperatures, a winter diapause in insects, lower rate of metabolism in the marmot, and burrows and galleries in the snow in many mammals and birds.

Several species can still survive in the Himalayas above 5000 metres. Some insects, bristle-tails and collembola, feed on plant debris deposited by the wind, and these detritivorous insects form the prey of several predators, including salticid spiders and carabid beetles. Several birds, including the corvid *Pyrrhocorax graculus* and a snow goose *Eulabeai indica* and mammals, including the wild goat *Pseudois nahura*, the yak and the rodent *Ochotona ladacensis*, can live at these altitudes.

3   FORESTS

As the result of man's activities, forests have been replaced by grasslands in many places; under natural conditions, forests would occupy a much larger area of the earth's surface than they do at present. The presence of a more or less dense tree cover creates the particular kind of microclimate that was described in chapter 2.

The forest habitat provides a good example of an ecosystem arranged in successive layers, thus allowing the maximum use of solar energy as well as greater diversification of ecological niches. The following layers or strata can be distinguished in a forest: a tree layer, shrub layer, field or herbaceous layer, and a ground or moss layer.

Animal populations vary in the different layers. The soil fauna is very rich, and some of its members move up on to dead tree trunks, helping to break these down into humus. The fauna on the soil surface does not contain many mammals (cf. p. 310 for data relating to temperate forests), this being true for both temperate and tropical forests, and large species are particularly scarce. There is, however, a well developed fauna of insects, mammals and birds in the trees, and these show a number of adaptations to the arboreal life. At the tops of the trees, the canopy is inhabited by a specialised group of species which rarely come down to the ground.

3.1   Temperate deciduous forests

These are well developed in Europe, consisting of oak and beech forest; and in North America, where they consist of maple, oak and beech, but of different species from the European ones.

The fauna is very varied. Arboreal vertebrates include squirrels and dormice amongst the mammals, and tree-creepers and woodpeckers amongst the birds. Terrestrial mammals include red deer, roe deer and wild boar, together with numerous rodents (wood mice) which form the prey of small carnivores like the fox and weasel. Bears are occasionally found, while insectivorous birds and owls are abundant. The niches of birds are separated because of their different vertical distributions, each species feeding and nesting in a fixed zone. Turcek (1951) found, in an oak forest in Czechoslovakia, that fifteen per cent of the bird species nested on the ground, twenty-five per cent in grass and bushes, thirty-one per cent in or on tree trunks, and twenty nine per cent in the foliage. However, fifty-two per cent of the species searched for their food on the ground. The variety of birds seems related, in deciduous forests, to the height and density of the foliage (MacArthur, 1961).

Xylophagous insects are common, and attack in succession either living trees or mainly diseased and dead trees (cf. p. 261, table 11.1). In this succession beetles play an important part, followed by flies and hymenopterans. Large numbers of insects make use of the dead leaf litter on the ground (cf. p. 313). High humidity often allows the development of mosses with a specialised fauna adapted to this habitat (darkness, abundance of decomposing organic matter and constant temperature). Beetles form a large part of this fauna. Lignicolous bracket fungi (*Polyporus, Pleurotus*) which attack dead wood harbour an extensive fauna of beetles and flies.

There is a marked seasonal cycle in temperate forests. In winter invertebrates find shelter in the soil and litter, many moving during summer into field and shrub layers.

3.2 Coniferous forests: the taïga

The taïga forms a belt of forest bordering the southern edge of the tundra in the northern hemisphere. It consists entirely of conifers: pine, fir spruce and larch. The climate of the taïga is cold with long winters, and it has a restricted fauna. Large mammals are represented by deer, the wapiti, *Cervus canadensis* and elk or moose. They feed on buds, bark and lichens. Carnivores include the bear, wolf, fox, wolverine, marten and mink. Fur mammals are numerous and are trapped. There are a few arboreal species, and they include several squirrels and the Canadian porcupine *Erethizon dorsatus*, which climbs trees and then strips the bark from them. The few resident birds are mainly seed eaters (capercaillie, crossbill), many other birds migrating southwards for the winter. Population fluctuations are particularly marked in the taïga.

Amongst invertebrates, xylophagous insects are abundant. The caterpillar of *Choristoneura fumiferana* attacks buds of *Abies balsamea* in Canada, where it causes considerable damage.

3.3   Mediterranean forest

This is formed of evergreen trees, including holm oak and cork oak, *Arbutus*, various pines (Aleppo pine, stone pine) and cedar of Lebanon. Much of this Mediterranean forest has been destroyed by felling. The fauna is rather different from that of temperate forests. The eyed and green lizards and colubrids are common, but vipers are rare. Birds and mammals are less numerous, the red and roe deer being absent. Insects, on the other hand, are well represented, some families like the buprestid and cetonid beetles and some orthopterans taking on an importance which is more indicative of tropical forests. Flies are scarce because of the dry conditions, and are replaced by a rich endogenous fauna, including many beetles restricted to the Mediterranean region, having been eliminated from northern regions by glaciations.

3.4   Tropical forests

This type of forest is particularly well represented in the Amazon basin, in tropical Africa and in Indo-Malaysia. Temperature is uniform and humidity high. Many species of trees are present (600 species on the Côte-d'Ivoire, 2000 in Malaysia), represented by widely spaced individuals. The ground vegetation is sparse, formed from shade-loving species like ferns. Lianes or woody climbers and epiphytes are numerous.

The fauna is abundant and varied as a result of the great antiquity of these communities (which have existed since the tertiary era without change in climate), abundance of food resources, and the great diversity of habitats, which allows the formation of numerous ecological niches (cf. p. 250). Many of the animals are arboreal species which rarely come down to the ground. Among mammals are monkeys, lemurs, rodents (squirrels and arboreal porcupines), some anteaters, sloths, opossums, and carnivores (jaguar and sun bear). The birds are often brightly coloured (toucans, parrots), and many are fruit-eating species. Many of the reptiles and amphibians are arboreal forms, and include boas, pythons, chamaeleons and frogs. Many arboreal mammals have powerful claws like the sloths, or prehensile tails (platyrrhine monkeys, opossums), or even a gliding membrane along the side of the body between the anterior and posterior limbs (flying squirrel, flying lemur). These membranes are also found in reptiles (the flying dragon) and in amphibians (the flying frog).

Mammals living on the ground include antelope of the genus *Cephalophus* okapi, peccaries, agouti and dwarf hippopotamus. There is convergence in form between the mammal fauna of African tropical forests and that of American forests providing evidence of similar environmental conditions. Some African species of *Cephalophus* have, as their equivalents, two species of American deer of the genus *Mazama*;

the African antelopes *Neotragus batesi* and *Cephalophus monticola*
resemble the American rodents *Dasyprocta* and *Myoprocta*; the tragulid
Hyemoschus aquaticus resembles the American rodent *Cuniculus paca*,
and finally, *Hippopotamus amphibius* resembles *Hydrochoerus hydro-
chaeris*, an aquatic American rodent.

The invertebrates are often large and coloured. The genus *Achatina*
are African snails that may reach one kilogramme in weight. Scarabid
beetles *Goliathus* (Africa) and *Dynastes* (America) are among the
largest insects known. Butterflies like *Ornithoptera* reach thirty
centimetres in wing span. Ants and termites have an important role on
the ground because of their abundance. There are many species of
mosquitos, each one living at a fixed height (fig. 2.35).

The tropical forest fauna is mainly active at night. There is not, in
general, a seasonal cycle, although some species reproduce during the
season when the rains are least abundant.

4   GRASSLAND FORMATIONS: STEPPE AND SAVANNA

There are two main types of grassland formations: steppe, which is
characterized by the predominance of grasses adapted to dry conditions, and savanna, a mixed tropical formation of tall grasses with
scattered trees and shrubs.

4.1   Steppe

Steppe develops in areas where the climate includes periods of prolonged drought. In the Old World they form a continuous band from
the Ukraine as far as Mongolia. In North America, they form the
prairie and in South America, the pampas. In Russia the steppe
develops on a particular soil type, the chernozem or black earth, which
may reach 1.50 m in depth. The characteristic grasses are mainly of the
genera *Stipa, Koeleria* and *Festuca*, having well developed roots which
allow them to obtain water from deep in the soil. In North America,
the relatively well watered tall grass prairie is coverd by *Andropogon*,
which reaches a height of two metres, while the short grass prairie in
drier regions is characterized by *Buchloe dactyloides* (buffalo grass) and
Bouteloua gracilis.

The fauna is rich in large herbivores living in herds, and includes
gazelle, saïga antelope and wild horse in the Old World, and buffalo and
pronghorn antelope in North America. Burrowing mammals, mainly
rodents, are common, and include the hamster and bobak marmot in
Eurasia, and prairie dog in America. Predators include the wolf in
Eurasia and the coyote and foxes in America. The birds of the steppes
are usually ground-living forms. Amongst the insects, migratory locusts
are the most noteworthy. Coprophagous beetles are numerous, feeding
on mammal faeces.

4.2   Savanna

Savanna occurs in tropical regions in Africa, Asia, America and Australia. The herbaceous or field layer consists of grasses of the genera *Andropogon, Pennisetum* and *Imperata*. The scrub layer is represented by *Acacia*, baobab and palms in Africa, cacti in America and *Eucalyptus* in Australia

The fauna of the savanna consists mainly of large herbivores (antelope, gazelle, zebra, giraffe, elephant, rhinoceros, etc.) and carnivores (lion, leopard and cheetah) in Africa. Flightless birds include the ostrich in Africa, rhea in America and emu in Australia. Monkeys and birds occur in the trees, while the dominant insects on the ground are termites, ants and locusts.

5   DESERTS

Deserts (arid zones) have been defined in chapter two. The vegetation is sparse, and consists partly of annual plants which grow rapidly, flower and fruit after the rare wet spells, and partly of perennial drought-resistant plants. There are few large vertebrates in the fauna, and these are mainly antelopes like the addax of the Sahara. Rodents are fairly abundant, and lead a subterranean life. The birds are largely flightless species. Tenebrionid beetles are the most important insects, while mantids of the genus *Eremiaphila* are found in the more arid regions of the Sahara.

Adaptations of the fauna are numerous, and are concerned with protection against drought and heat. Some have already been examined in chapter two. The main adaptations are the following. Firstly, the prevention of desiccation by the reduction of sweating and urine production, the use of metabolic water, the limited consumption of free water and the choice of foods with a high water content. Secondly, protection against heat and exposure to the sun, either by seeking shade, by leading a nocturnal or underground life, or by restricting activity to cooler periods. Adaptations for running and jumping are common, allowing for the increased activity necessary to compensate for the scarcity of food and other individuals. The tympanic bulla of some rodents is hypertrophied, and this is thought to provide an improved method for hearing individuals of the same species which are often some distance away. Animals are often pale coloured and so homochromous with their habitat. This homochromy is due to an arrest in oxidation of melanin precursors, and does not appear to have any protective value. At Hoggar, the bird *Ammomanes deserti* is represented on dark soils by a dark form and on lighter soils by a light form. In the Sahara, there are thirty-nine homochromous and only eleven non-homochromous mammals, but only eighteen out of forty-seven species of birds are homochromous (Heim De Balzac, 1936).

II   MARINE BIOMES

Only a brief summary can be given here because of the extent of the subject.

1   GENERAL FEATURES

The seas and oceans cover 363 million km^2 of the earth's surface, more than twice the area of dry land. The mean depth of the oceans is about 3800 metres, while the mean altitude of the continents is only 875 metres (fig. 2.15). The continents are only inhabited at or near the soil surface, while life is found at all depths in the seas, even in the deepest trench known (Mariana Trench, 11 034 m).

1.1   Abiotic factors

The main abiotic factors are the following:

(i) *Hydrostatic pressure* increases by approximately one bar for every ten metres in depth. Pressure variations are therefore much more important in the marine environment than on land (fig. 2.15).

(ii) *Light penetration* decreases rapidly, and euphotic and aphotic zones can be distinguished (cf. p. 64).

(iii) *Temperature* is characterised by thermal stratification, with a seasonal thermocline near the surface and permanent thermocline in the depths.

(iv) *The dissolved salts, oxygen and carbon dioxide contents* of the water are very important environmental factors (cf. p. 123).

1.2   Major ecological groupings of marine animals

Marine organisms show a wide range of life forms, and these may be divided into three main groups.

a   Benthos
This is composed partly of organisms that are fixed to the substrate (sessile benthos), and partly of mobile organisms (mobile benthos) whose movement is restricted to their immediate vicinity. Sessile benthos is made up of plants (algae and, very occasionally, higher plants like *Zostera* and *Posidonia*) and a wide range of animals, including cnidarians, bryozoans and protochordates. Mobile benthos includes crustaceans, fish and echinoderms. Also included in the benthos are animals that burrow in mud and sand, like bivalves and holothurians, together with micro-organisms from the mud.

b   Plankton
These are floating organisms whose movements are entirely dependent on water currents. They are, however, capable of vertical movements.

Plankton is made up of plants and animals. The plants consist of unicellular algae including diatoms, dinoflagellates, coccolithophores, silicoflagellates and occasionally Xanthophyceae and Myxophyceae. The animals can be divided into two groups, temporary plankton or meroplankton (eggs and larvae of benthic and nectonic species), and permanent plankton or holoplankton (foraminiferans, some coelenterates including siphonophores and medusae, rotifers, chaetognaths, crustacea and appendicularians). The meroplankton includes larvae of polychaetes, molluscs and echinoderms, and also fish fry. The components of the plankton may be classified according to their size. Ultraplankton consists of organisms less than five μm in size (bacteria and some small flagellates), and nannoplankton of unicellular organisms between five and fifty μm in size (bacteria and other protozoa). Ultraplankton and nannoplankton are too small to be retained by even the finest net. Microplankton measures between fifty μm and one millimetre, and forms the major part of the holoplankton. Mesoplankton varies between one and five millimetres, and macroplankton contains those organisms larger than five millimetres. The mesoplankton and macroplankton contain various crustaceans, molluscs, jellyfish and siphonophores.

c   Necton
These are swimming organisms living in open water and able to resist the movement of water currents. Necton consists mainly of pelagic fishes, marine mammals, cephalopods and various decapod crustacea.

d   Less important categories
The *neuston* describes those animals that move over the surface of the water. The marine neuston is almost restricted to Hemiptera of the genus *Halobates*. The *pleiston* contains siphonophores like *Physalia* and *Velella* which have some form of sail projecting from the water to catch the wind. The term *seston* is used to describe all the organic matter, dead and alive, that floats in the water. Plankton forms the living component and the dead component is the *tripton*.

2   THE MAJOR SUBDIVISIONS OF THE MARINE COMMUNITY

Three main divisions can be distinguished, the pelagic, open water and benthic regions. Each of these regions can be sub-divided vertically into different zones according to depth. The sea bed can be divided into:

(i)   the *continental shelf*, more or less following the land masses, with a gradual slope (0.5%) down to a mean depth of about 200 metres;

(ii)   the *continental slope*, with mean slope about 5%;

(iii)   the *abyssal plain*, going down to about 6000 metres and forming the larger part of the sea bed;

(iv)   the *hadal zone*, or ultra-abyssal, corresponding to the very deep areas.

The relative areas of these different regions is shown below:

Continental shelf	(0–200 m)	7.6%
Continental slope	(200–2000 m)	8.1%
Abyssal plain	(2000–6000 m)	82.2%
Hadal zone	(more than 6000 m)	2.1%

The term *neritic zone* is applied to the shallow, turbulent waters covering the continental shelf. These waters are rich in dissolved nutrients and suspended material and have a high productivity. The term *oceanic zone* is applied to the remaining waters with depths greater than 200 metres.

2.1 Subdivisions of the benthic region

The boundaries used here are those of Peres and Picard (1949), who defined the zones on the basis of ecological criteria and not simply on changes in depth.

The *supralittoral zone* is above the level of spring tides and is reached only by spray. The *mediolittoral zone* is situated within the tidal range, and organisms living there are subjected to regular cycles of immersion and emersion. The lower limit of the *infralittoral zone* is the point at which there is just sufficient light to allow the development of Zosteraceae and those algae with more exacting light requirements. This limit is situated at about fifteen metres in higher latitudes, thirty to forty metres in the Mediterranean and eighty metres in tropical regions. The *circalittoral zone* extends as far as the limit of algal growth at about 200 metres depth, i.e. the edge of the continental shelf. The group of four zones described above forms the *phytal system* characterised by the presence of benthic green plants.

The *bathyal zone* occupies the continental slope and the less steep areas at its base, down to a depth of about 2000 to 3000 metres depending on the criteria used. According to Bruun *et al.* (1956), the lower limit corresponds to the four degree Centigrade isotherm, which varies with latitude. It has also been suggested that the lower limit corresponds to the significant change in fauna which occurs a little below 3000 metres. The *abyssal zone* corresponds to the large abyssal plain extending to depths of 6000 and 7000 metres. The fauna is impoverished by the disappearance of nearly all the eurybathic species of the continental shelf. The *hadal zone* occupies the very deep trenches beyond 6000 to 7000 metres depth. It is characterised by the paucity of its fauna, from which a number of major systematic groups are absent (sponges, asteroids, ophiuroids and decapods), by several endemic groups and by barophilic bacteria adapted to high pressures. The group of bathyl, abyssal and hadal zones forms the *aphytal system.*

2.2 Subdivisions of the pelagic region

A similar system to that used for the benthic region is used here to subdivide the pelagic region.

The *epipelagic* or euphotic zone is the upper light zone. It is restricted to between fifty and one hundred metres in depth, according to the region, by a compensation zone where photosynthesis is just balanced by respiration. The *mesopelagic zone* from 50–100 metres to 200 metres, is faintly illuminated, but not sufficiently for phytoplankton to survive. Seasonal temperature changes are still felt in this zone. The *infrapelagic zone*, between 200 and 500–600 metres, is rich in species as it contains both species that move down during the day from zones above and species that move up at night from zones below it. This zone is not affected by seasonal changes in temperature and copepods are dominant here. The *bathypelagic zone*, from 500 to 2000 metres, contains copepods, jelly-fish of the genera *Atolla* and *Crossota,* amphipods, decapods and the nemertine *Pelagonemertes.* The lower limit of this zone corresponds almost to the four degree Centigrade isotherm in intermediate latitudes. The *abyssopelagic zone*, between 2000 and 6000 metres, is dominated not by copepods but by chaetognaths and crustaceans (mysids and decapods). The *hadopelagic zone*, which is faunistically poor, contains amphipods, ostracods and copepods.

3 MAJOR ECOLOGICAL GROUPINGS: THE BENTHOS

3.1 General

The total biomass of benthic organisms decreases rapidly with depth. Some estimates on a world scale were given on p. 300. At the local level this decrease is also clearly shown. It has been shown for rocky shores in California occupied by the large brown alga *Macrocystis pyrifera*, that the biomass at 1.5 m depth was 4792 g/m^2 (4667 for algae and 125 for animals), while at 22 m depth it was only 983 g/m^2 (606 for algae and 377 for animals).

Faunal diversity in shallow waters means that rocky substrates become almost entirely covered by organisms. The resulting competition for food has produced a wide range of feeding mechanisms and diets: phytophagous algal feeders (gastropods, crustaceans and fish), suspended particle feeders (bryozoans, polychaetes, ascidians and bivalves), burrowers, mud-feeders, etc.

Benthic animals may be either sessile, i.e. fixed to the surface of the substrate, like sponges, bryozoans, ascidians and barnacles, or 'rooted' in soft substrates, for example some polychaetes and cnidarians. Other animals are free living but remain on the substrate, and these include gastropods, echinoderms and decapods. Holothurians and molluscs burrow into soft substrates, while animals like *Teredo, Pholas* and sponges of the genus *Cliona* bore into harder substrates.

3.2 Description of benthic communities

a Supralittoral zone

The supralittoral zone of rocky shores is covered by organisms of which the most characteristic are black lichens of the genus *Verrucaria*. The fauna consists of gastropods (littorinids) and isopods like *Ligia*, which can withstand prolonged exposure. On sand beaches, amphipod crustaceans like *Talitrus* and *Orchestia* are dominant. They are found together with a number of insects, including carabid and tenebrionid beetles and flies.

b Mediolittoral zone

The mediolittoral zone, on rocky shores of seas with an appreciable tidal range, is covered by superimposed horizontal bands of brown algae, forming clearly visible zones on sheltered shores where the wave action is not too severe. On very exposed shores the algae are less important (fig. 14.2). Many invertebrates, including gastropods, hydrozoans, polychaetes and bryozoans, are found in the algal zone. The

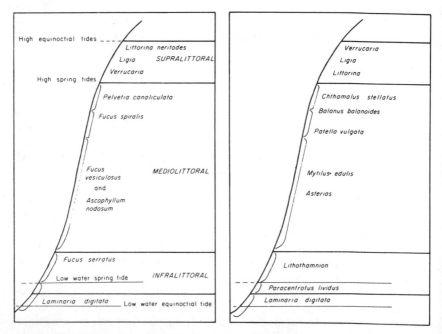

Figure 14.2   Communities of rocky shores of the English Channel. Left: sheltered shore; Right: exposed shore.

lower part of the mediolittoral zone is often occupied by populations of mussels and oysters with a biomass reaching up to 80 kg/m^2, which form the prey of asteroids and carnivorous gastropods (*Nucella, Murex*).

On soft substrates, the fauna consists partly of large species living on the surface and rarely burrowing into the substrate (echinoderms, crustaceans and fish), and partly of psammophilous or sand-dwelling species which live permanently in the sand. The species of the endopsammon burrow in the sand, constructing burrows and galleries, while those of the mesopsammon live in the interstices between sand grains. The latter form a marine interstitial fauna made up of small (one millimetre or less) elongated forms which are able to creep among the sand grains (cf. p. 134). The components of this interstitial fauna are mainly ciliates, gastrotrichs, annelids, nematodes and crustaceans (Delamare Deboutteville, 1961).

In non-tidal seas, the upper mediolittoral zone is inhabited by barnacles (*Balanus, Chthamalus*), and the lower mediolittoral zone by red algae of the Melobesieae. The latter often forms a cornice, the 'pavement' formed by the calcareous alga *Lithophyllum tortuosum* on the Mediterranean coast (Provence, Albères). This is inhabited by both marine invertebrates (e.g. crabs) and others of terrestrial origin (e.g. the spider *Desidiopsis* and collembola).

c Infralittoral zone

This is occupied by animals that remain continually immersed except at the equinoctial tides. Rocky shores are colonised by either photophilic algae or reef building, madreporarian corals. Photophilic algae are mainly brown algae, for example the laminarians of the coasts of Brittany. The corals give rise to coral reefs with a great variety of species, large biomass and high productivity. They only develop in warm waters that are clean, well oxygenated and with a high salinity. The structure of coral reefs has been described on p. 302.

Swards of the angiosperms *Zostera* and *Posidonia* develop on soft substrates. These plants form a new habitat, occupied by numerous animals living on or in the leaves and rhizomes.

d Circalittoral zone

The circalittoral zone is occupied by red calcareous algae of the genus *Lithothamnion*. In the Mediterranean, this zone is characterized by coralligenous formations consisting of encrusting algae, forming about eighty per cent of the total calcareous mass, together with an extensive fauna of sponges, gorgonians, alcyonarians, madreporarians, polychaetes and bryozoans which live mainly in cavities in the surface of the algae. The abundance of the red coral, *Coralium rubrum*, in this formation gives it the name coralligenous, but it is in no way related to coral reefs.

Lithothamnion also grows over loose substrates formed from detritus of gravel and coarse sand, and in Brittany these form a marl used as fertiliser. The fauna here consists largely of bivalves and echinoderms, while on muddy substrates polychaetes are dominant.

e Bathyal zone

In the North Atlantic this is occupied by a community formed from large white corals, mainly madreporarians of the genera *Amphelia* and *Lophohelia*, associated with many other cnidarians and with echinoderms, polychaetes, crustaceans, fish, etc. The muds of this zone are colonised by small bivalves (*Yoldia, Nucula, Chlamys*), scaphopods, sponges and various decapods and echinoderms.

f Abyssal zone

Rocky substrates are very sparsely populated here. On soft substrates (abyssal muds), the fauna is more diverse, and includes holothurians, asteroids, bivalves and polychaetes.

g Hadal zone

This is characterized by holothurians, anemones, polychaetes, molluscs, and a few isopod and amphipod crustaceans. The Pognophora reach their maximum abundance in the hadal zone, and barophilic bacteria are also characteristic (Bruun, 1956).

The benthos of the great depths shows a number of characteristic features. The gigantism of some species has been mentioned already (p. 83). There is a marked reduction in supporting structures. Carnivores are uncommon, as the main food resources are bacteria and organic debris. The biomass of the benthos at great depths is very small, being only 0.1 g/m^2 at a depth of about 5000 m. There are, however, more favourable areas, especially near the continents.

4 MAJOR ECOLOGICAL GROUPINGS: THE PLANKTON

4.1 The buoyancy problem

This has been discussed in chapter 3 (p. 119).

4.2 The problem of feeding

Planktonic animals are either filter feeders (like the Appendicularia, the crustacean Euphausiacea and many copepods) collecting microscopic organisms and debris in suspension in the water, or predators like most Cnidaria and some annelids.

4.3 . Vertical distribution of the plankton

Phytoplankton is clearly restricted to the upper light zone. Zooplankton, however, is more widely distributed and reaches the greatest depths, but its biomass decreases rapidly (fig. 14.3). Similar large

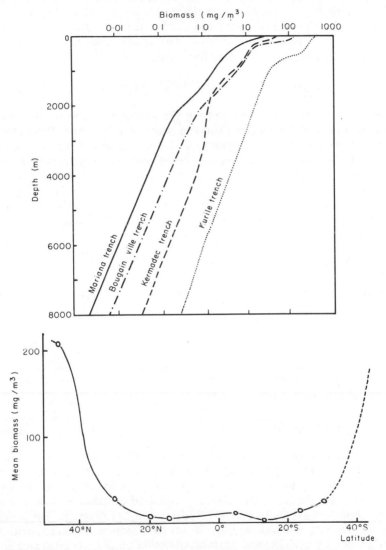

Figure 14.3   Above: the biomass of plankton in mg/m³ in relation to depth for four Pacific trenches.
Below: the mean biomass in mg/m³ for the 0 to 4000 m zone in relation to latitude in the Pacific Ocean (after Vinogradov).

variations in biomass can also be observed in relation to latitude. Well developed diurnal vertical migrations take place, and these may cover several hundreds of metres and show several modes. In general, however, animals move upwards during the day and descend at night (fig. 2.51). These migrations are partly due to changes in light intensity. Vinogradov (1962) showed that these vertical migrations at different levels, according to the species, make up a series of relays that gradually transport nutrients down into the depths for the abyssal fauna. This explains why, as depth increases, the relative importance of filter feeders decreases, and that of predators increases.

5   MAJOR ECOLOGICAL GROUPINGS: THE NECTON

Fish play an important role among the animals of the necton. Some species feed directly on the plankton, thus forming short food chains. This is the case for many clupeids (sardines, anchovies). The herring has a diet of plankton when young, and larger prey as it becomes adult. The whalebone whales are also plankton feeders. Food is collected by filtration, using branched gill-rakers in fish and baleen in whales. A single rorqual can consume one tonne of plankton (mainly euphausids) per day. Other pelagic fish like the tunny are predators, and so are seals and toothed whales.

III   FRESH WATER

The physico-chemical properties of lakes and rivers have been described in chapters 2 and 3.

1   RUNNING WATERS

The main environmental factors acting on animals and plants are current speed, form of substrate, temperature, oxygenation and chemical composition of the water.

1.1   The inhabitants of springs

Spring waters provide a stable environment where temperature is fairly uniform, and are usually occupied by stenothermal species. Cold springs support a sparse flora (algae, mosses like *Fontinalis*), and their fauna includes flatworms (*Planaria alpina*), amphipod crustaceans (*Gammarus*) and isopods (*Asellus*), several Hydracarina, phryganeid larvae and beetles (*Helmis, Riolus*). Animals of subterranean waters are sometimes carried out by springs, and examples include amphipod crustaceans of the genus *Niphargus*. Another type of spring is the thermal spring, which harbours stenothermal animals restricted to high temperatures (cf. p. 71).

1.2 The inhabitants of streams and rivers

Following the hydrobiologists of central Europe, it is possible to distinguish in watercourses, depending on their slope and width of channel, several zones, each characterized by a dominant fish. This scheme is somewhat arbitrary and does not always correspond to reality, but forms a convenient summary.

(i) *Mountain torrents and brooklets* correspond to the *trout zone*. The waters are turbulent and rich in oxygen. Plankton is absent, but benthos is abundant. It consists of forms fixed to stones (green algae of the genus *Cladophora*, red algae of the genus *Lemanea*, mosses, sponges and bryozoans), and forms that creep on the substrate (flatworms and molluscs like *Ancylus fluviatilis*). Insects are mainly Ephemeroptera, Plecoptera, Trichoptera, some Coleoptera and Diptera (simulids and blepharocerids). These animals show many modifications to withstand the current (cf. p. 122), and negative rheotropism is common. The fish of this zone are strong swimmers and have bodies that are rounded in cross section. They include the trout, bullhead and minnow.

(ii) *The zone of the grayling (Thymallus thymallus)* follows the trout zone. It corresponds to the zone where the channel of the river begins to enlarge, and where the bottom is covered with sand and gravel. In the south of France the grayling is replaced by *Telestes soufia*, which has the same requirements. The chub (*Leuciscus cephalus*) also occupies this zone. The fauna consists of a number of species that are less stenothermal than those of the previous zone. These include Ephemeroptera, Trichoptera and Hydracarina. These animals do not possess organs of attachment.

(iii) *The barbel and Chondrostoma nasus zone* corresponds to the slow flowing rivers of the plain. Vegetation is more abundant, with numerous flowering plants along the banks. In the muddy substrate are found bivalves, including *Unio, Anodonta* and *Pisidium*, oligochaetes and chironomid larvae. An extensive plankton (potamoplankton) may develop in the larger rivers, and includes diatoms, rotifers and copepods.

(iv) *Estuaries* show a progressive increase in salinity and the appearance of euryhaline marine species. These include the crab *Carcinus maenas*, bivalve *Cardium edule* and polychaete *Nereis diversicolor*, while fish become more numerous. (Many authors include an extra zone, *the zone of the bream*, before the estuary. Macan and Worthington (1972) outline the British classification, which differs from that used on the continent.)

A specialised environment is produced when the thickness of the water film is reduced to only two or three millimetres, and when this film is continually renewed. A characteristic fauna, the *madicolous* fauna (Vaillant, 1956), develops in this film, living either on rocky substrates or on mosses and liverworts attached to them. It consists of some ubiquitous species and some specialised species or eumadicoles, characterized by their small size, flattened form, less numerous gills in species with aquatic respiration, and dorsal spiracles in air-breathing insects. This fauna consists mainly of Diptera (psychodids, stratiomyids), some beetles and caddis larvae of the genus *Stactobia*.

2   STILL WATERS: LAKE COMMUNITIES

The physico-chemical features of the lacustrine environment have been described on page 66. It is possible to distinguish several zones as the depth increases from the shore.

(i) *The littoral zone*, which is the most shallow, contains many plants. Eutrophic lakes have a rich flora with successive zones of reeds (*Phragmites*), *Scirpus*, water lilies and pond weeds (*Potamogeton*) to a depth of a metre or more, and then a zone of stoneworts (Charales) to a depth of several metres. In oligotrophic lakes, vegetation is less abundant, and includes *Isoetes, Littorella* and stoneworts. The fauna of the littoral zone is rich and varied, including crustaceans (gammarids and *Asellus*), annelids and many insects. The psammon or interstitial fauna of the sand includes protozoa, rotifers, tardigrades, nematodes and copepods, all with similar adaptations to those that live in marine sands.

(ii) *The sublittoral zone* is one of transition, and extends to a depth of about thirty metres. It is mainly occupied by bivalves and larvae of chironomid flies, the latter becoming increasingly important.

(iii) *The profundal zone* occurs only in larger lakes, Lakes Geneva, Ohrid (Yugoslavia), Baikal and Tanganyika being among the more well known ones. Plants are absent in the profundal zone. Well-oxygenated oligotrophic lakes contain chironomid larvae of the genus *Tanytarsus*, which has a high oxygen requirement. Eutrophic lakes which are poorer in oxygen contain the chironomid genera *Chironomus* and *Tendipes*, which are less exacting and can live under semi-anaerobic conditions (cf. p. 125). The benthos of the profundal zone also includes crustaceans (*Asellus*) and oligochaetes (*Tubifex·*).

Lake plankton, which is less diverse than marine plankton, consists mainly of unicellular and filamentous algae, protozoa, rotifers, and crustaceans including Cladocera, Copepoda and Ostracoda. Vertical migrations similar to those in marine plankton take place. The neuston is better developed than in the marine environment, and insects are well represented. The characteristic fish of lake necton in Europe are salmonids, including the char *Salvelinus alpinus* and various species of the genus *Coregonus*, the latter being plankton feeders. Species like the perch, gudgeon, minnow, carp and tench live in the littoral zone.

Chapter 15

ECOLOGY, EVOLUTION AND ADAPTATION

One of the fundamental aims of ecology is to explain the interaction between the environment and living organisms, and to show how the latter are adapted to their mode of life. Only the capacity of living organisms for adaptation to changing environmental conditions can ensure the survival of a species and allow for the extension of its geographical range and evolution.

I THE ORIGIN AND EVOLUTION OF LIVING ORGANISMS

The environmental conditions on this planet many millions of years ago were very different to those which we know today, and it was under these circumstances that life first originated. Living organisms have gradually changed their own environment in the course of their progressive evolution from that time. Berkner and Marshall (1964) have put forward a theory that explains these changes and suggests that evolution through geological time can be seen as the result of a series of interactions between the oxygen produced by living organisms and the organisms themselves. Critical periods arising from significant variations in environmental conditions were immediately followed by adaptive responses on the part of living organisms.

The earth is about four and a half thousand million years old. There was, at first, a 'pneumatosphere' before the hydrosphere and atmosphere had become separated. This separation was made about 3.8 thousand million years ago. The primitive atmosphere was a reducing one, containing hydrogen, ammonia, water vapour, methane and carbon dioxide. The absence of oxygen allowed the entry of ultra-violet radiation. These conditions favour the synthesis of substances such as amino acids and their bases (e.g. pyrimidine), which are the basic constituents of living matter. Some 'abiogenic' organic material accumulating in suitable areas of the primitive oceans, for example on the shores, on contact with catalysts, such as clays, could give rise to more complex molecules like enzymes and, above all, chlorophyll. These events would have taken place about 2.7 thousand million years ago, when traces of oxygen were probably already present as the result of the photodissociation of water vapour by ultra-violet radiation. With the formation of chlorophyll, photosynthesis would begin gradually to enrich the atmosphere with oxygen. At this time ultra-violet radiation could easily penetrate the atmosphere and reach the surface of the earth; life would, therefore, be confined to shallow waters, and known fossils are rare. The latter include some schizophytes like *Eobacterium* from South Africa and *Guntflintia* from Ontario. In the upper atmosphere oxygen was gradually converted to ozone, forming a layer which gradually became so dense that it absorbed most of the short-wave ultra-violet radiation (240 to 290 nm) which is harmful to life. Living organisms began to spread slowly through the oceans and to become more varied. The first known eucaryote cells date from about 1.2 thousand million years ago. When the oxygen content had increased to one per cent of the present level in the atmosphere, the 'Pasteur effect' became evident. This is the process by which bacteria, which may be living either as aerobes or anaerobes, become aerobic whenever the oxygen level is sufficiently high. Aerobic respiration now becomes

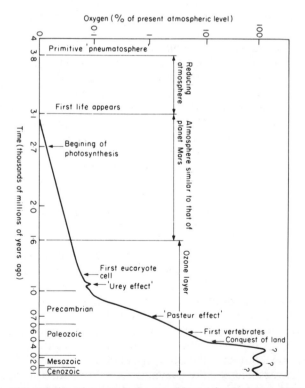

Figure 15.1   The evolution of oxygen in the atmosphere (after Berkner and Marshall, 1964).

344

the main metabolic process. This first critical level was accompanied by a rapid spread of life throughout the oceans. At the time that this occurred, during the Pre-cambrian 0.7 thousand million years ago, the Cnidaria and coelomates were evolving rapidly. A second critical level was reached when the oxygen content of the atmosphere rose to about ten per cent of its present level. The ozone layer was by then completely formed, and life began to develop on land, since ultra-violet radiation could no longer reach the earth. This stage was reached during the Ordovician period. Evolution then became much more rapid as a result of the abundant plant life, the present level of oxygen being reached in the Carboniferous period. Since that time it has undergone only slight variations (figure 15.1).

II THE ROLE OF COMPETITION IN EVOLUTION

One remarkable feature of evolution is the replacement of entire animal groups, in the course of time, by others which occupy the same ecological niche. The disappearance of many reptile groups in the Mesozoic era and their replacement by mammals provides an example of this phenomenon. The cryptograms were replaced in a similar way

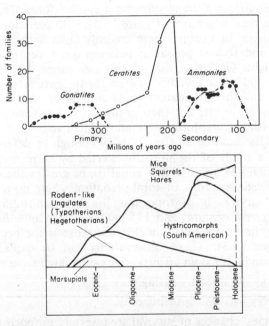

Figure 15.2   Examples of faunal succession in geological time.
   Above: The evolution of the three main groups of Ammonoidea. The successive appearance of each of the three groups is the result of competition.
   Below: The succession of rodent-type herbivores in the South American fauna.

345

by the flowering plants. The surviving groups are those which have avoided competition. The evolution of the Ammonoidea (Cephalopoda) during the Paleozoic and Mesozoic eras is a good example. *Goniatites* appeared first, and their extinction corresponds with the development of *Ceratites*, which reached a maximum at the end of the Paleozoic era and then disappeared abruptly, leaving in their place the *Ammonites* of the Mesozoic era. The invasion of South America by mammals shows the same sequences of replacement due to competition. In the case of small rodent-like herbivores which occupy clearly defined niches, this position was held during the Eocene by marsupials of the family Polydolopidae. They were replaced by eutherian ungulates similar to rodents, particularly *Typotheria* and *Hegetotherium*, which were in turn succeeded by hystricomorph rodents. The latter faced strong competition from invading species from North America (squirrels, mice and rabbits) at the end of the Pliocene, when a land bridge was established between North and South America (figure 15.2).

III THE EFFECT OF PREDATION ON SIZE OF MAMMALS

Studies on the relationship between the size of mammals and their ecology have shown that size is not randomly determined, graphs for species from different areas showing a similar form. There are more small species than larger ones (figure 15.3), with predators occupying a median position. In Europe there are only eight herbivore species, no insectivores and sixteen predators between thirty centimetres and one metre in height. On either side of this size range there are fifty-four prey species and three predators below thirty centimetres, and thirteen prey species and three predators above one metre. All the species that exceeded the mean size for their group show defensive modifications, and those with no adaptations are all of small size. Large prey species are adapted for running or else are large enough to defend themselves directly. As a result of the pressure exerted by carnivores, the size of herbivores increases so that it is equal to, or greater than, that of the carnivores. There is a zone of total predation where no terrestrial prey species can survive. In Europe this lies between 62.5 cm (crested porcupine *Hystrix cristata*) and 115 cm (roe deer *Capreolus capreolus*), and the only herbivore found in this size range is the beaver, which is aquatic (figure 15.3). These observations can be explained on bio-energetic grounds. If I is an affinity index calculated from

$$I = \frac{\text{energy obtained by eating prey}}{\text{energy used in capturing prey}}$$

the prey species' chances of survival are inversely proportional to I. The more energy a predator must use to capture its prey, the less the benefit gained from the prey. The development of a means of defence, for example a subterranean, arboreal or aquatic life, provides a way of decreasing the value of I.

346

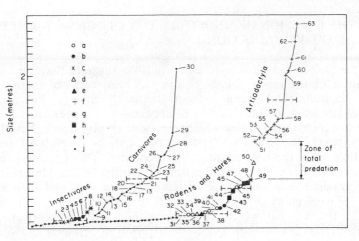

Figure 15.3 · Graphs showing variation in size of European mammals, excluding bats. The horizontal dotted line shows the mean size of each taxonomic group. There is a zone of total predation where no terrestrial prey species are represented.

a: semi-burrowing;
b: burrowing;
c: semi-aquatic;
d: aquatic;
e: semi-arboreal;
f: arboreal;
g: armed with spines;
h: lagomorph-type runners;
i: ungulate-type runners;
j: generalised animals with no particular adaptations.

Key to species:

1: *Neomys anomalus*	22: *Vulpes vulpes*	43: *Oryctolagus cuniculus*
2: *N. fodiens*	23: *Meles meles*	44: *Lepus capensis*
3: *Desmana pyrenaica*	24: *Lutra lutra*	45: *Marmota bobak*
4: *Talpa romana*	25: *Gulo gulo*	46: *M. marmota*
5: *Talpa coeca*	26: *Canis aureus*	47: *Lepus europaeus*
6: *Talpa europaea*	27: *Lynx pardina*	48: *L. timidus*
7: *Aethechinus algirus*	28: *L. lynx*	49: *Hystrix cristata*
8: *Erinaceus europaeus*	29: *Canis lupus*	50: *Castor fiber*
9: *Mustela minuta*	30: *Ursus arctos*	51: *Capreolus capreolus*
10: *M. vulgaris*	31: *Pteromyscus volans*	52: *Capra aegagrus*
11: *M. erminea*	32: *Glis glis*	53: *Rupicapra rupicapra*
12: *Vormela peregusna*	33: *Mesocricetus auratus*	54: *Saiga tatarica*
13: *Lutreola lutreola*	34: *Arvicola terrestris*	55: *Capra pyrenaica*
14: *Putorius putorius*	35: *A. sapidus*	56: *C. ibex*
15: *Martes martes*	36: *Rattus rattus*	57: *Dama dama*
16: *M foina*	37: *Citellus citellus*	58: *Sus scrofa*
17: *Mustela furo*	38: *C. suslicus*	59: *Rangifer tarandus*
18: *Genetta genetta*	39: *Sciurus vulgaris*	60: *Cervus elaphus*
19: *Herpestes ichneumon*	40: *Cricetus cricetus*	61: *Ovis musimon*
20: *Alopex lagopus*	41: *Spalax leucodon*	62: *Alces alces*
21: *Felis sylvestris*	42: *S. microphthalmus*	63: *Bison bison*

347

IV THE INVASION OF UNOCCUPIED ECOLOGICAL NICHES: ADAPTIVE RADIATION

All the palaeontological evidence available suggests that the differentiation of the major animal groups has been achieved in the same way. Environmental changes (most frequently climatic, but also the appearance of new land) or the disappearance of pre-existent groups, leave unoccupied territories which represent empty niches. New forms soon appear and colonise these niches and, as a result of natural selection, only the best adapted forms survive. This process is called adaptive radiation. The proliferation of reptiles in the Mesozoic era, their presence in a wide range of habitats, their diets and their varied sizes, provide an example of successful adaptive radiation. This is also true for the mammals which are replacing them at the present time.

Adaptive radiation corresponds to a period of accelerated evolution which is only possible when new habitats appear or when the absence of competition allows the diversification of new forms. In Australia, in the absence of placental mammals, the marsupials had a wide range of habitats at their disposal, and show almost as much ecological diversity as other mammals. *Thylacinus* resembles the wolf, *Myrmecobius* feeds like an anteater, *Notoryctes* burrows like a mole, *Petaurus* glides like the flying lemur, kangaroos are herbivores, etc.

A familiar example since Darwin's time is the Geospizinae of the Galapagos Islands. These volcanic islands are situated far out in the Pacific Ocean, over a thousand kilometres from the coast of South America, and appeared over a million years ago. They are inhabited by birds that have arrived by chance from the mainland, and, finding available ecological niches, have occupied them. The ancestral geospizine, which was a seed-eating ground-finch, has given rise, through evolution, to fourteen species distributed throughout the different islands. Figure 15.4 shows the main lines of evolution of the Geospizinae. They are birds whose size varies from that of a tit to that of a hawfinch. Their plumage is dull, usually black in the male and greyish in the female. Their songs are very similar, and so are their nests and eggs. The main specific difference is in the form of the beak, which has evolved in relation to the diet. The genus *Geospiza* contains six terrestrial species, which are mainly seed-eaters. In *G. magnirostris*, the beak is enormous and adapted for feeding on hard seeds and nuts. In *G. Scandens* the beak is long and conical, enabling it to feed on cactus flowers. The genus *Platyspiza* contains only one species, which is arboreal and feeds on buds and fruit, having a parrot-like beak. The genus *Camarhynchus* comprises four insectivorous tree-finches. One interesting species, *Cactospiza pallida*, occupies the woodpecker niche, feeding on xylophagous insects, which it extracts with the aid of a cactus spine since its beak is not sufficiently narrow. This is one of the rare examples of the use of tools by an animal. Finally, two other species have the thin beak of

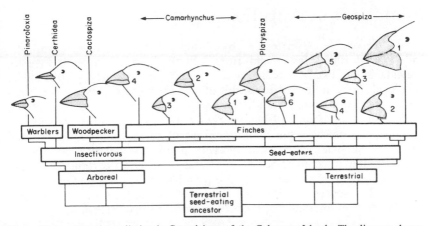

Figure 15.4   Adaptive radiation in Geospizinae of the Galapagos Islands. The diagram shows the evolution of this group from a seed-eating, ground-dwelling ancestor. Genus *Geospiza*; 1. *G. magnirostris*; 2. *G. fortis*; 3. *G. fuliginosa*; 4. *G. difficilis*; 5. *G. scandens*; 6. *G. conirostris*. Genus *Platyspiza*: a single species *P. crassirostris*. Genus *Camarhynchus*: 1. *C. psittacula*; 2. *C. pauper*; 3. *C. parvulus*; 4. *C. heliobates*. Genus *Cactospiza*: a single species *C. pallida*. Genus *Certhidea*: a single species *C. olivacea*. Genus *Pinaroloxia*: a single species *P. inornata*. Three morphological forms exist: finches, woodpeckers and warblers.

a warbler and are arboreal insectivores. These are *Certhidea olivacea* and *Pinaroloxias inornata*, the latter being rather isolated, both morphologically and geographically, since it is restricted to the island of Cocos, situated several hundred kilometres to the north of the main group of islands (figure 15.4).

An adaptation may also be of a physiological or biochemical nature. Several examples have been described already (cf. p. 12). One well-known example is that of mountain grasshoppers (Marty, 1968). Marty found that the haemolymph protein content increased with altitude for species common in the Pyrenees, like *Stenobothrus lineatus*. For example, for a female there was a protein content of 2.6 per cent at 200 metres, and 4.3 per cent at 1100 metres. In addition, species confined to mountains have haemolymph which is always richer in protein than that of lowland species. For example, the protein content of the haemolymph of *Cophopodisma pyrenaea*, which rarely occurs below 1800 metres, is five per cent, while that of lowland species is only three per cent. The increased protein content is necessary because, at high altitudes, rapid maturation of the ovaries is essential owing to the brief favourable season. Widespread species also show this characteristic in mountainous areas. The haemolymph with its proteins forms a reservoir which assists ovarian maturation. It is possible to distinguish proteins that are affected by environmental factors (*ecoproteins*) from *ecostable proteins*, which are an expression of the genotype and not affected by the environment.

349

1 THE RATE OF EVOLUTION

Some reasonable estimates can be obtained from well known examples of isolated island fauna and flora. On the Henderson atoll (Pitcairn Islands in the Pacific Ocean), which is only about thirty metres above sea level, there are endemic species of plants, molluscs, insects and birds which can be no more than a few million years old, i.e. the age of the atoll. In the Hawaii Islands, there are twenty-three species of Lepidoptera of the genus *Omiodes*. Five of these species are restricted to the banana, which was unknown in the islands before its introduction by man about 800 years ago. It seems certain that these five species have developed over the past 800 years (or less) from one or more other species of *Omiodes* living on grasses, legumes or Liliaceae.
Two further examples of recent evolution deserve mention.

2 INDUSTRIAL MELANISM

Industrialisation in Great Britain during the second half of the nineteenth century greatly changed the natural environment. Trees, especially, became more and more frequently covered with black soot, which restricted the development of lichens. Certain moths, particularly *Biston betularia*, evolved rapidly. These moths, when at rest, place their wings flat against a tree trunk, and are almost invisible as a result of their coloration, which has a white background covered with dark speckles. In 1848 a dark form (form *carbonaria*) appeared in the vicinity of Manchester, rare at first, but becoming increasingly common until it almost completely replaced the light form. More than seventy species show this phenomenon, which is known as industrial melanism. The dark form is almost invisible on trees blackened by smoke, while the light form can be more easily seen. Kettlewell (1961) was able to explain the mechanism for this rapid evolution. It takes the form of selective predation on light forms by birds, especially tits, which ignore individuals that are more or less camouflaged by their colour. The form *carbonaria* is a dominant mutation that has been able to establish itself rapidly, because natural selection has not favoured the light form. Selective predation by birds has been confirmed by release and recapture experiments under natural conditions.

3 RESISTANCE TO INSECTICIDES

Another example of rapid adaptation is the development of resistance to insecticides by many insects, and here the resistance to DDT in particular is well known. It is possible, in the space of about ten generations, to select a race of the house fly for which the LD 50 (the dose killing fifty per cent of the individuals in a test) is increased by 100 times. According to many authors, the use of DDT has resulted in the selection by man of naturally resistant individuals, already present in those wild populations with a varied genetic composition. 'The formation of resistant strains probably provides the best proof of the effectiveness of natural selection' (Dobzansky, 1947).

4 THE EFFECT OF ISOLATION

The genetic variability of natural populations has been firmly established. The great majority of species studied are subdivided into geographical races, and these species are polymorphic. Isolation appears to be the only possible mechanism for speciation, many authors agreeing with Mayr (1963) that speciation is impossible amongst sympatric forms This isolation may be geographic, different populations being separated by stretches of sea, mountain ranges or valleys; or it may be ecological, as in several species of frog of the genus *Rana* which reproduce at different times of the year. Some fish of the genus *Coregonus* can spawn in deep water, while others use shallow water, running water, or still water.

Evolution resulting from geographical isolation is especially well marked on islands, as certain characteristics of insular ecosystems tend to accentuate the process. The first of these characteristics is a reduction in the number of individuals in a population, and this increases the effect of isolation. The second is the simplification of communities and ecosystems resulting from the uniformity of habitats, which limits the number of niches and thus the number of species per niche. Large predators are usually absent from small islands, or they may be represented by races of smaller individuals than those found in similar habitats on the mainland. The simplification of these ecosystems is accompanied by a reduction in the intensity of factors such as competition and predation, and this permits, for example, the appearance of flightless birds. Small shrews (Insectivora) of the genus *Crocidura* have developed giant races on some Atlantic Islands in the absence of small carnivores. Similarly, voles have become larger, smaller size being a means of defence against predators which are absent from these islands. The increased variability of island species is probably a result of the decrease in predation pressure (cf. p. 249).

Geographical isolation also operates on the mainland. In the Pyrenees, there are about thirty species and sub-species of cave-dwelling beetles belonging to the genus *Aphaenops*, each restricted to a particular grotto or stream basin. Various mountain ranges contain many endemic races of animals and plants. On a much larger scale, the tit *Parus major* occurs as the typical form in Europe, as the form *cinereus* in India and in part of Central Asia, and as the form *minor* in China. Overlapping zones exist for the races *major* and *cinereus* and for the races *major* and *minor*, but in these areas where races are sympatric they do not hybridise, but behave as different species (figure 15.5).

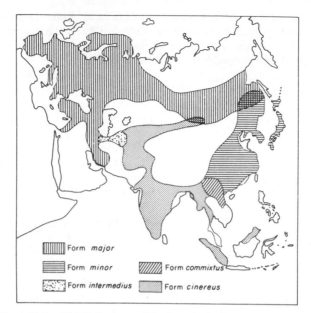

Figure 15.5   Distribution map of the forms in the Great Tit, *Parus major*.

One example of the selection of a particular genotype by environmental factors is that of sickle-cell anaemia in man. This abnormality, which is found mainly in dark races, is characterised by the abnormal appearance of the erythrocytes, which appear more or less crescent shaped. It is determined by the semi-dominant gene Hb^S, which is located on one of the autosomes. The disease occurs either as the severe form in homozygotes (Hb^S/Hb^S) where it is usually fatal, or as the milder form in heterozygotes (Hb^S/Hb^A). The higher frequency of heterozygotes found in some parts of East Africa and India is explained by the natural immunity developed by the heterozygotes (Hb^S/Hb^A) to malaria which occurs in these regions. The disadvantage of the gene Hb^S is counterbalanced by the resistance to malaria, and explains the survival of this gene.

V   GENETIC PLASTICITY – A FACTOR IN EVOLUTION

An example has already been given (cf. p. 160) of the variety in form which develops as a result of competition in Darwin's finches. This diversity has been described for many animals, including birds (partridges of the genus *Alectoris*, Geospizinae), fish (gambusids), mammals (*Peromyscus*), amphibians (*Hyla* and *Microhyla*), insects *Drosophila, Erebia, Lasius*), and has also been recorded for plants (*Phlox* and *Gilia*). Among ants of the genus *Lasius*, the species *L. flavus* can be easily separated from *L. nearcticus* by at least eight taxonomic characters when the two species occur together in the eastern United States. In the west, however, where *L. nearcticus* is rare or absent, *L. flavus* resembles this species much more closely. Behaviour may also be similarly modified, and may play a part in preventing hybridisation. The two warblers *Phylloscopus trochilus* and *P. collybita* have very different songs in those areas of France where they are sympatric. In Spain, however, where only *P. collybita* occurs, this species produces a song that is intermediate in form, probably similar to the primitive form of their common ancestor. The process of divergent evolution may, in this instance, indicate a close relationship.

The last example leads to a discussion of the evolution of species in parallel with the evolution of ecosystems. It has been suggested that the greater part of animal behaviour is simply an adaptation to the behaviour of other species living in the same habitat. For example the warning and avoidance behaviour of prey is an adaptation to the attacking behaviour of predators and vice versa. A good example is that of the Calao *Tropicranus albocristatus*, a bird which is associated with troops of cercopithecid monkeys in African equatorial forests. This bird appears to feed solely on large grasshoppers which are disturbed by the movements of the monkeys. The structure of the Calao is adapted to this social behaviour. It differs from related species in having a long layered tail, allowing a zig-zag flight among the trees in pursuit of its prey. This example shows how closely the evolution of a species is related to that of the ecosystem.

VI PLANT LIFE FORMS

The life form represents the morphological adaptation of plants to their environment. The most frequently used classification is that of Raunkiaer (1934), who proposed it mainly for plants in temperate regions, where the unfavourable season is the cold one. It can, however, be extended to countries where the unfavourable season is the dry one. In equatorial regions, life forms should also be considered from the point of view of interspecific competition and exploitation of the habitat. This is true for tropical rain forests, where conditions are favourable all the year round, and where the existence of chamae-phytes, geophytes and even therophytes in the undergrowth allows the best exploitation of a particular microclimate. The main life forms are:

(i) *Phanerophytes* ('visible plants'). These have their buds more than fifty centimetres from the soil, unprotected by the winter snow in northern Europe. They include trees, small trees, shrubs and woody climbing plants like ivy and *Clematis.*

(ii) *Chamaephytes* ('dwarf plants') have their buds above the soil surface, but below fifty centimetres in height. In this way they are protected by snow in winter. Some are woody (heather, bilberry), others herbaceous (*Vinca*, cabbage). In tropical regions the buds are protected by plant debris, which covers the soil during the dry season. In the savanna especially, the dense grass tussocks provide chamaephytes with a more humid microclimate.

(iii) *Hemicryptophytes* ('half-hidden plants') have their winter buds at soil level, enclosed in a rosette of leaves that often persist (dandelion, *Bellis*) or protected by scales (*Urtica dioica*).

(iv) *Cryptophytes* ('hidden plants') have no vegetative structures visible during the unfavourable season. They are perennial plants with buds hidden either in the soil, in which case they are called *geophytes* (rhizomes of Solomon's seal, tubers of *Orchis*, bulbs of *Endymion*); in the mud, when they are known as *helophytes*, or marsh plants (*Phragmites, Typha*); or in water, called *hydrophytes* or water plants (water lilies, *Elodea*).

(v) *Therophytes* ('fine season plants') are annuals which spend the unfavourable season as seeds. In deserts like the Sahara, there are therophytes capable of very rapid growth, called *ephemerophytes*, which take only a few weeks (sometimes only fourteen days) to germinate, flower and form seeds, and which only appear after the infrequent rains (for example *Convolvulus fatmensis, Schismus barbatus*).

As a general rule, phanerophytes, the plants least protected against extremes of climate, are most abundant in warm countries, where the climate favours continuous growth, and especially in warm and humid tropical regions. Where there is a dry or cold season, the proportion of phanerophytes decreases, and they are absent from arctic regions and high mountains. Chamaephytes are especially common in areas with a well defined dry season, while hemicryptophytes are characteristic of

TABLE 15.1   The distribution of life forms for four very different climatic regions.

LOCALITIES	NUMBER OF SPECIES	PERCENTAGE OF DIFFERENT LIFE FORMS				
		Ph	Ch	H	G	Th
Seychelles, 5° S.	258	*61*	6	12	5	16
Rwindi-Rutshuru plain, Zaire. 1° S.	464	25	*27.6*	14.4	11.2	21.8
Argentario, Italy, 42° N.	866	12	6	29	11	*42*
Sologne, France, 47° N.	760	8.2	5.6	*45.1*	24.8	16.3

Ph: phanerophytes,
H: hemicryptophytes,
Ch: chamaephytes,
G: geophytes,
Th: therophytes

temperate or cold regions. Geophytes occur where the climate includes a severe long dry season, and therophytes in warm dry countries.

The biological spectrum (table 15.1), the percentage of different life forms present in a region, is a remarkable mirror of the sum of environmental factors operating in a given habitat. Some plant associations often present very different biological spectra, and these show in a striking way how microclimatic and edaphic differences may distinguish two neighbouring habitats.

Chapter 16

APPLIED ECOLOGY

The practical applications of ecology are becoming increasingly numerous as the science of ecology develops. Only a few of the many examples of applied ecology are described here.

I BIOLOGICAL CONTROL

This term is applied to those methods which employ living organisms to reduce or prevent the damage caused by animal pests to crops or livestock. The object of the method is to modify the population equilibrium in a particular ecosystem at the expense of the pest species. The ecosystem may be either a natural one, or, more frequently, one which has been modified by cultivation. Although already well established, interest in biological control has increased over the past twenty years, particularly as the result of the increased ineffectiveness of insecticides which, in addition, have shown many side effects (cf. chap. 17). It must be emphasised that the object of biological control is not the total elimination of a species, but simply the reduction of pest numbers to a level at which damage is negligible and where control methods would be uneconomic. The total removal of a single species by any method other than a chemical one is virtually impossible.

Biological control is applied mainly to insect pests. Research has, however, been carried out against rodents, weeds, etc. The methods used are varied, and are becoming more numerous as our knowledge of the ecology of pest species increases. Sometimes the technique used is simple. For example, caterpillars of the tortrix *Adoxophyies reticulana*, a fruit tree pest, cannot begin their diapause, which is essential if they are to survive the winter, if there is a very brief interruption of the dark period. Floodlighting the trees for a two minute period, sixteen and a half hours after sunrise prevents eighty per cent of caterpillars from entering diapause, and they eventually die.

Biological control often makes use of insects which are predators or parasites of pest species. An insect parasite, the braconid wasp, *Opius concolor*, from North Africa, provides an example of this. The larva develops in that of the olive fly *Dacus oleae*, a serious pest in the Mediterranean region, causing the loss of 800 000 tonnes of olives in some years. The difficulty has been to find a method of rearing the parasite in the laboratory, since it was not possible to acclimatise *Opius* in some areas, especially the French Mediterranean region, because of unfavourable winter conditions and the absence of hosts which would allow it to continue its life cycle through the bad weather. A substitute host was found for *Opius concolor* in the laboratory, this being the fruit fly *Ceratitis capitata*, which produces thousands of progeny when reared on a mixture based on carrot pulp. It is now possible to obtain *Opius concolor* at any season, and they can be released in the olive groves at a suitable time. Another important pest is the San José scale *Quadraspidictus perniciosus*, a scale insect which attacks fruit trees. It may be effectively controlled by the chalcid wasp *Prospaltella perniciosi*. The San José scale can be cultured on watermelons, which means that *Prospaltella* is available throughout the year. The wasps are released in orchards by suspending the melons from strings tied to the trees. The acclimatisation of *Prospaltella* has been successful because of the synchronisation between the parasite and its host, the diapause of the latter causing a quiescent stage in the wasp, and ensuring that the two species develop simultaneously.

The use of predators has, perhaps, had the greatest success in biological control. Since the nineteenth century, a ladybird beetle, *Rodolia cardinalis*, which originated from Australia but is now raised on a world wide scale, has held the cottony-cushion scale *Icerya purchasi* in check. This scale has, since its accidental introduction into many countries, ruined plantations of orange and lemon trees, and also mimosas.

Another source of biological control lies in the use of pathogenic micro-organisms, usually chosen because of their specificity. A bacterium, *Bacillus thuringiensis*, produces a toxin which is harmless to vertebrates and most insects other than Lepidoptera. Thus the many beneficial insects are not harmed. This selectivity appears to be due to the mode of action of the toxin, the proteinaceous crystals of which it is formed acting only in a strongly alkaline medium which is rarely found except in the intestine of caterpillars, where the pH may exceed 9. Preparations based on *B. thuringiensis*, and effective against caterpillars, are commercially produced in many countries. Viruses are also used as weapons of biological control. The most spectacular large scale treatment was made in 1958 and 1959 over 500 hectares of pine devastated by processionary moth caterpillars on the north-west slopes of Mount Ventoux between 300 and 900 metres altitude. The virus used, *Smithiavirus pityocampae*, causes a disease characterised by polyhedric cytoplasmic inclusions. Large quantities were obtained in the laboratory in 1959 from 500 000 caterpillars. A powder containing

20 million polyhedra per gramme was dusted on the trees from machines on the ground and from helicopters. There was a considerable decrease in numbers of the processionary moth caterpillars, and in 1961 only eighteen egg batches were found in an area of twenty-two hectares.

Myxomatosis, a virus disease specific to the rabbit (although hares do very rarely contract this disease), has been used effectively to control this pest. In France, the disease was introduced on a voluntary basis, and has spread rapidly over lower ground but more slowly in hilly areas. The virus is transmitted from rabbit to rabbit by several arthropods, including the mosquito *Anopheles maculipennis*, which are not found on higher ground. A sufficiently large rabbit population is essential for the development of an epidemic and, if an area contains only small groups of isolated individuals, the disease will spread slowly or, since rabbits are sedentary and the range of mosquitos limited, will cease. The effects of myxomatosis have been spectacular in France, where mortality reached ninety-nine per cent in some instances. In the game reserves at Sologne, the number of rabbits killed reached nearly 60 000 in 1952–3, and then fell to only a few dozens in 1953–54. The numbers were then maintained at a much lower level than before through the development of resistant strains. There has been a marked effect on the vegetation. In woodland, there has been a spontaneous regeneration of trees where previously this was prevented by the selective grazing of rabbits. The same changes have been observed in Australia, where rabbits were formerly a serious pest. There have been marked and rapid changes in the diets of predators of the rabbit. The fox now includes birds, mice and even fish in its diet. The buzzard, a bird-of-prey which fed largely on rabbits, has shown a reduced fecundity, and is unable to breed in some areas.

Biological control also makes use of genetic methods, which involve the release of individuals which are sterile but still capable of mating. Males are usually sterilised, and females with which they mate will become sterile, especially if they normally mate only once. Calculations show that these self-destructive methods have a theoretical advantage over insecticides. Table 16.1 summarises the data relating to the development of an isolated population, beginning with one million individuals of each sex, among which two million sterile males are released each generation.

The most striking success of radiation sterilization was obtained initially in the island of Curacao, and then in southern U.S.A., where the fly *Cochliomyia hominivorax* has been eliminated. This fly spends its life cycle on cattle, and even on man, causing large sub-cutaneous abscesses. Fifty million sterile male flies were released every week. These flies had been irradiated with cobalt-60 at a dosage which was sufficient to cause lethal mutations in their spermatozoids, while not interfering with their sexual behaviour. Total success was achieved

TABLE 16.1   Population data for the radiation sterilisation method of biological control

GENERATIONS	ESTIMATED NUMBER OF MALES AND OF FEMALES FORMING EACH GENERATION	NUMBER OF STERILE MALES RELEASED	RATIO OF STERILE MALES/ FERTILE MALES
1	1 000 000	2 000 000	2:1
2	333 333	2 000 000	6:1
3	47 619	2 000 000	42:1
4	1 107	2 000 000	1 807:1

GENERATIONS	PERCENTAGE OF FEMALES FERTILISED BY STERILE MALES	THEORETICAL NUMBER OF MALES AND OF FEMALES PRODUCED
1	66.7	333 333
2	85.7	47 619
3	97.7	1 107
4	99.95	<u><1</u>

through this technique, and the cost of the operation, estimated at 10 million dollars, was significantly less than the annual loss due to the fly, estimated at 20 million dollars.

Biological control of weeds has been introduced to avoid the problems which have been associated with chemical herbicides. For example, the eradication of weeds from cereal fields has almost completely eliminated red poppies and cornflowers, but creeping grasses like black-grass, wild oats and bent have not been affected. The systematic use of simazine as a herbicide favours the spread of bindweed and *Cynodon dactylon*, species resistant to this chemical. The biological control of weeds has had three marked successes. In the islands of Hawaii, the plant *Lantana camara*, originally introduced as an ornamental from Mexico, has spread throughout the country, overrunning thousands of hectares of grassland, which have become unproductive. It has been possible to control this weed effectively by the introduction of insects which feed on *Lantana* in Mexico, and it has not succeeded in recolonising those areas from which it has been eliminated after 1962. Several phytophagous insects have been involved, including the tortrix moth, *Crosidosema lantanae*, and the agromyzid fly, *Agromyza lantanae*. A similar success has been recorded in Australia against the cactus *Opuntia*, which was introduced in 1842. By 1920, almost 20 million hectares were overrun, and the invasion was extending at a rate of 400 000 hectares per year. The moth *Cactoblastis cactorum*, originally from central America, has now effectively controlled the cactus, two million eggs of this insect having been produced in 1929 for cactus control. St. John's Wort, *Hypericum perforatum*, is a relatively unimportant weed in Europe, but its introduction into Australia was catastrophic. This plant, freed from natural enemies, invaded tens of thousands of hectares. It was eventually controlled by the introduction of two beetles from Europe, a chrysomellid, *Chrysolina*, and a buprestid, *Agrilus*.

A change in cultural practice is often sufficient to control a pest species. The improvement of reclaimed land in the polders of the Zuyder Zee in Holland was hindered by the invasion of two weeds, a thistle (*Cirsium arvense*) and coltsfoot (*Tussilago farfara*), whenever any dry ground appeared. Ecological research produced a new method for controlling these weeds. Since the thistle can only disperse over a short distance using its achenes, it can be controlled by destroying those plants which appear at the edge of the polder during regeneration. Coltsfoot cannot tolerate shade during the early stages of growth and, if seeds of the reed *Phragmites communis* are sown by plane as new ground appears, a dense cover is produced, in whose shade coltsfoot cannot grow. These reeds assist the drying out of the ground, and they may be cropped in the following year; after a number of years the ground can be cultivated.

In regions where the Hessian fly, *Mayetiola destructor*, is a pest, the late sowing of wheat breaks the life cycle by preventing oviposition in early spring. On the other hand, the early sowing of spring oats prevents attack by the chloropid fly, *Oscinella frit*. In southern areas of the U.S.S.R., the quick, early harvesting of cereals maintains numbers of *Eurygaster integriceps* at a sufficiently low level for damage to be negligible. This insect completes its development when the wheat ripens, building up fat reserves to enable it to overwinter and achieve high fecundity the following spring. Early harvest prevents the build up of food reserves, and so reduces fecundiy.

INTEGRATED CONTROL

The idea of integrated control is due largely to American research workers (Stern *et al.*, 1959). The integrated control of a pest is the co-ordination of all known cultural, biological, ecological and chemical methods in such a way as to obtain the maximum total benefit, and especially to minimise harmful side effects that may result from exclusive use of chemical pesticides (Milne, 1965). Thus integrated control sets out to replace pesticides by biological control methods wherever possible. In the U.S.S.R., for example, advantage is taken of the fact that the toxicity of *Bacillus thuringiensis* to caterpillars of the apple tree ermine moth can be increased by the addition of small quantities of DDT, although the effect of DDT alone on these caterpillars is almost negligible. In France traditional winter washes against the fruit tree scale, *Pseudaulacaspis pentagona*, were also found to kill its predator *Prospaltella berlesei*, and an alternative wash at a lower rate of application, which is less harmful, is now used. The success of a programme of integrated control depends on a thorough knowledge of the insect fauna of a particular habitat, including their identity and changes in numbers. This emphasises the fundamental interest of studies of taxonomy and population dynamics.

II   THE MAINTENANCE OF NATURAL EQUILIBRIA

It has been shown that, in the absence of human intervention, ecosystems tend to become more mature, that is, they evolve towards increased stability and complexity. The action of man, by creating arable land, which is relatively low in species diversity, produces communities which are not mature and, in consequence, are subject to considerable population fluctuations. The maintenance of natural equilibria is thus one of the aims of applied ecology.

Where parasite control measures are carried out carelessly, they often seriously affect the balance of nature. Attempts to eliminate quickly a particularly troublesome pest often result in 'pest resurgence' i.e. an increase in the pest population, and so in the damage done. A second species, unaffected by the control measures, may replace the eliminated pest species in the now vacant niche, and this may result in a secondary pest outbreak. This substitution may also occur through the immigration of a species from outside. This phenomenon has been noted in olive groves, where the excessive use of chemical sprays, while producing the required reduction in numbers of the olive fly *Dacus oleae*, also caused an increase in the scale insect *Saissetia oleae*, which, up to that time, had been of little importance as a pest. Furthermore, the simultaneous presence of an introduced parasite, *Opius concolor*, and of an indigenous parasite of *Dacus oleae*, such as *Eupelmus urozonus*, had the unexpected result of a larger proportion of olive flies surviving the winter than would have survived if *Opius concolor* had been present by itself.

NATURAL RESERVOIRS

These examples show that a great deal of care must be taken when modifying the environment. The conservation of natural reservoirs plays an important part in preserving the balance of nature. The term reservoir can be applied to areas of varying size where natural conditions are maintained, sheltered from the influence of man. Virgin forest, alpine grassland, river banks, coastal habitats, hedgerows and even game reserves form natural reservoirs (Balachowsky, 1951). These shelter a reserve of useful species, such as insect parasites and various vertebrate predators, which play an important part in the regulation of pest populations. These natural reservoirs also represent ecosystems which are more diverse and, therefore, more stable. They may separate crops and so interfere with the dispersal of insect pests. Under monoculture systems there are no natural barriers to the dispersal of insect pests. The olive groves of Crete, noted for their oil, receive little attention. As a result, a dense undergrowth develops under the olive trees, especially of the composite, *Inula viscosa*, and this serves as an overwintering site for hymenopterous parasites of the olive fly.

The removal of hedges and ditches from the countryside of western France, which accompanies the re-allocation of the land, provides a typical example of the harmful effects of removing natural reservoirs. Reptiles, insectivores, birds-of-prey, and beneficial insects such as coccinellids and syrphids, were numerous there. Parasitic hymenoptera collect the pollen and nectar required for egg formation from wild plants growing in the hedgerows. Insect pollinators, such as bumble-bees and solitary wasps, nest in the banks. If the west of France is a region where 'insect outbreaks' are uncommon, it is due partly to the abundance of these beneficial animals, and some provision should be made for their conservation. Coccinellids and syrphids control aphid numbers, for example. In Russia and in central European countries, wooded areas are preserved so that entomophagous insects and other useful animals may maintain their numbers.

There are numerous examples of the way in which the balance of nature has been affected by ill-timed interventions. In Poland, the otter was accused by fishermen of eating fish and, in consequence, this mammal was exterminated through over-hunting. Fish numbers, however, continued to decline. Otters, in fact, tend to take mainly diseased fish, which are more easily caught, and so help to maintain healthy fish stocks. The destruction of the otter allowed an increase in disease among the fish, and resulted in higher losses. Now that the lesson has been learnt, the otter is protected, and stocks of this beneficial mammal have been increased through rearing. In 1872 the mongoose, a carnivorous mammal, was introduced into Jamaica to control rats, which damaged the sugar cane crop. Within ten years the mongoose had increased to such an extent that it destroyed, not only most of the rats, but also a large part of the island fauna, including mammals, birds, reptiles, crabs, etc. The surviving rats became adapted to an arboreal life and continued to damage the sugar cane.

Upsetting the balance of nature also has important consequences in the medical field. The introduction of cattle into the tropical savanna of America resulted in an increase of blood-feeding vampire bats, which act as vectors of rabies. Irrigation of the African savanna has resulted in an increase in the extent of schistosomiasis. Systematic deforestation has frequently resulted in the transfer to man of arboviruses normally restricted to the forest canopy, where they are responsible for relatively mild infections in monkeys and arboreal rodents. These diseases (yellow fever, dengue, kyasanur forest disease) can, however, be dangerous in man. Similarly, the eradication of tropical diseases, if this is even possible, presents long-term problems. In East Africa, some populations are protected from malaria by the possession, in the heterozygous condition, of abnormal erythrocytes. When the disease disappears, however, the unfavourable effects of this genetic adaptation persist (cf. p. 352).

It is sometimes possible, through careful management, to re-establish the natural balance or to create a new one. In Australia, where cattle and sheep were introduced by man, the rare indigenous coprophagous scarab beetles feed only on the dung of marsupials or flightless birds. Yet it has been estimated that domestic mammals produce thirty-three million tonnes of dung (dry weight) annually, much of which remains on the surface, hardens and dries out, losing its volatile and unstable nitrogenous components. The dry dung may remain on the ground for five years, although in other countries it disappears within days or even hours. Thus the Australian pastures, which receive no additional fertilisers, are also deprived of an important source of plant nutrients. The dried dung covers a large area, and more than 100 000 hectares of grassland are lost each year because the vegetation developing in these spoilt areas is not palatable to cattle. To solve this problem, Bornemissza (1960) introduced several species of dung beetle into Australia, and these have helped to improve the pastures in that country. Some data are available for the activity of dung beetles. A common French species such as *Geotrupes niger* buries 725 g of dung in each larval nest, while *Copris lunaris* buries 300 g. In Kansas (U.S.A.), *Dichotomus carolinus* moved 150 kg of earth and buried 25 kg of dung per hectare during the observation period in September. Six species have already been introduced successfully into Australia and, in addition to their beneficial action on the soil, it seems probable that they reduce parasite infestations in cattle by burying in the soil those stages in the life cycle of parasitic worms which pass through dung, making them less accessible to cattle and sheep. Many flies are unable to lay their eggs in the dung, and so their numbers are reduced. It is interesting to note that those parts of the world which have the richest mammal fauna also have the most diverse dung fauna. Africa, with numerous mammals, possesses several thousands of coprophagous species of scarab beetle.

III INDICATOR SPECIES AND ECOLOGICAL DIAGNOSIS

Species with very restricted ecological requirements may be used to describe particular habitats. These species are 'indicator species'. Dragonflies are restricted to particular types of waters. The carnivorous diet of the larva means that it must find abundant prey to complete its

363

development, and so the abundance of dragonflies over a water body is a good indicator of the trophic state of that habitat. In polluted waters, the extent of pollution can be expressed in terms of saprobic zones, each defined by the presence of certain species (cf. chap. 17). The disappearance of lichens from tree trunks indicated an increased level of sulphur dioxide in the air. Overgrazing, leading to the gradual deterioration of natural grassland, may be diagnosed by examining the vegetation. In the Ossau valley (Pyrenees), where wandering flocks are numerous in summer, rich pastures with several species of clover, and grasses like *Nardus stricta*, are gradually transformed into poor swards invaded by *Asphodelus albus* and *Festuca spadicea*. Nitrogen-loving plants (Chenopodiaceae, nettles), avoided by animals, slowly invade those areas used by sheep and cattle for defecation.

IV ECOLOGICAL STUDIES AS THE BASIS FOR INTERVENTION

In conclusion, some examples are given to show the importance of preliminary ecological studies before any kind of human interference in an ecosystem.

In coastal salt marshes in the south of France, ceratopogonid flies of the genus *Leptoconops* are the likely vectors of arboviruses and filariasis. It has been possible to restrict the extent of chemical spraying against the larvae of these flies through an accurate knowledge of the characteristic soil and flora of the larval habitats. The larval habitat of *L. irritans* is restricted to base-rich saline muds and, even in these habitats, only those muds supporting *Arthrocnemum glaucum* and *Salicornia fruticosa* associations provide suitable sites for adult emergence. The larval habitats of *L. kerteszi* occur only in saline areas of dunes at the edges of temporary pools, where adult emergence can take place as the water level recedes and while the water table is still within 0.20 m of the surrounding area.

Mosquito control along the Mediterranean coast of France is now based on ecological principles, avoiding the unnecessary use of large quantities of insecticides. Careful study of the different mosquito species has shown that plant cover is a good indicator of potential breeding sites. *Juncus maritimus* and *Salicornia fruticosa* salt marshes, for example, harbour large numbers of eggs, while few are found among adjacent plant communities. A map can be drawn, using a scale of 1/5000, showing the distribution of the main plant formations under four categories: sansouires with saline soil, reed beds with fresh or slightly brackish water, areas of rushes and, finally, the slightly saline dry meadows. Potential breeding sites can be located from this map, and their areas estimated. These maps also give information about the type of soil. A group of *Schoenus nigricans*, for example, is an indication of a well-drained sandy soil, while *Molinia caerulea* or *Salicornia fruticosa* suggests a heavy, impermeable soil, where drainage is impeded and where there may be opportunities for rice culture (Rioux, 1967). A similar plant mapping technique enables oviposition sites of the migratory locust *Schistocerca gregaria* to be located.

The collaboration of ecologists could be very useful to those engaged in land management. In Upper-Volta, where there is a large water deficit during the dry season, dams have been constructed to provide water reserves. Unfortunately, large numbers of *Simulium damnosum* larvae, a vector of human onchocerciasis, appear in the spillways of reservoirs which are already full. In some instances, this disease has now extended into areas where it was previously almost unknown, and this can be correlated with the construction of dams. *Simulium* larvae are restricted to running water, and the normal types of spillway, in the form of stairways, provide ideal sites. It should be possible to eliminate the larva by changing the design, and a group of ecologists found that the problem could be solved by using a system of sluices and siphons. However, this arrangement increases the cost of the scheme and requires the supervision of watchmen. At the present time, therefore, it is only possible to continue with the installation of dams in areas where simulids do not occur, to avoid the spread of onchocerciasis. In this instance the ecologist faces a problem which is out of his province.

Chapter 17

APPLIED ECOLOGY AND CONSERVATION

The biosphere is endowed with a marked stability with respect to external influences, that is a considerable plasticity. This plasticity represents an important asset for man, as it enables him to use and transform the components of the biosphere to a large extent, according to his needs. This transformation, however, cannot be carried beyond certain limits, as otherwise it may imperil the established dynamic equilibrium of the biosphere. In some large areas of the globe, these limits have already been transgressed, resulting in the deterioration of considerable parts of the biosphere, as well as in the disappearance of numerous plant and animal species, freshwater basins and soils. Man and his society live within, and constitute an important part of, the biosphere, and use the resources of the biosphere. The protection of the biosphere is of great importance to him. (Utilisation and conservation of the biosphere, UNESCO, 1970, p. 13).

The close relationship which exists between the individual components of an ecosystem and, on a much larger scale, the components of the biosphere, is a fundamental ecological truth. Man, in common with other living organisms, forms part of the biosphere, and this has important consequences. It is true to say that 'the conservation of nature is ultimately and essentially the conservation of man' (Michel Hervé Julien).

I MAN'S EFFECT ON THE BIOSPHERE

1 THE CONSUMPTION OF NATURAL RESOURCES

The world population is rapidly increasing in an 'explosive' fashion, and the following estimates have been made for past and future populations:

Neolithic (9000 years ago)	10 millions
Beginning of present era (2000 years ago)	160 millions
Year 1900	1 617 millions
Year 1960	3 010 millions
Year 1966	3 370 millions
Year 2000 (anticipated)	6 thousand millions

The population doubling time has decreased from 2500 years in the Neolithic, to 100 years in 1900, and thirty-five years in 1960. The productivity of the biosphere, however, remains relatively low (cf. chap. 13). Much of the land surface is desert (cf. figure 2.8), and increased crop yields do not keep pace with population growth. Natural resources are also being depleted at an ever increasing rate in the twentieth century, and unfortunately there have been many examples of the misuse of these resources. Forest fires (deliberate or accidental) each year destroy two million tonnes of organic material. To this may be added the large number of trees which are felled to supply paper for a multitude of publications, many of which are probably never read. The burning and clearing of tropical rain forests leaves in their place, after a few years of cropping, a laterite desert. Nine tenths of the island of Madagascar has been ruined as the result of this kind of activity. The monoculture of sugar cane, coffee and other tropical crops exhausts the soil, and the impoverished land is then abandoned.

Overfishing on a large scale has gradually reduced stocks of many marine fish, and the tonnage landed decreases annually despite the effects of more efficient fishing techniques (figure 17.1). Whales have been similarly exploited, and existing stocks of some species are now so low that some countries have discontinued their whale fisheries. The Biscayan whale has almost disappeared, and protection has come too late. The blue whale is seriously threatened, and is likely to become extinct.

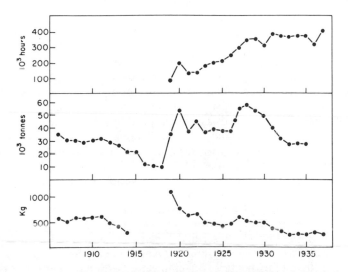

Figure 17.1   An example of how overfishing can almost destroy the existing stocks. Above: the length of time spent fishing for cod in the seas of Iceland. Centre: fluctuations in catches of fish landed. Below: changes in catch of fish taken per unit of effort. The tonnage decreases despite increased fishing, a sure sign of over-exploitation.

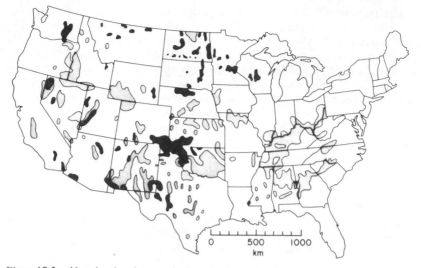

Figure 17.2   Map showing the spread of erosion in the U.S.A. Black: erosion by wind, dots: erosion by rain.

The deliberate destruction of food may be added to this list of misuse of resources. In 1942 eight million tonnes of wheat were destroyed in Argentina. The fact that an increasing amount of arable land disappears each year through urbanisation and road building must be a cause for concern to the ecologist.

Land erosion and drainage may be included among those processes playing an important part in the reduction of available resources. In many regions extensive erosion removes the soil, resulting in the progressive formation of desert. Man is responsible for this erosion as he destroys the protective plant cover by poor agricultural practices, burning forests and overgrazing with sheep and goats. 'Man has probably already destroyed as much productive land as there is in existence today' (Duvigneaud, 1967). Land in the Near East, Greece and North Africa, which was formerly covered by forest, is today semi-desert. In China, a quarter of the land surface has lost its fertility, and 2.5 thousand million tonnes of fertile yellow earth is removed each year through erosion. In the U.S.A., the central Great Plains became covered in about 1935 by clouds of dust (dust bowls) formed by wind erosion as a result of monoculture and dry farming (figure 17.2). This striking phenomenon was described by Steinbeck in the first few pages of his novel *The Grapes of Wrath.* Over the past 150 years, in that country, 120 million hectares of land have been damaged by erosion, and 2000 km² of good soil have been lost each year. It was estimated that 700 million dollars were lost in 1950 through a reduction in crop yields resulting from the impoverished soils.

Over the world's land surface as a whole, five million km^2 of arable land have been lost through man's action. In Italy, 50 000 km^2 have been subjected to erosion, mainly in the Apennines. The Arno alone carries away 2.6 million tonnes of earth each year. In France an area of five million hectares is threatened by erosion, especially in the south-west. Destruction of the plant cover causes the soil to dry out at an ever increasing rate. This is accelerated through the deliberate drainage of wetlands. The water table is lowered more quickly as the demand for water increases daily. The requirements of industry are especially great; 250 m^3 of water are required to manufacture a tonne of paper, and 600 m^3 for a tonne of fertilizer. A water shortage has become noticeable in many areas over the last ten years, and this problem will increase if variations in rainfall are taken into account. In Italy, at Milan, the level of the water table has fallen by twenty metres in twenty years, and seventy wells have been closed during the past ten years, because the water produced was polluted to such an extent that it was dangerous. At Bologna the water table, twelve metres below ground level in 1945, has now fallen to thirty-five metres.

The systematic drainage of wetlands in temperate zones is a serious mistake. These marshes act as sponges, regulating the level of the water table by supplying water through the summer and absorbing water after heavy rains, thus reducing flooding. The productivity of these wetlands is very high, reaching or even exceeding that of the most productive crop plants (cf. p. 309). They also provide reserves for many plants and animals that are disappearing elsewhere, and may have some economic importance (e.g. wildfowling). Drainage of wetlands is expensive and often yields little in return. Marshes in the bay of Aiguillon were drained at a cost of seven million francs, and cereal crops grown on the reclaimed land in spite of the fact that France already produces a surplus of cereals.

2   POLLUTION OF THE BIOSPHERE

There are many different types of pollutants, and they will be considered here under three headings: industrial effluents, pesticides and radioactive wastes.

2.1   *Pollution by industrial effluents*

The atmosphere has become contaminated by harmful gases emitted by factories and vehicles, the most important being carbon dioxide, carbon monoxide and various compounds of sulphur, chlorine and nitrogen. It has been estimated that sixty per cent of atmospheric pollution is derived from vehicles, but another important source is the combustion of coal and, increasingly, of oil. Oxygen, derived from photosynthesis, comprises twenty-one per cent of the atmosphere. A single aeroplane uses thirty-five tonnes of oxygen in crossing the Atlantic. It seems reasonable to ask, therefore, whether the industrial consumption of this gas will eventually exceed production, especially when the rate of destruction of the plant cover, especially forest, is taken into consideration. The carbon dioxide content of the atmosphere is gradually increasing and, because this gas absorbs infra-red heat re-radiated from the earth's surface, the temperature of the biosphere could eventually increase. This might cause a melting of the polar ice cap and a catastrophic rise in sea level.

Atmospheric pollution reaches a maximum level in large towns, where trees sometimes lose their leaves prematurely. Smog, a toxic fog containing high levels of sulphur dioxide and smoke, caused the deaths in London of 4000 people in a single day, the 5th of December 1952. In Paris, vehicles produce fifty million cubic metres of carbon dioxide per year. A large power station produces 500 tonnes of sulphur compounds and dust each day, and ten per cent of these pollutants may be carried over a distance of five kilometres. It has been shown in Los Angeles that the exhaust fumes of cars contain harmful oxides of nitrogen. Atmospheric pollutants produced in Texas have been detected at Cincinnati (Ohio) over 1600 km away. The atmospheric levels of tetraethyl lead, a compound added to petrol as an antiknock agent, are rapidly increasing. Each vehicle produces about one kg of lead per year, and the tissues of people living near motorways have a higher lead content than those from elsewhere. There is a correlation between frequency of lung cancer and population density in the Netherlands, cancer being twice as numerous in towns with 500 000 or more inhabitants. There is a similar increase in mortality due to lung cancer in Great Britain, if rural areas are compared with towns of over 100 000 inhabitants.

Water pollution has reached unhealthy levels in many countries. In some areas the further development of industry is hindered because water is either too saline or scarce. Pollution occurs when the quantity of sewage effluent is too large to be broken down by natural processes. These effluents also contain non-biodegradable detergents which produce masses of white foam, often more than a metre high, which

TABLE 17.1   The quantities of several pollutants in different countries (in millions of tonnes)

TYPE OF POLLUTANT	U.S.A.	FRANCE	GREAT BRITAIN	FEDERAL REPUBLIC OF GERMANY
Agricultural wastes	1300	560	340	470
Spoil heaps from mining	1000	240	350	380
Household wastes and industrial effluents	400	80	120	130
Derelict cars	17	3	4	6
Atmospheric pollutants	150	37	44	55

floats on rivers. Industrial effluents also play a significant part in water pollution. In the U.S.A. twenty million tonnes of chemical effluent are discharged annually into Lake Superior, while Lake Erie has been transformed into a sewer. Fish have become increasingly scarce, and salmon rivers will soon become only a memory. It is interesting to recall that, at the time of Philip Augustus in the thirteenth century, people visited Paris in order to collect drinking water from the Seine. The water reserves of the water table have now become contaminated themselves because water percolating through has carried with it pesticides and other harmful substances. The water table may also be polluted directly: dredging of the river bed of the Saône has damaged the impermeable clay lining of the river so that polluted river water has entered the underlying water table, which supplies the drinking water for the region. Organic wastes from paper mills, dairies, sugar refineries and sawmills are oxidised by bacteria, and this reduces the dissolved oxygen content of the water to a point where living organisms can no longer survive.

The Rhine is probably the most seriously polluted river in Europe. In its upper reaches (valley of Grisons) the water contains between thirty and one hundred micro-organisms per millilitre. This value is increased to 2000 by the time it enters Lake Constance; at Kembs it is 24 000, and by the time it reaches the mouth of the river, the water contains between 100 000 and 200 000 micro-organisms per millilitre. This water also carries away 30 000 tonnes of mineral salts per day which have not been removed by purification plants. It is also necessary to include the effect of 11 000 litres of oil wastes that are discharged into the river, as well as accidental spillages like the recent examples involving a toxic insecticide.

The extent of pollution can be estimated by determining the biochemical oxygen demand (B.O.D.) of a sample of water. This is measured by the ability of a given volume of polluted water to absorb oxygen over a period of five days at a temperature of eighteen degrees centigrade. The biological assessment of pollution depends on the use of *saprobien zones*. The distribution of aquatic organisms is determined by several factors, but particularly by the dissolved oxygen and organic matter contents of the water. Organisms capable of living in saprobic (i.e. rich in organic matter) habitats are known as saprophiles. It is possible to assess the extent of pollution along a river exposed to organic pollution by recording the saprophile associations present, providing the several indicator species are considered rather than a single species.

(i) Clean, unpolluted waters, with a blue colour, near springs and upstream of any other water courses, correspond to the *oligosaprobic zone*. Characteristic organisms include the moss *Fontinalis antipyretica*, the limpet *Ancylus fluviatilis*, the flatworm *Planaria gonocephala*, and mayfly larvae such as *Oligoneuriella rhenana*.

(ii) The water of the *polysaprobic zone* is grossly polluted, often with a reddish coloration and an unpleasant smell. Because of the lack of oxygen, sulphur in these waters is usually converted to hydrogen sulphide and not to sulphates. Characteristic organisms include sulphur bacteria (*Beggiatoa* and *Thiothrix*) and a fungus, *Leptomitus,* which has a gelatinous appearance and forms black, filamentous tufts attached to stones. Animals include oligochaetes (*Tubifex*), chironomid larvae, leeches and the isopod, *Asellus aquaticus.*

(iii) The *mesosaprobic* β zone corresponds to mildly polluted waters, with a greenish coloration, where ammonium compounds are nearly all converted to nitrates. They occur more frequently in sparsely populated areas, and contain some insects (corixids, notonectids, nepids and dytiscids), crustaceans (Cladocera) and green unicellular algae (*Scenedesmus* and *Spirogyra*).

(iv) The *mesosaprobic* α zone is situated at some distance downstream from a pollution source, and shows some evidence of recovery. These waters are poorly oxygenated, yellowish in colour and still contain nitrites and ammonium salts. The fauna includes species characteristic of the polysaprobic zone, such as *Tubifex*, chironomids, leeches and *Asellus* and, in addition, insect larvae such as *Sialis lutaria*, and the bivalve *Sphaerium corneum.*

The sea forms an enormous depository where man deposits much of his waste: oil, mineral, radioactive and others. The oceans and beaches are becoming increasingly polluted by oil residues, and it has been estimated that five million tonnes of oil were deposited in the sea in 1968 despite regulations that banned this practice. The well known *Torrey Canyon* disaster is, unfortunately, only one example of this form of permanent pollution. Over 100 tonnes of oil waste were found in a single day on an island off the French coast in the English Channel. Many birds had been affected, and especially razorbills, guillemots and puffins of the Sept Iles reserve; there is concern for the future of these and other birds like the grebes, divers and many gulls. It also seems likely that oysters, mussels and other edible shellfish will eventually become unfit for human consumption. Fishing nets have been damaged, and fish stocks will probably decline through the destruction of eggs and fry and interference with food chains, especially through a reduction in the plankton. There is an urgent need for legislation, against those responsible for discharging oil into the sea, in order to check a situation that could end in catastrophe.

2.2   Pesticide pollution

Pesticides include all those chemical compounds designed for use against animals and plants that are considered to be pests. They include insecticides, herbicides, fungicides, etc., these compounds being applied in large quantities to most ecosystems. The annual world production

372

for a single insecticide, DDT, is estimated at 100 000 tonnes. Three million hectares in Great Britain were treated with herbicides in 1963, while insecticides are probably the most widely used pesticides. Synthetic organic insecticides, especially organochlorine compounds, have become increasingly important since 1940. They have been used with marked success in, for example, the arrest of a typhus epidemic in Naples in 1943, and have been responsible for a significant reduction in malaria in many countries. Unfortunately these results have encouraged the belief, over a period of some years, that insecticides can solve every insect pest problem. Most chemists employed in the development of insecticides appear unable to appreciate the possible ecological consequences of their widespread use. In fact, the ultimate effect of insecticides scattered over the countryside is just as important as their effect on insect pests at the time of application.

The indirect effect of these insecticides can be considered under three headings. Firstly, the destruction of harmless and often beneficial species and a reduction in species diversity in many habitats. Secondly, the development of resistant strains which have become increasingly difficult to control; and, thirdly, the accumulation of residues in ecosystems which may persist over many years, as in the case of the persistent organochlorine insecticides. Estimates show that there are about one million tonnes of DDT in the soil at the present time.

Insecticides are as toxic to beneficial species as they are to pest species, honey bees and other insect pollinators having been killed by BHC used as a spray on rape crops. Fish are very susceptible to DDT and BHC, and in 1956 over 800 000 salmon and trout died through the aerial spraying of forests in Canada with DDT to control spruce budworm. Bird populations, especially birds-of-prey, have been significantly reduced through poisoning by organochlorine compounds. In the Scottish Highlands, the proportion of breeding pairs of the golden eagle was reduced from seventy-two per cent of the total population during the period 1937–1960, to only twenty-nine per cent for the period 1961–1963. The bald eagle has been similarly threatened in the U.S.A., and fatal doses were found in twenty-five out of twenty-six specimens analysed. Deaths due to insecticide poisoning are especially noticeable among carnivores, as residues may be concentrated along food chains; the mechanism for this is, as yet, not clear.

Clear Lake in California has become a classic example of the misuse of insecticides. The lake was sprayed in 1949, 1954 and 1957 with TDE (a compound related to DDT) to control a small gnat of the genus *Chaoborus*. This fly does not bite, but was considered a nuisance as it formed dense swarms in the vicinity of water. Although the TDE was applied at a rate less than 0.014 parts per million (p.p.m.), after spraying it was found that planktonic crustacea in the lake contained five p.p.m., fish feeding on these crustaceans even more, and catfish, which had fed on small fish, contained 22 to 221 p.p.m. of TDE in muscle tissue and 40 to 2400 p.p.m. in their fat reserves. At the top of the food chain, the western grebe, *Aechmophorus occidentalis*, which is exclusively piscivorous, showed a dramatic reduction from 1000 pairs to only thirty pairs which appeared to be sterile. Dead birds were found to contain between 1500 and 2500 p.p.m. of TDE in their body fat, showing that the insecticide had been concentrated by a factor of nearly 100 000. (The idea that organochlorine insecticide residues are concentrated along food chains is, however, not entirely correct. Reference should be made to Moriarty (1972).)

Mussels and oysters cultured in water containing DDT can concentrate this substance by more than 70 000 times. Even low concentrations of organochlorine compounds may inhibit photosynthesis in phytoplankton. A concentration of one p.p.m. lowers productivity in the unicellular algae *Platymonas* and *Dunaliella* by seventeen per cent. If heptachlor is used, there is a ninety-five per cent reduction in productivity. If the waters over the continental shelf continue to be contaminated with organochlorine residues, there may eventually be a decrease in the biomass of food available for human consumption. The L.D. 50 (concentration causing 50% mortality in a given time) of insecticides for three aquatic animals is compared with the minimum recommended concentrations of these compounds for mosquito control in Florida salt marshes in Table 17.2. It can be seen that all the insecticides, with the possible exception of lindane, although this is harmful to prawns, are used at rates from 20 to 300 times greater than the L.D. 50. The mosquito control programmes using these chemicals will probably cause a considerable reduction in the fauna of the areas treated.

When elm trees in the U.S.A. were sprayed with DDT to protect

TABLE 17.2   The effect of some insecticides used to control mosquito larvae in Florida, on three aquatic animals

INSECTI-CIDES	L.D. 50 in p.p.m. 96 HOURS AFTER CONTACT OYSTER (*Crassostrea virginica*)	48 HOURS AFTER CONTACT PRAWN (*Penaeus* sp.)	FISH (*Mugil curema*)	DECREASE IN PRODUCTIVITY OF PLANKTON EXPOSED TO 1 p.p.m. FOR FOUR HOURS (as a %)	MINIMUM CONCEN-TRATION RECOM-MENDED AGAINST MOSQUITOS (in p.p.m.)
DDT	7×10^{-3}	10^{-3}	6×10^{-4}	77	3×10^{-2}
Aldrin	5×10^{-3}	—	5.5×10^{-3}	—	—
Chlordane	7×10^{-3}	4.4×10^{-3}	5.5×10^{-3}	94	2×10^{-2}
Heptachlor	2.7×10^{-2}	2.5×10^{-4}	3×10^{-3}	95	6×10^{-2}
Dieldrin	3×10^{-2}	5×10^{-3}	7.1×10^{-3}	85	3×10^{-2}
Lindane	0.45	4×10^{-4}	3×10^{-2}	29	6×10^{-2}

them against beetles carrying Dutch elm disease, some of the insecticide reached the soil and was assimilated by earthworms. These were eaten by American robins (*Turdus migratorius*), and nearly eighty per cent of these birds were then killed by a paralysis. Man is equally exposed to this poisoning. In the U.S.A., a mean value of 925 mg per person of organochlorine residues was recorded in 1961, mainly in the adipose tissue; while in France, a value of 370 mg per person has been estimated. These levels continue to increase, and there is a danger that repeated small doses may result in an insidious chronic poisoning which would be difficult to detect. Insecticides could act in this way like other cumulative poisons, for example fluorine, some carcinogenic hydrocarbons and radioactive substances.

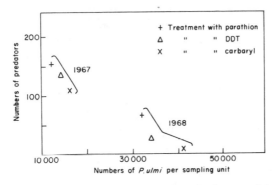

Figure 17.3   Effect of insecticides applied to apple orchards on predators of the mite pest *Panonychus ulmi*.

The effect of insecticides on the fauna may disturb the natural balance. The moth, *Lymantria dispar*, a pest of the oak, was successfully controlled in Spain by the use of insecticides. However, many other insects were also killed, including parasites of other oak pests like the green oak-roller, *Tortrix viridana*. This moth, in the absence of parasites, increased rapidly, causing considerable damage, and this new problem has still not been solved. A similar problem has developed in several countries through spraying DDT on fruit trees to control insect pests of apples and pears. The red spider mite, *Panonychus ulmi*, has shown a rapid increase due partly to a lack of predators and partly to increased fecundity caused by the insecticide (figure 17.3).

The development of resistant strains is another unexpected consequence of the widespread use of artificial insecticides. By 1964 at least 140 species had acquired a resistance to various insecticides, eighty of these, like the mosquito, being vectors of diseases.

·2.3   Radioactive pollution

Radioactive pollution posed an important threat to the future of man and the whole biosphere before the cessation of nuclear tests. Data relating to explosions carried out between 1945 and 1962 are summarized in table 17.3.

There are three ionizing radiations of ecological importance. Alpha (α) particles are formed from the nucleus of the helium atom, and are positively charged. They do not penetrate far and, in the case of α particles from thorium C, can be stopped by a water film fifty μm thick, or a layer of air eight cm thick. Beta (β) particles are high-speed electrons and are negatively charged. They penetrate further, travelling up to eight cm in living tissue. The non-particulate gamma (γ) rays are electromagnetic radiations similar to X-rays. As they have no electrical charge, they are deep penetrating.

The effect of radiation depends on its energy and intensity, the latter being determined by the number of particles emitted in unit time. Radiation intensity is measured in *curies* (symbol Ci). The curie is the amount of material in which 3.7×10^{10} atoms disintegrate each second. Smaller units used in biology include the *millicurie* (mCi = 10^{-3} Ci) and the *microcurie* (μCi = 10^{-6} Ci). The radiation dose received by an organism is expressed as the amount of energy absorbed by a given mass of an organism. Units used include:

— the *roentgen* (R) is the radiation dose which results in the absorption of 10^{-2} J per litre of water, or 8.33×10^{-3} J per kg of air. It is now being replaced by the S.I. unit, coulomb/kilogramme.

— the *rad* (radiation absorbed dose) is the amount of radiation which deposits 10^{-2} J/kg of energy in living tissue.

— the *rem* (roentgen equivalent, man) is the amount of radiation which deposits in man an amount of energy equivalent to one roentgen of γ radiation.

The maximum dose to which man can be exposed without injury has been estimated at 0.3 rad per week for repeated exposure, and

TABLE 17.3   The total power of atomic explosions in the years 1945−62

| YEARS | POWER IN MEGATONNES | | |
	EXPLOSION BY FISSION	EXPLOSION BY FUSION	TOTAL
1945−51	0.5	0.8	1.3
1952−56	51.0	88.0	139.0
1957−58	40.0	85.0	125.0
1959−60	0.07	−	−
1961−62	101.0	337.0	438.0

TABLE 17.4   Average values for naturally occurring radiation (millirads per year)

RADIATION SOURCE	SEDIMENTARY ROCK AT SEA LEVEL	GRANITIC ROCK	
		at sea level	at 3300 m altitude
Cosmic rays	35	35	100
Potassium-40 in living tissues	17	17	17
Radionuclides in rocks and soil	23	90	90
Total	75	142	207

twenty-five rad for a single dose. Naturally occurring radiation varies with altitude and sub-soil and some values are given in table 17.4.

Radiosensitivity varies between one organism and another. The lethal dose is of the order of 10^6 rads for bacteria, 10^5 rads for insects, and 10^3 rads for mammals. Radiosensitivity varies with age, children under four years of age accumulating strontium-90 more rapidly than adults, and appearing to be more sensitive to radiation. This is the reason it was feared, before nuclear tests were discontinued, that the danger level would be reached within fifteen years.

The half-life of a radionuclide is the time taken for fifty per cent of its atoms to disintegrate. Some elements have a very brief half-life and are of little ecological interest. The more important radionuclides are those occurring in radioactive fallout which pollute the biosphere, and also those components of protoplasm that can be used as markers in metabolic studies (table 17.5).

Radioactive pollution arises from the disposal of industrial wastes and from the fall-out produced by testing nuclear weapons. The most important pollutants are strontium-90, iodine-131 and caesium-137, all of which occur in the human body. Iodine is found in the thyroid, while strontium accumulates in the skeleton because its chemical properties are similar to those of calcium. The damage caused by the accumulation

TABLE 17.5   Some data relating to the more important radionuclides

NUCLIDE	HALF-LIFE	TYPE OF RADIATION EMITTED
carbon-14	5568 years	β
potassium-42	12.4 hours	β,γ
zinc-65	250 days	β,γ
iodine-131	8 days	β,γ
strontium-90	28 years	β
caesium-137	33 years	β,γ
plutonium-239	2.4×10^4 years	α,γ
cobalt-60	5.27 years	β,γ

of radionuclides in the tissues may affect both the individual (e.g. development of cancer) and the species. Genetic changes may occur, resulting in an increased rate of mutation and the production of abnormal progeny. This danger is made even more serious by the concentration of radionuclides along food chains. The problem is especially important in the tundra of Lapland, where the ecosystem is fairly simple, the main food chain passing from lichens, through the reindeer, up to the Lapps themselves. Lichens, which form the main source of reindeer food in winter, have a slow growth rate, and obtain a large proportion of their nutrients from the air and rainwater. The Lapps, feeding on reindeer, consume up to two kg of meat per day. Fallout from extensive nuclear tests carried out in 1961 and 1962 in the Arctic contained caesium-137, polonium-210 and iron-59, caesium being the most abundant. The fate of these radionuclides in man has been monitored since 1961. The highest radiation dose recorded for Lapps was 1.5 μ Ci in 1965, and since that time there has fortunately been a gradual decline in radioactivity due to the cessation of nuclear tests. It has been calculated that the radiation dose due to caesium-137 fallout alone in 1968 was seventy-five rem. During the period 1955–1985, by the end of which time the isotope should have disappeared, it has been estimated that Lapps living in the north of Finland will receive a total radiation dose of one rem, while Finns from the south of the country, not dependent on the reindeer, will receive only twenty-five mrem (about forty times less) (figure 17.4).

When radioactive phosphorus was added to a pond, insects present were found to concentrate this element by 500 times, while ducks increased the concentration by 7500 times. The concentration of radioactive phosphorus in duck eggs was 200 000 times the original concentration.

Radioactive pollution also results from industrial wastes. Disposal of wastes in the sea in water-tight barrels is a dangerous practice, since

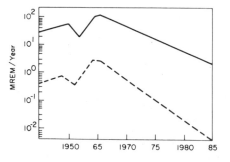

Figure 17.4   Caesium-137 levels in the body of
(a) Finns living in the south of Finland (dotted line)
(b) Lapp reindeer herdsmen at Inari in the north of Finland (continuous line).

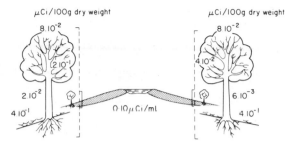

Figure 17.5   Ruthenium-106 levels (in μCi) in the leaves, branches, bark and litter of trees 60–80 m from a waste pit (Odum, 1959).

there is a risk that corrosion by sea water will liberate radioactive wastes before radioactivity has fallen to a safe level. The radioactivity of industrial wastes in the sea had reached two million curies by 1957. Burying this material in the soil would not solve the problem, however. Figure 17.5 shows how, in the case of the isotope ruthenium-106 placed in a waste pit, the isotope can move through the soil into oak trees located sixty to eighty metres away. These trees concentrated the radionuclide in the leaves and branches, and levels were particularly high in the leaf litter produced.

The effect of radiation on two populations of animals is illustrated in figure 17.6. Insects appear to be more tolerant, a dose of 3000 roentgens producing a lower mortality than in the control experiment. It is difficult to explain this unexpected result, but it seems possible that the action of radiation on the culture produces growth-stimulating substances.

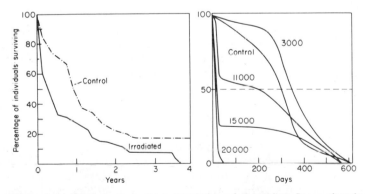

Figure 17.6   Left: Survival curves for a population of the rodent *Perognathus formosus* subjected to a continuous radiation dose of between 1 and 2 R per day.
Right: Survival curves for populations of *Tribolium* receiving doses of gamma radiation of 3000, 11000, 15000 and 20000 roentgens respectively, plus a control (after Cork, 1957).

3 THE EXTINCTION OF SPECIES AND DESTRUCTION OF
   ENTIRE ECOSYSTEMS

Man's activities in the biosphere have resulted in a reduction in numbers, or even extinction, of many species of animals and plants. Some detailed information is available for mammals and birds. One hundred and sixty-two species and sub-species of bird have become extinct since 1600 as the result of man's activities, and a further 381 species are threatened. About one hundred species of mammal have disappeared, and 255 species are on the verge of extinction. Forty-two per cent of Australian marsupial species are probably extinct or severely threatened. The time scale for these extinctions is available, and four familiar examples are given below:

1627: death of the last aurochs (*Bos taurus primigenius*), the ancestor of domestic cattle. This European ruminant still survived in France in the Middle Ages, and the last animal was killed in Poland.

1681: extinction of the dodo from the island of Mauritius. The Mascarene Islands (Mauritius, Reunion and Rodriguez) harbour birds that have lost the power of flight because there are no indigenous mammals and no predators. Since the early seventeenth century, when the islands were first colonised by man, the fauna of these islands has been significantly reduced, and twenty-four out of twenty-eight species of bird have disappeared. The most spectacular were the larger species weighing up to twenty kilogrammes and belonging to the pigeon family. They included the dodo (*Raphus cucullatus*) on Mauritius, the solitaire (*R. solitarius*) on Reunion, and the Rodriguez solitaire (*Pezophaps solitarius*). These defenceless birds were massacred by sailors who landed on the islands.

1870–1880: two species of South African zebra, Burchell's zebra and the quagga, were exterminated by the Boers.

1914: death in captivity at the Cincinnati Zoo of the last living passenger pigeon, *Ectopistes migratorius*. This North American bird formed enormous colonies, the population exceeding several thousand millions of birds in 1810. The pressure from hunting was so great that this species had disappeared from the wild by 1909.

Some reference must also be made to the long list of threatened species. The American bison and European bison have both been saved just in time. The Asiatic lion survives only in one Indian forest, where numbers have been reduced to only 150 individuals. The brown bear and birds of prey are becoming scarcer every day in France as the result of hunting, and measures should be taken to save them. Some mention must also be made of the large numbers of invertebrates threatened by pesticides, and the destruction of their habitats through cultivation.

Whole ecosystems are also threatened. Coastal sand dunes and beaches, with their characteristic animals and plants, are being replaced by buildings. Wet lands are being drained. Ancient forests, containing unique associations of insects and fungi, are clear-felled and replaced by coniferous woodland where life appears to be almost absent. At the present time, two-thirds of the tropical African forests have been destroyed. Four hundred years ago there were 170 million hectares of forest in the U.S.A., and now only eight million remain. The rate at which forests are destroyed by felling and burning is increasing in the Mediterranean region. The clearing of forest can have serious consequences. The vegetation that protects against erosion is removed. A hectare of beech may hold between 3000 and 5000 m³ of water, and loses 2000 m³ through transpiration. Catastrophic floods, such as those in Florence in 1968, are the direct result of a centuries old policy of deforestation in the surrounding catchment areas, and an attempt should be made to replant where this is possible. The planting of softwoods, a relatively recent development, is not always beneficial, conifers tending to acidify the soil, which gradually loses its fertility. These trees also burn easily and provide little shelter to animals. Deciduous trees would be more suitable in most instances.

II CONSERVATION

It seems fairly clear from the first section that the conservation of nature is one of the most important tasks facing the biologist at the present time.

1 SOIL CONSERVATION

This is an important consideration, and there are a number of possible methods. They include the reafforestation of waste arable land to prevent erosion, and carefully planned cropping programmes to avoid harmful practices like extended monoculture or dry farming. Overgrazing must be avoided, and goats, which tear up vegetation thus preventing regeneration, should be removed from those areas, such as the Mediterranean, where they continue to cause extensive damage and where there is a delicate balance. Itinerant agriculture, practised too frequently in tropical areas, can only be justified where the population is not too dense. In central Africa, for example, the cultivated area of ground is small and is surrounded by bush. After three or five years of cultivation, the areas are abandoned and rested for up to forty years. The original vegetation is re-established and the soil recovers. As the population density increases, however, this ground is put back into cultivation too quickly and the soil deteriorates. Itinerant agriculture is the scourge of central America and northern South America, where the shortage of available land and a rapidly increasing population create a difficult situation. The erosion that results has a long-term effect because it interferes with the water balance in the area.

2   THE CONSERVATION OF SPECIES AND OF ECOSYSTEMS

It soon becomes clear that a single species alone cannot be protected, but that the whole ecosystem, of which it is a part, must be conserved. The conservation of a single species may be justified for several reasons. Firstly, on aesthetic grounds. The protection of attractive countryside and the animals and plants living there can be justified for exactly the same aesthetic reasons as the conservation of ancient monuments. Secondly, for scientific and practical purposes. The diversity of living organisms, the result of a long evolution, is one of the most important factors contributing to the stability of the biosphere over a period of time (cf. p. 227). A reduction in the numbers of individuals or species in an ecosystem reduces its stability and biogeochemical activity.

Natural areas require protection because they may act as reservoirs for the improvement of domesticated plant and animal stocks. They may also contain chemical substances suitable for the control of pests, for example plant extracts like pyrethrins and rotenone, and for medicinal purposes. It is possible that compounds more effective than pyrethrum or penicillin will disappear without being discovered if ecosystems containing millions of species, known and unknown, are destroyed. The Nevada Indians have long used an extract from plants of the genus *Larrea* to prevent butter from becoming rancid. In 1950, Reichstein was awarded the Nobel prize in medicine for his work on the synthesis of cortisone from *Strophanthus* seeds. Other workers have since shown that the Mexican yam provides an even better source of cortisone. A new disease appears from time to time in rice or wheat or in cattle. One of the most effective ways of controlling this is to breed resistant strains and, in almost every instance, wild stock of the species must be used to take advantage of the variation in natural populations and so provide a greater chance of discovering the required resistant individuals. Wild stocks cannot be maintained in zoos, where the attention they receive results in a rapid loss of variability. The lemurs of Madagascar and some African monkeys are threatened with extinction. These species are of considerable scientific interest, so efforts are being made to rear some species on 'monkey farms'; but how many other species are likely to disappear?

It may also be necessary to revise our ideas about pest species in the interests of conservation, because few species are genuine pests. This is especially true of birds-of-prey, whose numbers in France, in common with the whole of western Europe, have been subjected to catastrophic reductions. This is due partly to unrestricted hunting (justified in terms of unreasonable prejudices due to ignorance, and carefully nurtured by vested interests such as cartridge manufacturers), and partly to the poisoning of many species through the accumulation of pesticide residues in their bodies. The accusation that birds-of-prey take large numbers of game birds is completely unjustified. These birds are only responsible for small losses, since they feed largely on small rodents (cf. p. 266). Game taken are usually diseased individuals, and in this way healthy stocks are maintained. Stories of ramblers attacked by eagles are usually due to unsubstantiated rumours or to journalists seeking sensational material.

The relative harmlessness of birds-of-prey with regard to game birds has been demonstrated by studies on the goshawk in Germany, and the diet of this bird was described on page 266, where the relative proportions of partridges, crows, hares, rabbits and pigeons taken as prey by a pair over a territory of 3700 hectares are recorded. Comparison of the graphs in figure 17.7 gives no grounds for assuming that the goshawk seriously interferes with numbers of partridges. The depredations of hunters are ten or twenty times more important. Analysis of captures made over the year shows a well-marked peak in March and April at the time of the nuptial displays. Between eighty and ninety per cent of partridges taken were males, as females were more shy and easily avoided the predators. Thus predation helps by

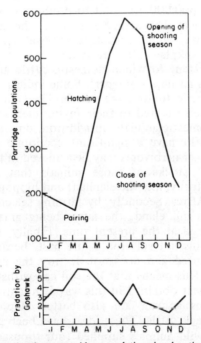

Figure 17.7   Effect of goshawk on partridge populations in a hunting reserve of 1700 ha near Hamburg.

Above: annual cycle of partridge.

Below: predation by goshawk (note that the scale on the abscissa is twenty times greater on the upper graph). Shooting is clearly the factor regulating numbers of partridges.

383

eliminating surplus males. Goshawks also play a useful role by attacking corvids, and poultry captures are so uncommon that farmers did not even comment on this. Domestic cats running wild probably cause much larger losses in farmyards.

Total stocks of the most threatened birds-of-prey in France include about fifty pairs of the golden eagle, between thirty and thirty-five pairs of Bonelli's eagle, about fifty booted eagles, about one hundred peregrines and about fifty eagle owls. These figures strongly support the law in France which seeks to protect all diurnal birds-of-prey larger in size than the buzzard, together with all nocturnal birds-of-prey, until such time as all birds are protected, as is the case in many neighbouring countries. Several other birds and mammals also receive the same amount of protection.

Careful measures have made it possible to save some species that are on the verge of extinction. The European bison is slowly increasing in numbers on the Bialowieza reserve at the frontier between Poland and Russia. The saiga antelope, which formerly ranged from Poland to Mongolia, appeared to be on the verge of extinction in 1920. Numbers are now maintained at about two million individuals, and this allows for the killing of 300 000 per year to produce about 6000 tonnes of excellent quality meat. This a good example of the rational exploitation of a species in a way which ensures its survival. Among those reasons put forward to explain the relatively low efficiency of meat production, the extent to which domestic cattle are adapted to their environment plays an important part. Some indigenous species, such as the reindeer of arctic tundra, the yak in Tibet and the llama in the Andes, are very well adapted to their environment, but this is not true for most other domestic animals. In addition, only sixteen species out of a total of 5000 have a significant economic role. The rational exploitation of larger herbivores may be achieved in three ways. Firstly, by the culling of stocks of those animals that are abundant and reproduce satisfactorily, like the elephant and hippopotamus in parts of central and east Africa. Secondly, by farming selected species like the gazelle, antelopes and eland, the latter being grazed with cattle in Kenya and Rhodesia at the present time. Finally, by setting up wild animal ranches. This last solution is the best, because the indigenous fauna is best adapted and does not destroy the natural grassland as quickly as cattle. This means that the soil can withstand drought more effectively. The meat obtained in this way is nutritious and has a good flavour. Indigenous animals are also better adapted to resist those diseases which attack cattle. These ideas have been put into practice in Rhodesia, where wild animals supplied four thousand tonnes of meat from thirty-three ranches in 1964. The wild fauna has also been exploited successfully in the Transvaal. In Tanzania, however, only five per cent of meat from the 3000 elephants killed annually as the result of control operations is eaten, and this represents a serious loss which could be avoided by better marketing methods.

In addition to the protection and conservation of threatened species, habitats that are particularly interesting can be given the special status of National Park or Nature Reserve to prevent further deterioration. Areas of outstanding scientific importance can be conserved in this way for future generations. Habitats can only be conserved after careful ecological studies, including the careful consideration of possible sources of interference and the prediction of their consequences. The unexpected changes in large mammal populations in the Albert National Park in Zaire after it was made into a reserve were described above (p. 3). The complete protection in Arizona of the mule deer, *Odocoileus*, after predators were destroyed, resulted in overpopulation. Widespread overgrazing occurred, followed by a rapid decline in numbers to well below the limiting capacity for the habitat (cf. p. 214). Red deer in the Swiss National Park became so numerous, in the absence of natural predators to check the population (brown bears and lynxes having been destroyed), that there was a danger that the plant cover would be destroyed through overgrazing. This problem has been solved by allowing the hunting of 250 animals at the edge of the park when the deer move down to their overwintering pastures.

These examples show how complex is the management of a reserve, particularly as man has almost entirely destroyed the natural balance in many places by eliminating predators. The conservation of plants is equally difficult. In California, where numbers of thousand year old *Sequoia sempervirens* grow, some trees growing on alluvial terraces were swept away in the floods of 1955. Total protection by barrages to control the floods would not be successful, however, as other forest species would grow in the absence of flooding and, by competition, replace the *Sequoia*. These trees are unique in their resistance to flood, wind and fire.

3 NATIONAL PARKS AND NATURE RESERVES

In many countries the problem of maintaining certain natural areas in a relatively undisturbed condition is achieved by creating National Parks and Nature Reserves. The Yellowstone National Park (U.S.A.) was created by an Act of Congress in 1872 to be an area where plants and animals would be sheltered from harm, and also as 'a pleasant land for the benefit and joy of the people'. Over 400 National Parks now exist in different parts of the world, and Nature Reserves cover 700 000 km² or 0.5% of the total land area. A Nature Reserve is briefly defined as an area where a particular habitat is completely protected, together with its flora and fauna. Human activity is reduced, and access is restricted to scientific workers. Examples in France include the Sept Iles Reserve and the Camargue. In Africa, the Garamba National Park in Zaire is a reserve of 491 000 hectares which is both inhabited and protected. National Parks are areas of some biological interest, but which are not prohibited to the public; however, industrial, commercial and other activities 'of a kind that would interfere with the natural development of the flora and fauna' are prohibited. This is the case in France, at least in theory, in the Vanoise Park and in the western Pyrenees. The Swiss National Park, which covers 15 000 hectares of Engadine, and the Grand Paradis Park (62 000 hectares) in Italy are two other examples. The conditions for the real protection of nature ought to be obvious. Is it reasonable to believe that everything possible has been done as soon as a small number of areas have been designated Nature Reserves or National Parks? It is already known that restricted populations tend to disappear quite quickly, and a small reserve which is enclosed, preventing free access to the surrounding area, would therefore provide only temporary protection to a species. The formation of a park around reserves, and of a peripheral zone round this park would partly solve the problem. Is it necessary to stop at this point? Unless the size of National Parks themselves is increased, they will become only temporary refuges, and in time the more interesting plants and animals will simply disappear. In order to be effective, the protection of nature needs to be extended to cover the whole country. There is little point in protecting certain areas if the remainder of the country is allowed to deteriorate. Conservation can be reconciled, for the majority of species, with the rational exploitation of the land. The attention of the authorities and the public generally must be drawn to this fact, and, if this were done effectively, then motorways would not be routed through large forests, such as Fontainebleau, against the advice of competent naturalists. Motorists would not ruin the countryside when they go out on a Sunday in search of fresh air and tear up plants to decorate their homes. Hunters would no longer see carnivores and birds-of-prey as rivals, but would, instead, understand that these animals have a part to play in maintaining the balance of nature. It will only be possible to conserve nature when the public as a whole becomes a determined guardian everywhere.

It is absolutely essential that all types of pollution should be controlled if the progressive poisoning of the biosphere is to be avoided. This is basically a technical and economic problem because, in most cases, the solution is either known or can easily be found. Techniques for the removal of carbon monoxide from car fumes, for cleaning smoke from factory chimneys, for purifying water contaminated by industrial effluents and sewage, exist but are not always employed. The use of biodegradable detergents and a reduction in the use of pesticides, through improved methods of control, are welcome developments.

Priority should be given in agriculture to improved cropping techniques to control the famine that persists in an endemic state in many parts of the world, where a high birth rate increases the extent of overpopulation daily.

It is possible to improve crop production somewhat by simply preventing waste and overcoming superstition. In India, where famine is perhàps most severe, there are herds of thousands of 'sacred cows' which may not be harmed in any way.

It is useful to refer back at this point to some data in chapter 13 relating to productivity in various ecosystems. A large part of the earth's surface has a low productivity, and one of the tasks of the scientist is to discover if this can be improved. Experience has shown that certain complex natural ecosystems, such as tropical rain forests, are very delicate formations and cannot be cultivated. The slightest interference causes them to disappear, and they are slowly replaced by desert.

The low ecological efficiency from one trophic level to the next is a strong argument in favour of adopting an entirely vegetarian diet. Research is aimed at selecting those varieties of plants which are rich in protein, and at finding ways of extracting this protein. Traditional fish farming in south China makes use of five or six species of fish having different diets, so that, by exploiting different trophic levels including detritivores, the highest level of efficiency can be reached (cf. p. 161).

It is more efficient to rear poikilotherms because homeotherms such as cattle use part of the energy they assimilate to maintain their body temperature. The large biomass of herbivores on the African savanna (cf. Table 6.1, p. 171) suggests that these animals might be used, since they represent a better resource than the traditional cattle, which are poorly adapted to this habitat.

The study of ecological succession (cf. p. 262) shows that, under natural conditions, there is a tendency for stable ecosystems to form, having a high biomass; i.e. the ratio

$$\frac{\text{gross productivity, } P}{\text{biomass, } B} \text{ is low.}$$

This is in conflict with man, who attempts to produce a high productivity and low biomass, i.e. with a high P/B ratio. This is the condition in young communities, of which arable fields are an example. Thus the situation that develops when man's efforts are applied systematically to the available land contravenes all ecological rules. The functions of mature ecosystems are ignored, including an aesthetic role (forests for example), a climatic role (regulation of the oxygen and carbon dioxide content of the atmosphere; stabilising effect with regard to climatic changes), and a geological role (water cycle, protection against erosion). It is essential that the diversity of nature is conserved and a mosaic of communities maintained at different stages of maturity so that man can become fully integrated with nature.

III CONCLUSION

The techniques of the nineteenth century are not necessarily suitable for the very different conditions at the end of the twentieth century. The task of the ecologist is to develop, on sound scientific principles, new and original techniques related to the total biosphere. Unless there is an urgent assessment of the size of the problem, the future of man is threatened. Hope for the rational management of our planet may be found in the International Biological Programme, which began in 1967. Nothing can be achieved, however, without the understanding and cooperation of all men.

It seems appropriate to conclude this chapter with a quotation from Professor J. G. Baer, the president of the special committee of the I.B.P.: 'At the present time, when vast sums of money are spent on the exploration of extraterrestrial space, one often tends to forget that biology, a science concerned with terrestrial phenomena, is indispensable to man, since it is due to biology that he is adequately fed, able to understand his illnesses and can improve his physical well-being . . . It is the earth that will decide the future of humanity.'

FURTHER READING

Many books have been published on ecology and related subjects. Some of the more useful textbooks are listed below for further reference, those marked with an asterisk having extensive bibliographies.

*Allee, W. C., Emerson, A. E., Park, O., Park, T. and Schmidt, K. P. (1949) *Principles of Animal Ecology*. W. B. Saunders Co., Philadelphia.

Andrewartha, H. G. (1970) *Introduction to the Study of Animal Populations*. 2nd Ed. Methuen & Co. Ltd, London.

*Andrewartha, H. G. and Birch, L. C. (1954) *The Distribution and Abundance of Animals*. University of Chicago Press, Chicago.

Bodenheimer, F. S. (1958) *Animal Ecology Today*. W. Junk, The Hague.

*Clark, L. R., Geier, P. W., Hughes, R. D. and Morris, R. F. (1967) *The Ecology of Insect Populations in Theory and Practice*. Methuen & Co. Ltd, London.

*Collier, B. D., Cox, G. W., Johnson, A. W. and Miller, P. C. (1973) *Dynamic Ecology*. Prentice-Hall, New Jersey.

Dempster, J. P. (1975) *Animal Population Ecology*. Academic Press, London and New York.

Ehrlich, P. R. and Ehrlich, A. H. (1972) *Population, Resources, Environment*. 2nd Ed. W. H. Freeman & Co., San Francisco.

Elton, C. (1927) *Animal Ecology*. Reprint 1966. Methuen & Co. Ltd, London.

Elton, C. (1958) *The Ecology of Invasions by Plants and Animals*. Methuen & Co. Ltd, London.

*Elton, C. (1966) *Pattern of Animal Communities*. Methuen & Co. Ltd, London.

*Geiger, R. (1957) *The Climate Near the Ground*. Harvard University Press, Cambridge, Mass.

*Greig-Smith, P. (1967) *Quantitative Plant Ecology*. 2nd Ed. Butterworths, London.

*Kershaw, K. A. (1973) *Quantitative and Dynamic Ecology*. 2nd Ed. Edward Arnold, London.

Klopfer, P. H. (1959) *Behavioural Aspects of Ecology.* Prentice-Hall, New Jersey.

*Krebs, C. J. (1972) *Ecology, The Experimental Analysis of Distribution and Abundance.* Harper & Row, New York and London.

Lewis, T. and Taylor, L. R. (1967) *Introduction to Experimental Ecology.* Academic Press, London and New York.

Macan, T. T. and Worthington, E. B. (1972) *Life in, Lakes and Rivers.* 2nd Ed. Collins, London.

Mellanby, K. (1972) *The Biology of Pollution.* Edward Arnold, London.

*Odum, E. P. (1971) *Fundamentals of Ecology.* 3rd Ed. W. B. Saunders Co., Philadelphia.

Owen, O. S. (1971) *Natural Resource Conservation: An Ecological Approach.* Collier-Macmillan Ltd, New York and London.

*Ricklefs, R. E. (1973) *Ecology.* Nelson, London.

Russell-Hunter, W. D. (1970) *Aquatic Productivity.* Collier-Macmillan Ltd, New York and London.

*Ruttner, F. (1963) *Fundamentals of Limnology.* 3rd Ed. University of Toronto Press, Toronto.

*Southwood, T. R. E. (1966) *Ecological Methods.* Methuen & Co. Ltd, London.

Wallwork, J. A. (1970) *Ecology of Soil Animals.* McGraw-Hill, London and New York.

Williamson, M. (1972) *The Analysis of Biological Populations.* Edward Arnold, London.

*Wynne-Edwards, V. C. (1962) *Animal Dispersion in Relation to Social Behaviour.* Oliver and Boyd, Edinburgh and London.

Yapp, W. B. (1972) *Production, Pollution, Protection.* Wykeham Publications Ltd, London.

The I.B.P. Handbooks (Blackwell Scientific Publications, Oxford) form an invaluable source of information on ecological methods. More than twenty volumes have been published to date in this series. The published symposia of the British Ecological Society (Blackwell Scientific Publications) contain many useful papers on a wide range of ecological topics. Two other series, *Advances in Ecological Research* and *Advances in Marine Biology* (Academic Press, New York and London) contain review articles summarizing specific areas of ecological research.

Ecological journals include the following:

Journal of Ecology
Journal of Animal Ecology
Journal of Applied Ecology
Oecologia
Limnology and Oceanography
Bulletin of the Marine Biological Association of the U.K.
Ecology

Ecological Monographs
Oikos
Freshwater Biology
In addition, popular articles relating especially to man and his environment frequently appear in *New Scientist, The Ecologist* and *Scientific American.*

REFERENCES

Adams, C. C. (1913) *Guide to the Study of Animal Ecology.* Macmillan, New York.

Adams, C. C. (1915) An Ecological Study of Prairie and Forest Invertebrates. *Bulletin of the Illinois State Laboratory of Natural History* 11, 33–280.

Allee, W. C. (1941) Integration of problems concerning protozoan populations with those of general biology. *American Naturalist* 75, 473–487.

Amanieu, M. (1969) Recherches ècologiques sur les faunes des plages abritées de la région d'Arcachon. *Helgolander Wissenschaftliche Mieresuntersuchungen* 19, 455–557.

Amiet, J. L. (1967) Les groupements de Coléoptères terricoles de la haute vallée de la Vesubie (Alpes-Maritimes). *Mémoires du Muséum nationale d'histoire naturelle* Série Zoologie. 46, 125–213.

Amiet, J. L. (1968) Variabilité, extension et synécologie de l'entomocenose à *Abax ater contractus* (Abacetum contracti) en Alpes-Maritimes. *Vie et Milieu* 19, 437–450.

Andrewartha, H. G. (1935) Thrips investigation No. 7. On the effect of temperature and food upon egg production and the length of adult life of *Thrips imaginis*, Bagnall. *Journal of the Council for Scientific and Industrial Research Australia* 8, 281–288.

Auer, C. (1961) Ergebnisse zwölfjahriger quantitativer Untersuchungen der Populationsbewegungen des Grauen Larchenwicklers *Zeiraphera griseana* Hb. im Oberengadin (1949–60). *Mitteilungen der Schweizerischen Anstalt für das forstliche Versuchswesen* 37, 175–263.

Augesse, M. P. (1957) La classification des eaux poikilohalines, sa difficulté en Camargue; nouvelle tentative de classification. *Vie et Milieu* 8, 341–365.

Bachelier, G. (1963) *La vie animale dans les sols.* Éditions de l'O.R.S.T.O.M., Paris.

Baker, J. R. (1938) The evolution of breeding seasons. In: *Evolution.* ed. G. R. de Beer. Clarendon Press, Oxford. 161–177.

Balachowsky, A. S. (1951) *La lutte contre les insectes. Principes, méthodes, applications*. Payot Éd., Paris.

Balogh, J. (1958) *Lebensgemeinschaften der Landtiere*. Akademie Verlag, Berlin.

Barley, K. P. (1961) Earthworms and salinity. *Advances in Agronomy* 13, 249–268.

Bates, M. (1945) Observations on climate and seasonal distribution of mosquitos in Eastern Colombia. *Journal of Animal Ecology* 14, 17–26.

Baudoin, R. (1955) La physio-chemie des surfaces dans la vie des arthropodes sereins surfaces. *Bulletin biologique de la France et de la Belgique* 85, 16–164.

Beauchamp, R. S. A. and Ullyott, P. (1932) Competitive relationships between certain species of fresh-water Triclads. *Journal of Ecology* 20, 200–208.

Berkner, L. V. and Marshall, L. C. (1964) The history of growth of oxygen in the earth's atmosphere. In: *The Origin and Evolution of Atmospheres and Oceans* (eds. D. J. Brancazio and A. G. W. Cameron). John Wiley & Sons, Inc., New York.

Bernard, F. (1951) *Les insectes sociaux du Fezzan. Comportement et biogéographie*. Mission scientifique du Fezzan (1944–1945). Institute pour Recherches sahariennes de l'Université d'Algerie.

Bertin, L. (1925) Recherches bionomiques, biochemiques et systématiques sur les Épinoches. *Annales de l'Institute d'Oceanographie, Monaco* 2, 1–204.

Bigot, L. (1965) Essai d'écologie quantitative sur les Invertébrés de la 'sansouire' camarguaise. *Mémoires de la Société Zoologique de France*.

Birch, L. C. (1945) The influence of temperature, humidity and density on the oviposition of the small strain of *Calandra oryzae* L. and *Rhizopertha dominica* Fab: (Coleoptera). *Australian Journal of Experimental Biology and Medical Sciences* 23, 197–203.

Blondel, J. (1964) L'avifaune nidificatrice des eaux saumâtres camarguaises en 1962 et 1963. *La Terre et la Vie* 309–330.

Blondel, J. (1965) Étude des populations d'oiseaux dans une garrigue méditerranéenne; description du milieu, de la méthode de travail et exposeé des premiers résultats obtenus à la période de reproduction. *La Terre et la Vie* 311–341.

Blondel, J. (1969) *Synécologie des Passereaux résidents et migrateurs dans un échantillon de la région méditerranéenne française*. C.D.R.P., Marseille.

Bodenheimer, F. S. (1926) Ueber die Voraussage der Generationenzahl von Insecten III. *Zeitschrift für Angewandte Entomologie* 12, 91.

Bodenheimer, F. S. (1928) Welche Factoren regulieren die Individuenzahl einer Insektenart in die Natur? *Biologisches Zentralblatt* 48, 714–739.

393

Bodenheimer, F. S. (1930) Uber die Grundlagen einer allgemeinen Epidemiologie der Insektenkalamitäten. *Zeitschrift für Angewandte Entomologie* 16, 433–450.

Bodenheimer, F. S. (1938) *Problems of animal ecology.* Clarendon Press, Oxford.

Bodenheimer, F. S. (1955) *Précis d'écologie animale.* Payot, Paris.

Bodenheimer, F. S. and Klein, H. J. (1930) Uber die Temperaturabhängigkeiten von Insekten II. Die Abhängigkeit der Aktivität bei der Ernteameise *Messor semirufus* E. Anderung von Temperature und anderen Faktoren. *Zeitschrift für vergleichende Physiologie* 11, 345–385.

Boness, M. (1953) Die Fauna der Wiesen unter besonderer Berucksichtigung der Mahd. *Zeitschrift für Morphologie und Okologie der Tiere* 42, 225–277.

Bonnet, L. (1964) Le peuplement thécamoebien des sols. *Revue d'écologie et de biologie du sol* 123–408.

Bornemissza, G. F. (1960) Could dung-eating insects improve our pastures? *Journal of the Australian Institute of Agricultural Science* 26, 54–56.

Bourlière, F. and Lamotte, M. (1962) Les concepts fondamentaux de la synécologie quantitative. *La Terre et la Vie* 16, 329–350.

Bourlière, F. and Verschuren, J. (1960) *Introduction à l'écologie des Ongulés du Parc National Albert.* Institut des Parcs Nationaux du Congo Belge, Bruxelles.

Braun-Blanquet, J. (1927) *Pflanzensoziologie.* Springer, Wien.

Bray, J. R. and Gorham, E. (1964) Litter production in forests of the ·world. *Advances in Ecological Research* 2, 101–157.

Brereton, J. le G. (1962) Evolved regulatory mechanisms of population control. Evolution of living organisms. *Symposia of the Royal Society of Victoria* 8, 81–93.

Brett, J. R. (1944) Some lethal temperature relations of Algonquin Park fishes. *University of Toronto Studies. Biological Series* 52, 201.

Bruun, A. R., Greve, S., Mielche, H. and Sparch, R. (1956) *The Galathea Deep Sea Expedition 1950–1952.* Trans. by R. Spink. The Macmillan Co., New York.

Buechner, H. K. and Golley, F. B. (1967) Preliminary estimation of energy flow in Uganda Kob (*Adenota kob thomasi* Neumann). In: *Secondary Productivity of Terrestrial Ecosystems.* Ed. K. Petrusewicz, Warszawa-Krakow. pp. 243–253.

Bullock, T. H. (1955) Compensation for temperature in the metabolism and activity of poikilotherms. *Biological Reviews* 30, 311–342.

Cailleux, A. (1953) *Biogéographie Mondiale.* Presse Universitaires, Paris.

Cain, A. J. and Sheppard, P. M. (1954) The theory of adaptive polymorphism. *American Naturalist* 88, 321–326.

Chapman, J. A. (1955) Towards an insect ecology. *Canadian Entomologist* 87, 172–177.

Chapman, R. N. (1928) The Quantitative Analysis of Environmental Factors. *Ecology* 9, 111–122.

Chapman, R. N. (1931) *Animal ecology with special reference to insects.* McGraw-Hill, New York.

Chauvin, R. (1957) Réflexions sur l'écologie entomologique. *Revue de zoologie agricole* 79p.

Chitty, D. (1967) The natural selection of self-regulatory behaviour in animal populations. *Proceedings of the Ecological Society of Australia* 2, 51–78.

Chopard, L. (1938) *Biologie des Orthoptères.* Paris.

Christian, J. J. (1961) Phenomena associated with population density. *Proceedings of the National Academy of Sciences of the United States of America* 47, 428–429.

Clarke, G. L. (1954) *Elements of Ecology.* Wiley, New York.

Clarke, G. L. and James, H. R. (1939) Laboratory analysis of the selective absorption of light by sea water. *Journal of the Optical Society of America* 29, 43–55.

Clarke, M. (1963) Economic importance of North Atlantic squids. *New Scientist* 17, 568–570.

Clements, F. E. (1916) Plant succession: analysis of the development of vegetation. *Publications of the Carnegie Institute, Washington* 242, 1–512.

Clements, F. E. and Shelford, V. E. (1939) *Bioecology.* Wiley, New York.

Coker, R. E. (1939) The problem of cyclomorphosis in *Daphnia. Quarterly Reviews of Biology* 14, 137–148.

Connell, J. H. (1961) The influence of interspecific competition and other factors on the distribution of the barnacle, *Chthamalus stellatus. Ecology* 42, 710–723.

Connell, J. H. and Orias, E. (1964) The ecological regulation of species diversity. *American Naturalist* 98, 399–414.

Corbet, P. S. (1957) The life-history of the Emperor dragonfly, *Anax imperator* Leach (Odonata: Aeshnidae). *Journal of Animal Ecology* 26, 1–69.

Cork, J. M. (1957) Gamma-radiation and longevity of the flour beetle. *Radiation Research* 7, 551–557.

Crombie, A. C. (1942) The effect of crowding upon the oviposition of grain-infesting insects. *Journal of Experimental Biology* 19, 311–340.

Crombie, A. C. (1943) The effect of crowding upon the natality of grain-infesting insects. *The Journal of Zoology* 113, 77–98.

Crossley, D. A. (1963) Consumption of vegetation by insects. In: *Radioecology.* Ed. V. Schultz and A. W. Klement. Reinhold Publishing Company, New York. pp. 427–430.

Crossley, D. A. (1964) Biological elimination of radionuclides. *Nuclear Safety* 5, 265–268.

Crossley, D. A. (1966) Radioisotope measurement of food consumption by a leaf beetle species, *Chrysomela knabi* Broun. *Ecology* 47, 1–8.

Dajoz, R. (1966) Ecologie et biologie des Coléoptères xylophages de la hêtraie. *Vie et Milieu* series C, 523–763.

Danilyevsky, A. S. (1957) Photoperiodism as a factor in the formation of geographical races in insects. *Entomologicheskoe obozrenie* 36, 5–27.

Darwin, C. (1859) *The Origin of Species.* Murray, London.

Darwin, C. (1881) *The formation of vegetable mould, through the action of worms.* Murray, London.

Davenport, C. B. (1903) *The animal ecology of the Cold Spring Harbor Sand spit with remarks on the theory of adaptation.* Decennial Publications of the University of Chicago, Chicago.

Davidson, J. (1938) On the ecology of the growth of the sheep populations in South Australia. *Transactions of the Royal Society of South Australia* 62, 141–148.

Davidson, J. (1944) On the relationship between temperature and rate of development of insects at constant temperatures. *Journal of Animal Ecology* 13, 26–38.

Deevey, E. S. (1947) Life tables for natural populations of animals. *Quarterly Review of Biology* 22, 283–314.

Deevey, E. S. (1950) The probability of death. *Scientific American* 182, 58–60.

Delamare Deboutteville, C. (1960) *Biologie des eaux souterraines littorales et continentales.* Hermann, Paris.

Dempster, J. P. (1960) A quantitative study of the predators on the eggs and larvae of the broom beetle, *Phytodecta olivacea* Forster, using the precipitin test. *Journal of Animal Ecology* 29, 149–167.

Dempster, J. P. (1961) A sampler for estimating populations of active insects upon vegetation. *Journal of Animal Ecology* 30, 425–427.

Dice, L. R. (1952) *Natural Communities.* University of Michigan Press, Ann Arbor.

Doane, J. R. (1961) Movement on the soil surface of adult *Ctenicera aeripennis destructor* Brown and *Hypolithus bicolor* Esch. (Coleoptera, Elateridae) as indicated by funnel pitfall traps, with notes on captures of other arthropods. *Canadian Entomologist* 93, 636–644.

Dobzhansky, T. (1947) Adaptive changes induced by natural selection in wild populations of Drosophila. *Evolution* 1, 1–16.

Dobzhansky, T. (1950). Evolution in the tropics. *American Science* 38, 209–221.

Dorst, J. (1963) Les techniques d'échantillonnage des populations d'oiseaux. *La Terre et la Vie* 180–202.

Dorst, J. (1967) Considérations zoogéographiques et écologiques sur les oiseaux des hautes Andes. In: *Biologie de l'Amérique Australe.* Documents biogéographiques, Editions du Centre National de la Recherche Scientifique, Paris. pp. 471—504.

Downes, J. A. (1965) Adaptations of insects in the arctic. *Annual Review of Entomology* 10, 257—274.

Dreux, P. (1962) Recherches écologiques et biogéographiques sur les Orthoptères des Alpes françaises. *Annales de Sciences Naturelles, zoologie* 325—766.

Dudich, E., Baloga, J. and Loksa, I. (1952) Produktionsbiologische Untersuchungen über die Arthropoden der Waldböden. *Acta Biologica Academiae Scientarum Hungaricae* 3, 295—317.

Dussart, B. (1966) *Limnologie. L'étude des eaux continentales.* Gauthicr-Villars, Paris.

Duvigneaud, P. (1967) *L'écologie, science moderne de synthèse.* Volume 2: *ecosystèmes et biosphère.* Ministère de l'Education Nationale et de la Culture, Bruxelles.

Dybas, H. S. and Lloyd, M. (1962) Isolation by habitat in two synchronized species of periodical cicadas (Homoptera: Cicadidae: Magicicada). *Ecology* 43, 432—444.

Edney, E. B. (1953) The temperature of woodlice in the sun. *Journal of Experimental Biology* 30, 331—349.

Edwards, C. A. and Heath, G. W. (1963) The role of soil animals in breakdown of leaf material. In: *Soil organisms.* Ed. J. Doeksen & J. van der Drift. North-Holland Publ. Co., Amsterdam.

Edwards, D. K. (1961) Activity of two species of *Calliphora* (Diptera) during barometric pressure changes of natural magnitude. *Canadian Journal of Zoology* 39, 623—635.

Elton, C. and Miller, R. S. (1954) The ecological survey of animal communities: with a practical system of classifying habitats by structural characters. *Journal of Ecology* 42, 460—496.

Emberger, L. (1955) Afrique du nord-ouest. *In* Plant ecology, reviews of research. Unesco, Paris. *Arid Zone Research* 6, 219—249.

Emlen, J. T. (1940) Sex and age ratios in survival of the California Quail. *Journal of Wildlife Management* 4, 92—99.

Errington, P. L. (1945) Some contributions of a 15-year local study of the northern bobwhite to a knowledge of population phenomena. *Ecological Monographs* 15, 1—34.

Errington, P. L. (1956) Factors limiting higher vertebrate populations. *Science* 124, 304—307.

Ferry, C. (1964) Un dénombrement d'oiseaux nicheurs dans 16 hectares en forêt de Citeaux, printemps 1963. *Le Jean le Blanc* 3, 4—9.

Ferry, C. and Frochot, B. (1958) Une méthode pour dénombrer les oiseaux nicheurs. *La Terre et la Vie* 85—102.

Fisher, J. and Vevers, H. G. (1944) The breeding distribution, history and population of the north Atlantic gannet (*Sula bassana*): Changes in the world numbers of the gannet in a century. *Journal of Animal Ecology* 13, 49–62.

Fisher, R. A., Corbet, A. S. and Williams, C. B. (1943) The relation between the number of species and the number of individuals in a random sample of an animal population. *Journal of Animal Ecology* 12, 42–58.

Flemming, R. H. and Laevastu, T. (1956) The influence of hydrographic conditions on the behaviour of fish. *FAO Fisheries Bulletin* 9, 181–196.

Forbes, E. (1844) Report on the Mollusca and Radiata of the Aegean Sea, and on their distribution considered as bearing on geology. *Report of the British Association for the Advancement of Science* 13, 130–193.

Forbes, S. A. (1887) The lake as a microcosm. *Bull. Sc. A. Peoria*, 77–87.

Fox, C. J. S. and Maclellan, C. R. (1956) Some Carabidae and Staphylinidae shown to feed on a wireworm, *Agriotes sputator* (L) by the precipitin test. *Canadian Entomologist*, 88, 228–231.

Fraenkel, G. S. and Blewett, M. (1944) The utilization of metabolic water in insects. *Bulletin of Entomological Research* 35, 127–137.

Frank, P. W. (1957) Coactions in laboratory populations of two species of *Daphnia. Ecology* 38, 510–518.

Friedrichs, K. (1930) Zur Epidemiologie des Kiefernspanners. *Zeitschrift für Angewandte Entomologie* 16, 197–205.

Fry, F. E. J., Brett, J. R. and Clausen, G. H. (1942) Lethal limits of temperature for young goldfish. *Revue canadienne de biologie* 1, 50–56.

Gaarder, T. and Gran, H. H. (1927) Investigations of the production of plankton in the Oslo Fjord. *Rapport et procés-verbaux des réunions. Conseil permanent international pour l'exploration de la mer* 42, 1–48.

Gadgil, M. and Solbrig, O. T. (1972) The concept of r- and K-selection: evidence from wild flowers and some theoretical considerations. *American Naturalist* 106, 14–31.

Gams, H. (1918) Prinzipienfragen der Vegetationsforschlung. Ein Begriffsklarung und Methodik der Biocoenologie. *Naturforschung Gesellschaft, Zurich, Vierteljahresschrift* 63, 293–493.

Gause, G. F. (1931) The influence of ecological factors on the size of population. *American Naturalist* 65, 70–76.

Gause, G. F. (1935) Vérifications expérimentales de la théorie mathématique de la lutte pour la vie. *Actualités Scientifiques et Industrielles* 277, 1–61.

Gillon, Y. and Gillon, D. (1965) Recherche d'une méthode quantitative d'analyse du peuplement d'un milieu herbacé. *La Terre et la Vie* 378–391.

Gisin, H. (1947) Analyses et synthèses biocénotiques. *Archives des sciences physiques et naturelles* 5, 42–75.

Gisin, H. (1949) Exemple du développement d'une biocénose dans un tas de feuilles en décomposition. *Mitteilungen der Schweizerischen entomologischen Gesellschaft* 22, 422.

Goethe, F. (1961) A survey of moulting Shelduck on Knechtsand. *British Birds* 54, 106–115.

Golley, F. B. (1960) Energy dynamics of a food chain of an old-field community. *Ecological Monographs* 30, 187–206.

Gounot, M. (1969) *Méthodes d'étude quantitative de la végétation.* Masson, Paris.

Govaerts, J. and Leclercq, J. (1946) Water exchange between insects and air moisture. *Nature, London* 157, 483.

Grasse, P. P. (1929) Étude écologique et biogéographique sur les Orthoptéres français. *Bulletin Biologique* 63, 489–539.

Grasse, P. P. (1951) *Traité de Zoologie: X. Insectes Supérieurs et Hemipteroides. Part II.* Masson, Paris.

Grasse, P. P. (1965) Les effets de groupe et les actions psycho-somatiques chez les insectes. *Proceedings of the XIIth International Congress of Entomology* 52–58.

Grasse, P. P. and Chauvin, R. (1944) *Revue scientifique* 82, 461–464.

Grüm, L. (1959) Sezonowe zmiany aktywnosci biegaczowatych (Carabidae). *Ekologia Polska* A 7, 255–268.

Guinochet, M. (1938) *Etudes sur la végétation de l'étage alpin dans le Bassin supérieur de la Tinée (Alpes-Maritimes).* Thèse de Doctorat ès-sciences. Lyon, Bosc. et Riou, and Communication no. 59, S.I.G.M.A.

Guinochet, M. and Casal, P. (1957) Sur l'analyse différentielle de Czekanowski et son application à la phytosociologie. *Bulletin du Service de la carte Phytogéographique*, Series B. 2, 25–33.

Haddow, A. J. (1947) The mosquitos of Bwamba County, Uganda. V. The vertical distribution and biting cycle of mosquitos in rain forest, with further observations on microclimates. *Bulletin of Entomological Research* 37, 301–330.

Haeckel, E. (1870) Ueber Entwickelungsgang u. Aufgabe der Zoologie. *Jenaische Zeitschrift* 5, 353–370.

Hairston, N. G., Smith, F. E. and Slobodkin, L. B. (1960) Community structure, population control and competition. *American Naturalist* 94, 421–425.

Hall, F. G. (1922) The vital limits of exsiccation of certain animals. *Biological Bulletin* 42, 31–51.

Hall, R. R., Downe, A. E. R., Maclellan, C. R. and West, A. S. (1953) Evaluation of insect predator-prey relationships by precipitin test studies. *Mosquito News* 13, 199–204.

Hamilton, A. G. (1950) Further studies on the relationship of humidity and temperature to the development of two species of African locusts – *Locusta migratoria migratorioides,* R. and F., and *Schistocerca gegaria,* Forsk. *Transactions of the Royal Entomological Society of London* 101, 2–56.

Hardy, A. C. (1924) The herring in relation to its animate environment. *Fishery Investigations. Ministry of Agriculture, Fisheries and Food Series 2* 7, no. 3.

Harker, J. E. (1956) Factors controlling the diurnal rhythmn of activity of *Periplaneta americana. Journal of Experimental Biology* 33, 224–234.

Hassell, M. P. and Varley, G. C. (1969) New inductive population model for insect parasites and its bearing on biological control. *Nature, London* 223, 1133–1137.

Hazard, T. P. and Eddy, R. E. (1950) Modification of the sexual cycle in the brook trout (*Salvelinus fontinalis*) by control of light. *Transactions of the American Fisheries Society* 80, 158–162.

Heatwole, H. and Davis, D. M. (1965) Ecology of three sympatric species of parasitic insects of the genus *Megarhyssa* (Hymenoptera: Ichneumonidae). *Ecology* 46, 140–150.

Heim de Balzac, H. (1936) Biogéographie des mammifères et des oiseaux de l'Afrique du Nord. *Bulletin biologique de la France et de la Belgique* 21, 1–446 (supplement).

Hinton, H. E. (1960) A fly larva that tolerates dehydration and temperatures of $-270°$ to $+102°C$. *Nature, London* 188, 336–337.

Hjort, J. (1926) Fluctuations in the year classes of important food fishes. *Journal du Conseil Permanent Internationale pour L'Exploration de la Mer* 1, 1–38.

Hock, R. J. (1964) Terrestrial animals in cold: reptiles. In: *Handbook of Physiology, Section 4, Adaptation to the Environment.* pp. 357–359. American Physiological Society, Washington D.C.

Holling, C. S. (1961) Principles of insect predation. *Annual Review of Entomology* 6, 163–182.

Holling, C. S. (1968) The tactics of a predator. *Symposia of the Royal Entomological Society of London* 4, 47–58.

Holme, N. A. (1950) Population dispersion in *Tellina tenuis* Da Costa. *Journal of the Marine Biological Association UK* 29, 267–280.

Howard, L. O. and Fiske, W. F. (1911) The importation into the United States of the parasites of the gipsy moth and the brown tail moth. *Bulletin of the United States Bureau of Entomology* No. 91.

Huffaker, C. B. (1958) Experimental studies on predation: dispersion factors and predator-prey oscillations. *Hilgardia* 27, 343–383.

Hutchinson, G. E. (1967) *A treatise on limnology.* Vol. II. Wiley & Sons, London.

Ibbotson, A. (1958) The behaviour of the frit fly in Northumberland. *Annals of Applied Biology* 46, 474–479.

Jaccard, P. (1912) The distribution of the flora in the alpine zone. *New Phytologist.* 11, 37–50.

Jenkins, D., Watson, A. and Miller, G. R. (1963) Population studies on red grouse, *Lagopus lagopus scoticus* (Lath), in northeast Scotland. *Journal of Animal Ecology* 32, 317–376.

Jespersen, D. (1924) On the frequency of birds over the high Atlantic Ocean. *Nature, London* 114, 281–283.

Johnson, D. W. and Odum, E. P. (1956) Breeding bird populations in relation to plant succession on the Piedmont of Georgia. *Ecology* 37, 50–62.

Jones, J. R. E. (1949) A further ecological study of a calcareous stream in the "Black Mountain" district of south Wales. *Journal of Animal Ecology* 18, 142–159.

Jourdheuil, P. (1960) Influence de quelques facteurs écologiques sur les fluctuations de population d'une biocénose parasitaire. Étude relative à quelques Hyménoptères parasites de divers Coleoptères des Crucifères. *Annales des Epiphyties* 224.

Juday, C. (1942) The summer standing crop of plants and animals in four Wisconsin lakes. *Transactions of the Wisconsin Academy of Science* 34, 103–135.

Kalela, O. (1949) Uber Fjeldlemming-Invasionen und andere irreguläre Tierwanderungen. *Annales zoologici Societatis zoologico-botanicae fennicae* 13, 1–90.

Kendeigh, S. C. (1961) *Animal Ecology*. Prentice Hall, New Jersey.

Kettlewell, H. B. D. (1961) The phenomenon of industrial melanism in Lepidoptera. *Annual Review of Entomology* 6, 245–262.

Kirmiz, J. P. (1962) *Adaptation de la gerboise au milieu désertique. Etude comparée de la thermorégulation chez la gerboise* (Dipus aegyptus) *et chez le rat blanc*. Societé Publicateur Egyptienne, Alexandria.

Klopfer, P. H. (1962) *Behavioural aspects of ecology*. Prentice Hall, New Jersey.

Klopfer, P. H. and MacArthur, R. H. (1961) On the causes of tropical species diversity: Niçhe overlap. *American Naturalist* 95, 223–226.

Kohn, A. J. and Helfrich, P. (1957) Primary organic productivity of a Hawaiian coral reef. *Limnology and Oceanography* 2, 241–251.

Kontkanen, P. (1957) On the delimitation of communities in research on animal biocoenotics. In: *Population Studies; Animal Ecology and Demography*. Ed. M. Demerec. Cold Spring Harbor Symposia on Quantitative Biology 22, 373–378.

Krebs, C. J., Gaines, M. S., Keller, B. L., Myers, J. H. and Tamarin, R. H. (1973) Population cycles in small rodents. *Science* 179, 35–41.

Krebs, C. J. and Myers, J. H. (1974) Population cycles in small mammals. *Advances in Ecological Research* 8, 267–399.

Kuhnelt, W. (1965) *Grundriss der Okologie*. Fischer, Jena.

Labeyrie, V. (1960) Contribution à l'étude de la dynamique des populations. I: influence stimulatrice de l'hôte *Acrolepia assectella* sur la multiplication d'un Hyménoptère Ichneumonidae (*Diadromus* sp.). *Entomophaga, Mémoires hors série* 9, 193.

Lack, D. L. (1945) Ecology of closely related species with special reference to cormorant (*Phalacrocorax carbo*) and shag (*P. aristotelis*). *Journal of Animal Ecology* 14, 12−16.

Lack, D. L. (1947) *Darwin's Finches.* Cambridge University Press, Cambridge.

Lack, D. L. (1966) *Population Studies of Birds.* Clarendon Press, Oxford.

Lamotte, M. and Bourlière, F. (1969) *Problèmes d'écologie. L'échantillonnage des peuplements animaux des milieux terrestres.* Masson, Paris.

Lees, A. D. (1953) Environmental factors controlling the evocation and termination of diapause in the fruit tree red spider mite *Metatetranychus ulmi* Koch (Acarina: Tetranychidae). *Annals of Applied Biology* 40, 449.

Lees, A. D. (1955) *The physiology of diapause in arthropods.* Cambridge University Press, Cambridge.

Lemee, G. (1967) *Précis de biogéographie.* Masson, Paris.

Leopold, A. (1943) Deer Irruptions. *Wisconsin Conservation Department Publications* 321, 1−11.

Levêque, R. (1957) L'avifaune nidificatrice des eaux saumâtres camarguaises en 1956. Essai de recensement suivi d'une première esquisse écologique. *La Terre et la Vie* 150−178.

Liebig, J. (1840) *Chemistry in its application to agriculture and physiology.* Taylor and Walton, London.

Lieth, H. (1962) *Die Stoffproduktion der Pflanzendecke.* Fischer, Verlag.

Lindeman, R. L. (1942) The trophic-dynamic aspect of ecology. *Ecology* 23, 399−418.

Lindroth, C. H. (1949) Die fennoskandischen Carabidae. Ein Tiergeographische Studie. *Vetensk. Samh. Handl. Goteborgs* 3.

Lotka, A. J. (1925) *Elements of Physical Biology.* Williams & Wilkins, Baltimore.

Lotka, A. J. (1934) Théorie analytique des associations biologiques. *Actualités Scientifiques et Industrielles* 187, 1−45.

MacArthur, R. H. (1955) Fluctuations of animal populations and a measure of community stability. *Ecology* 36, 533−536.

MacArthur, R. H. (1964) Environmental factors affecting bird species diversity. *American Naturalist* 98, 387−397.

MacArthur, R. H. and MacArthur, J. (1961) On bird species diversity. *Ecology* 42, 594−598.

MacFadyen, A. (1963) *Animal Ecology, Aims and Methods.* 2nd Ed. Pitman, London.

MacFadyen, A. (1966) Les méthodes d'étude de la productivité des

Invertébrés dans les écosystèmes terrestres. *La Terre et la Vie* 20, 361–392.

MacLulich, D. A. (1937) Fluctuations in the numbers of the varying hare (*Lepus americanus*). *University of Toronto Studies. Biological Series.* no. 43.

Maillet, P. (1959) Essai sur l'écologie des Jassides praticoles du Périgord noir. Contribution à l'étude des Homoptères Auchénorhynques II. *Vie et Milieu* 10, 117–134.

Malthus, T. R. (1798) *An essay on the principle of population.* Johnson, London.

Margalef, R. (1963) On certain unifying principles in ecology. *American Naturalist* 97, 357–374.

Martonne, E. de (1926) Aréisme et indice d'aridité. *Compte rendu hebdomadaire des séances de l'Academie des Sciences* 182, 1395–1398.

Marty, R. (1968) *Recherches écologiques et biochimiques sur les Orthoptères des Pyrénées.* Thèse, Toulouse.

Mayr, E. (1963) *Animal Species and Evolution.* Harvard University Press, Harvard, Mass.

Mellanby, K. (1939) Low temperature and insect activity. *Proceedings of the Royal Society of London*, B 127, 473–487.

Michal, K. (1931) Die Beziehung der Populationsdichte zum Lebensoptimum und Einfluss des Lebensoptimum auf das Zahlenverhältnis der Geschlechter bei-Mehlwurm und Stubenfliege. *Biologia generalis* 7, 631–646.

Miller, R. S. (1964) Ecology and distribution of pocket gophers (Geomyidae) in Colorado. *Ecology* 45, 256–272.

Milne, A. (1957) The Natural Control of Insect Populations. *Canadian Entomologist* 89, 193–213.

Milne, A. (1964) Biology and ecology of the garden chafer, *Phyllopertha horticola* (L) IX. Spatial distribution. *Bulletin of Entomological Research* 54, 761–795.

Milne, A. (1965) Pest ecology and integrated control. *Annals of Applied Biology* 56, 338–341.

Mobius, K. (1877) *Die Auster und die Austernwirtschaft.* Wiegumdt, Hempel & Parey, Berlin.

Monchadskii, A. S. (1958) [On the classification of environmental factors] *Zoologicheskii Zhurnal* 37, 680–692.

Monchadskii, A. S. (1961) [Concepts of factors in ecology]. *Zoologicheskii Zhurnal* 40, 1299–1303.

Moriarty, F. (1972) Pollutants and food chains. *New Scientist* 53, 594–596.

Morris, R. F. (1955) The development of sampling techniques for forest insect defoliators, with particular reference to the spruce budworm. *Canadian Journal of Zoology* 33, 225–294.

Morris,. R. F. (1957) The interpretation of mortality data in studies on population dynamics. *Canadian Entomologist* 89, 49–69.

Morris, R. F. (1960) Sampling insect populations. *Annual Review of Entomology* 5, 243–264.

Morris, R. F. (1963) The dynamics of epidemic Spruce Budworm populations. *Memoirs of the Entomological Society of Canada* 31, 1–332.

Moss, R. (1967) Probable limiting nutrients in the main food of red grouse (*Lagopus lagopus scoticus*). In: *Secondary Productivity of Terrestrial Ecosystems.* Ed. K. Petrusewicz. Warszawa-Krakow. pp. 369–379.

Murdie, G. and Hassell, M. P. (1973) Food distribution, searching success and predator-prey models. In: *The Mathematical Theory of the Dynamics of Biological Populations.* Ed. M. S. Bartlett and R. W. Hiorns. Academic Press, London and New York.

Murie, A. (1944) Dall Sheep. Ch. 3 In: *Wolves of Mount McKinley.* National Parks Service Fauna No. 5, Washington.

Murphy, P. W. (1962) *Progress in Soil Zoology.* Butterworths, London.

Murray, J. and Hjort, J. *Depths of the Ocean.* Macmillan, London.

Nef, L. (1957) Etat actuel des connaissances sur le rôle des animaux dans la décomposition des litières de fofets. *Agriculture* 5, 245–316.

Nicholson, A. J. (1950) Population oscillations caused by competition for food. *Nature, London* 165, 476–477.

Nicholson, A. J. (1954) An outline of the dynamics of animal populations. *Australian Journal of Zoology* 2, 9–65.

Nicholson, A. J. (1957) The self-adjustment of populations to change. *Cold Spring Harbor Symposia on Quantitative Biology* 22, 153–173.

Nicholson, A. J. (1958) Dynamics of insect populations. *Annual Review of Entomology* 3, 107–136.

Nicholson, A. J. and Bailey, V. A. (1935) The balance of animal populations. *Proceedings of the Zoological Society* 551–598.

Nielsen, C. O. (1955) Studies on the Enchytraeidae. 2. Field Studies. *Natura jutlandica* 4, 1–58.

Nikolsky, G. V. (1963) *The Ecology of Fishes* (translated from Russian by L. Birkett) Academic Press, New York and London.

O'Connor, F. B. (1957) An ecological study of the enchytraeid worm population of a coniferous forest soil. *Oikos* 8, 161–199.

Odum, E. P. (1959) *Fundamentals of Ecology.* 2nd Ed. Saunders, London.

Odum, E. P., Connell, C. E., and Davenport, L. B. (1962) Population energy flow of three primary consumer components of old-field ecosystems. *Ecology* 43, 88–96.

Odum, E. P. and Kuenzler, E. J. (1963) Experimental isolation of food chains in an old-field ecosystem with use of phosphorus-32. In: *Radioecology.* Ed. V. Schultz and A. W. Klement. Reinhold Publishing Company, New York.

Odum, H. T. (1957) Trophic structure and productivity of Silver Springs, Florida. *Ecological Monographs* 27, 55–112.

Odum, H. T. and Odum, E. P. (1955) Trophic structure and productivity of a windward coral reef community on Einwetok Atoll. *Ecological Monographs* 25, 291–320.

Paine, R. T. (1966) Food web diversity and species diversity. *American Naturalist* 100, 65–75.

Palmblad, I. G. (1968) Competition in experimental populations of weeds with emphasis on the regulation of population size. *Ecology* 47, 26–34.

Park, O. (1930) Studies on the ecology of forest Coleoptera. Seral and seasonal succession of Coleoptera in the Chicago Area, with observations on certain phases of hibernation and aggregation. *Annals of the Entomological Society of America* 23, 57–80.

Park, O. and Keller, J. G. (1932) Studies in nocturnal ecology II. Preliminary analysis of activity rhythm in nocturnal forest insects. *Ecology* 13, 335–346.

Park, T. (1941) The laboratory population as a test of a comprehensive ecological system. *Quarterly Review of Biology* 16, 274–293; 440–461.

Park, T. (1946) Some observations on the history and scope of population ecology. *Ecological Monographs* 16, 313–320.

Payne, N. M. (1927) Measures of insect cold hardiness, *Biological Bulletin* 52, 449–457

Pearl, R. (1925) *The biology of population growth*. Knopf, New York.

Pennak, R. W. (1955) Comparative limnology of eight Colorado mountain lakes. *University of Colorado Studies. Biology Series* 2, 1–74.

Pères, J. M. and Deveze, L. (1961, 1963) *Océanographie biologique et biologie marine*. Tome I: la vie benthique, tome II: la vie pélagique. Presses Universitaires de France, Paris.

Pères, J. M. and Picard, J. (1949) Notes sommaires sur le peuplement des grottes sousmarines de la région de Marseille. *Compte rendu sommaire des séances de la Société de la biogéographie* 26, 42–46.

Perttunen, V. (1960) Seasonal variation in the light reactions of *Blastophagus piniperda* L. (Col. Scolytidae) at different temperatures. *Annales entomologici fennici* 26, 86–92.

Petrides, G. A. (1950) The determination of sex and age ratios in fur animals. *American Midland Naturalist* 43, 355–382.

Petrides, G. A. and Swank, W. G. (1965) *Estimating the productivity and energy relations of an African elephant population.* Report presented to the Ninth International Grasslands Congress, Sao Paulo, 1965.

Petter, F. (1961) Répartition géographique et écologie des Rongeurs désertiques. *Mammalia* 1–222.

Pfeffer, P. (1967) Le mouflon de Corse. Position systématique, écologie et éthologie comparées. *Mammalia* supplement pp. 262.

Pianka, E. R. (1966) Latitudinal gradients in species diversity: a review of concepts. *American Naturalist* 100, 33–46.

Pianka, E. R. (1970) On *r*- and *K*-selection. *American Naturalist* 104, 592–597.

Pierre, F. (1958) *Écologie et peuplement entomologique des sables vifs du Sahara nord-occidental.* C.N.R.S., Paris.

Pimentel, D. (1958) Alteration of microclimate imposed by populations of flour beetles (*Tribolium*). *Ecology* 39, 239–246.

Pinet, J. M. (1967) Observations sur le vol crépusculaire d'*Oligoneuriella rhenana* Imhoff. (Ephem. Oligoneuriidae). Importance de la température et de la luminosité lors de l'émergence. *Bulletin de la Société entomologique de France.* 72, 144–155.

Pittendrigh, C. S. (1960) Circadian rhythms and the circadian organisation of living systems. *Cold Spring Harbor Symposia on Quantitative Biology* 25, 159–184.

Pourriot, R. (1963) Utilisation des algues brunes unicellulaires, pour l'élèvage des rotifères. *Compte rendu hebdomadaire des séances de l'Académie des Sciences* 256, 1603–1605.

Prenant, M. (1934) *Adaptation, écologie et biocoenotique.* Actualités scientifiques et industrielles. Hermann, Paris.

Rabaud, E. (1937) *Phénomène social et sociétés animales.* Alcan Editions, Paris.

Rapp, M. (1970) *Contribution à l'étude du bilan et de la dynamique de la matière organique et des éléments minéraux biogènes dans les écosystèmes à chêne vert et à chêne kermès du midi de la France.* Thèse, Montpellier.

Raunkiaer, C. (1928) Dominansareal, Artstaethed og Formationsdominanter. *Biologiske Meddelelser* 7, 1.

Raunkiaer, C. (1934) *The life forms of plants and statistical plant geography.* Clarendon Press, Oxford.

Reaumur, R. A. F. (1734–1742) *Mémoires pour servir à l'histoire des insectes.* Paris (Impr. Royale), 1–6.

Reichle, D. E. (1970) *Analysis of Temperate Forest Ecosystems.* Chapman & Hall, London.

Renaud Debyser, J. (1963) Recherches écologiques sur la faune interstitielle des sables. Bassin d'Arcachon, Île de Bimini, Bahamas. Supplement to *Vie et Milieu.*

Reynoldson, T. B. (1966) The distribution and abundance of lake-dwelling triclads – towards a hypothesis. *Advances in Ecological Research* 3, 1–71.

Ricou, G. (1967) Etude biocénotique d'un milieu 'naturel', la prairie permanente pâturée. *Annales des épiphyties* 18, supplement.

Riley, G. A. (1944) The carbon metabolism and photosynthetic efficiency of the earth as a whole. *American Scientist* 32, 132–134.

Riley, G. A. (1957) Phytoplankton of the north central Sargasso Sea. *Limnology and Oceanography* 2, 252–270.

Roth, M. (1963) Comparaisons de méthodes de capture en écologie entomologique. *Revue de pathologie végétale et d'entomologie agricole de France* 42, 177-197.

Sacchi, C. F. (1952) I Raggruppamenti di Molluschi terrestri sul litorale italiano. Considerazioni e richerche introdutire. *Bollettino della Società veneziana di storia naturale e del Museo Civico de storia naturale* 7, 1–51.

Saint-Girons, H. and Saint-Girons, M. C. (1956) Cycle d'activité et thermorégulation chez les reptiles (Lézards et Serpents). *Vie et Milieu* 7, 133–226.

Saint Hilaire, G. (1854) *Histoire générale des règnes organiques.*

Sapin Jaloustre, J. (1960) *Écologie du Manchot Adélie.* Hermann, Paris.

Satchell, J. (1955) Some aspects of earthworm ecology. In: *Soil Zoology.* Ed. D. K. McKevan. Butterworths, London.

Savely, H. E. (1939) Ecological relations of certain animals in dead pine and oak logs. *Ecological Monographs* 9, 321–385.

Schmidt-Nielsen, K. (1964) *Desert Animals. Physiological Problems of Heat and Water.* O.U.P., London.

Schmidt-Nielsen, B., Schmidt-Nielsen, K., Houpt, T. R. and Jarnum, S. A. (1956) Water balance of the camel. *American Journal of Physiology* 185, 185–194.

Scholander, P. F. (1955) Evolution of climatic adaptation in homeotherms. *Evolution* 9, 15–26.

Scholander, P. F., Hock, R., Walthers, V., Johnson, R. and Irving, L. (1950) Heat regulation in some arctic and tropical mammals and birds. *Biological Bulletin, Marine Biological laboratory, Woods Hole, Mass.* 99, 225–236.

Schröter, C. and Kirchner, O. (1896, 1902) Die Vegetation des Bodensees. *Verhandlungen der Gesellschaft des Bodensees Umgebung* 25, 1–119; 31, 1–86.

Schultz, A. M. (1964) The nutrient-recovery hypothesis for arctic microtine cycles. II. Ecosystem variables in relation to the arctic microtine cycles. In: *Grazing in Terrestrial and Marine Environments.* Ed. D. Crisp, Blackwells, Oxford. pp. 57–68.

Schwerdtfeger, F. (1941) Uber die Ursachen des Massenwechsels der Insekten. *Zeitschrift für Angewandte Entomologie* 28, 254–303.

Schwerdtfeger, F. (1953, 1968) *Okologie der Tiere.* Vol. I: Autokologie, Vol. II: Demokologie. Ed. P. Parey.

Shelford, V. E. (1911) Physiological animal geography. *Journal of Morphology* 22, 551–618.

Shelford, V. E. (1913) Animal communities in temperate America. *Bulletin of the Geographical Society of Chicago* 5, 1–368.

Shelford, V. E. (1927) An experimental investigation of the relations of the codling moth to weather and climate. *Bulletin of the Illinois State Natural History Survey* 16, 307–440.

Shelford, V. E. (1929) *Laboratory and Field Ecology*. Williams & Wilkins, Baltimore.

Shorrocks, B. (1970) The distribution of gene frequencies in polymorphic populations of *Drosophila melanogaster*. *Evolution* 24, 660–669.

Simpson, G. G. (1964) Species diversity of North American mammals. *Systematic Zoology* 13, 57–73.

Slobodkin, L. B., Smith, F. E. and Hairston, H. G. (1967) Regulation in terrestrial ecosystems, and the implied balance of nature. *American Naturalist* 101, 109–124.

Smalley, A. E. (1960) Energy flow of a salt marsh grasshopper population. *Ecology* 41, 672–677.

Smith, H. S. (1935) The role of biotic factors in the determination of population densities. *Journal of Economic Entomology* 28, 873–898.

Solomon, M. E. (1957) Dynamics of insect populations. *Annual Review of Entomology* 2, 121–142.

Solomon, ME. (1962) Ecology of the flour mite, *Acarus siro* L. *Annals of Applied Biology* 50, 178–184.

Sorensen, T. (1948) A method of establishing groups of equal amplitude in plant sociology based on similarity of species content and its application to analyses of the vegetation on Danish commons. *Biologiske Skrifter*. N.S. 5, 1–34.

Springer, L. M. (1950) Aerial census of interstate antelope herds of California, Idaho, Nevada and Oregon. *Journal of Wildlife Management* 14, 295–298.

Steeman-Nielsen, E. (1952) The use of radioactive carbon (C^{14}) for measuring organic production in the sea. *Journal du Conseil Permanent Internationale pour l'Exploration de la mer.* 18, 117–140.

Steeman-Nielsen, E. and Jensen, E. A. (1957) Primary oceanic production. In: *Galathea Report*, Vol. I. Copenhagen. pp. 49–136.

Stern, V. M., Smith, R. F., van den Bosch, R. and Hagen, K. S. (1959) The integrated control concept. *Hilgardia* 29, 81–101.

Strecker, R. L. and Emlen, J. T. (1953) Regulatory mechanisms in house-mouse populations: the effect of limited food supply on a confined population. *Ecology* 34, 375–385.

Sverdrup, H. U., Johnson, M. W. and Fleming, R. H. (1942) *The Oceans: their physics, chemistry and general biology*. Prentice Hall, New Jersey.

Sweetman, H. L. (1938) Physical ecology of the firebrat, *Thermobia domestica*, Packard. *Ecological Monographs* 8, 285–311.

Symposium on the Classification of Brackish Waters. Venice, 8–14 April, 1958. *Archivio di Oceanografica e Limnologia* 11, supplement. (1959).

Taber, R. E. and Dasmann, R. (1957) The dynamics of three natural populations of the deer, *Odocoileus hemionus columbianus*. *Ecology* 38, 233–246.

Tamarin, R. H. and Krebs, C. J. (1969) *Microtus* population biology. II. Genetic changes at the transferrin locus in fluctuating populations of two vole species. *Evolution* 23, 183–211.

Tamiya, H. (1957) Mass culture of algae. *Annual Review of Plant Physiology* 8, 309–334.

Tansley, A. G. (1935) The use and abuse of vegetational concepts and terms. *Ecology* 16, 284–307.

Teal, J. M. (1957) Community metabolism in a temperate cold spring. *Ecological Monographs* 27, 283–302.

Teal, J. M. (1962) Energy flow in the salt marsh ecosystem of Georgia. *Ecology* 43, 614–624.

Thallenhorst, W. (1950) Die Koenzedenz als gradolozisches Problem. *Zeitschrift für Angewandte Entomologie* 32, 1–48.

Thienemann, A. (1939) Grundzüge einer allgemeinen Oekologie. *Archiv für Hydrobiologie* 35, 267–285.

Thomas, N. D. and Hill, G. R. (1949) Lucerne field production. In: *Photosynthesis in Plants*. Ed. J. Franck and W. E. Coomis. Iowa State University Press, Ames, Iowa.

Thompson, W. R. (1923) La théorie mathématique de l'action des parasites entomophages. *Revue générale des sciences pures et appliquées* 34, 202–210.

Thompson, W. R. (1929) On natural control. *Parasitology* 21, 269–281.

Thompson, W. R. (1956) The fundamental theory of natural and biological control. *Annual Review of Entomology* 1, 379–402.

Thornthwaite, C. W. (1931) The climate of North America according to a new classification. *Geographical Review* 21, 633–655.

Thornthwaite, C. W. (1940) Atmospheric moisture in relation to ecological problems. *Ecology* 21, 17–28.

Thornthwaite, C. W. (1948) An approach towards a rational classification of climate. *Geographical Review* 38, 55–94.

Tinbergen, L. (1946) De Sperwer als roofvijand van zangvogels. *Ardea* 34, 1.

Tinbergen, L. (1960) The natural control of insects in pine woods I. Factors influencing the intensity of predation by songbirds. *Archives néerlandaises de zoologie* 13, 266–336.

Tischler, W. (1949) *Grundzüge der terrestrischen Tierökologie*. Braunschweig.

Trave, J. (1963) Ecologie et biologie des Oribates (Acariens) saxicoles et arboricoles. *Vie et Milieu* supplement.

Turcek, F. J. (1951) On the stratification of the avian population of the *Querceto-carpinetum* forest community in Southern Slovakia. *Sylvia* 13, 71–86.

UNESCO (1970) *Use and conservation of the biosphere*. UNESCO, Paris.

Utida, S. (1955) Fluctuations in the interacting populations of host and parasite in relation to the biotic potential of the host. *Ecology* 36, 202–206.

Uvarov, B. P. (1931) Insects and climate. *Transactions of the Royal Entomological Society of London* 79, 1–247.

Vaillant, F. (1956) Recherches sur la faune madicole (hygropétrique s.l.) de France, de Corse et d'Afrique du Nord. *Mémoires du Muséum nationale d'histoire naturelle* n.s. A 11, 1–258.

Vandel, A. (1965) *Biospeleology* (translated B. E. Freeman). Pergamon Press, London.

Vannier, G. (1964) Extracteur automatique de microfaune du sol à programmation pour études écologique. *Revue d'écologie et de biologie du sol* 1, 421–441.

Vannier, G. (1967) Définition des rapports entre les microarthropodes et l'état hydrique des sols. *Compte rendu hebdomadaire des séances de l'Académie des sciences* 265, 1741–1744.

Vannier, G. and Cancela da Fonseca, J. P. (1966) L'échantillonnage de la micro faune du sol. *La Terre et la Vie* 77–103.

Varley, G. C. and Gradwell, G. R. (1960) Key factors in population studies. *Journal of Animal Ecology* 29, 399–401.

Verhulst, P. F. (1844) Recherches mathématiques sur la loi d'accroissement de la population. *Mémoires de l'Académie Royale de Belgique* 20, 1–52.

Vibert, A. and Lagler, F. (1961) *Pêches continentales. Biologie et aménagement.* Dunod, Paris.

Vinogradov, M. E. (1962) Quantitative distribution of deep-sea plankton in the western Pacific and its relation to deep-water circulation. *Deep Sea Research* 8, 251–258.

Volterra, V. (1926) Variazioni e fluttuazioni del numero d'individui in specie animali conviventi. *Memorie dell' Accademia pontificia dei Nuovi Lincei* 2, 31–113.

Volterra, V. (1931) Variation and fluctuations of the number of individuals in animal species living together. In: *Animal Ecology* R. N. Chapman, McGraw-Hill, New York.

Waloff, N. (1948) Development of *Ephestia elutella*, Hb. (Lep. Phycitidae) on some natural foods. *Bulletin Entomological Research* 39. 117–130.

Waterhouse, F. L. (1950) Humidity and temperature in grass microclimates with reference to insolation. *Nature, London* 166, 232–233.

Wautier, J. (1952) *Introduction à l'étude des biocénoses.* Imprimerie Ferréol, Lyon

Weis-Fogh, T. (1948) Ecological investigation of mites and collembola in the soil. *Natura jutlandica* 1, 135–270.

Wellington, W. G. (1964) Qualitative changes in populations in unstable environments. *Canadian Entomologist* 96, 436–451.

Went, F. W. (1957) *The Experimental Control of Plant Growth.* Chronica Botanica Co., Waltham.

Whittaker, R. H. (1952) A study of summer foliage insect communities in the Great Smokey Mountains. *Ecological Monographs* 22, 1–144.

Wiegert, R. G. (1965) Population energetics of meadow spittlebugs (*Philaenus spumarius L.*) as affected by migration and habitat. *Ecological Monographs* 34, 217–241.

Woodruffe, L. L. (1912) Observations on the origin and sequence of the protozoan fauna of hay infusions. *Journal of Experimental Zoology* 12, 205–264.

Wright, J. C. (1954) The hydrobiology of Atwood lake, a flood-control reservoir. *Ecology* 35, 305–316.

Zenkevitch, L. (1963) *Biology of the seas of the USSR.* Allen & Unwin, London.

Zwolfer, W. (1931) Studien zur Oekologie und Epidemiologie der Insekten. I. Die Kieferneule, *Panolis flammea*, Schiff. *Zeitschrift für Angewandte Entomologie* 17, 475–562.

INDEX

In general the scientific names of organisms are not indexed.